Power Systems

Pavlos S. Georgilakis

Spotlight on Modern Transformer Design

With 121 figures

Springer

Pavlos S. Georgilakis, Asst. Prof.
Department of Production Engineering
and Management
Technical University of Crete
University Campus
731 00 Chania
Greece
pgeorg@dpem.tuc.gr

ISSN 1612-1287
ISBN 978-1-84882-666-3   e-ISBN 978-1-84882-667-0
DOI 10.1007/978-1-84882-667-0
Springer Dordrecht Heidelberg London New York

British Library Cataloguing in Publication Data
A catalogue record for this book is available from the British Library

Library of Congress Control Number: 2009928676

© Springer-Verlag London Limited 2009
Apart from any fair dealing for the purposes of research or private study, or criticism or review, as permitted under the Copyright, Designs and Patents Act 1988, this publication may only be reproduced, stored or transmitted, in any form or by any means, with the prior permission in writing of the publishers, or in the case of reprographic reproduction in accordance with the terms of licences issued by the Copyright Licensing Agency. Enquiries concerning reproduction outside those terms should be sent to the publishers.
The use of registered names, trademarks, etc. in this publication does not imply, even in the absence of a specific statement, that such names are exempt from the relevant laws and regulations and therefore free for general use.
The publisher makes no representation, express or implied, with regard to the accuracy of the information contained in this book and cannot accept any legal responsibility or liability for any errors or omissions that may be made.

Cover design: deblik, Berlin, Germany

Printed on acid-free paper

Springer is part of Springer Science+Business Media (www.springer.com)

This book is dedicated to my family

Foreword

Power transformer advantages of high efficiency and reliability have certainly contributed to the domination of alternating current in power networks since the beginning of the last century. From early times, their design has been a major concern and has been the subject of extended research. The first efforts were based on conveniently adapted analytical solutions enabling one to optimize their construction and to take advantage of the improvements in magnetic and electric material properties.

During recent decades the development of the philosophy of transformer design has been a logical extension of the use of computers and numerical tools enabling one to model accurately the geometrical complexities as well as the nonlinear material characteristics for problem analysis. In addition, optimization algorithms have been very successfully combined with numerical techniques to represent the electromagnetic and thermal phenomena developed in power transformers, resulting in very powerful composite computational methodologies. In particular, artificial intelligence algorithms incorporated in such techniques have dramatically enhanced the speed and capability for achieving detailed optimum designs.

With this book Professor Pavlos Georgilakis contributes to the diffusion of composite numerical methodologies for power transformer design based on the combination of standard design techniques for transformers with advanced numerical methods such as the finite element method, and efficient optimization algorithms such as sequential quadratic programming, the branch-and-bound technique, genetic algorithms, decision trees and artificial neural networks. The proposed approach to the subject creates a proper link between the various methodologies implemented and their particular contribution to this field. The important problem of transformer selection criteria is systematically treated by using total owning cost considerations and external environmental cost issues.

The author's involvement in research both in the design office of a transformer construction company and the Technical University of Crete has given him a wide experience of the subject. His previous industry experience is reflected in the book by many references to actual practices. His academic background and the number of papers he has published in refereed journals ensure that a thorough theoretical treatment is given to important topics.

An important advantage of this work is that all methodologies presented are illustrated through detailed practical examples, concerning general power transformer construction, including also the shell type core transformer case. The proposed examples cover all features of power transformer design and have been worked out in a rigorous and coherent manner. The introductory detailed presentation of the fundamental topics and phenomena involved enables the implementation of a textbook for teaching step by step the mysteries of transformer design both at undergraduate and postgraduate level. Certainly, it constitutes an excellent reference for researchers in the field, practicing electrical engineers and transformer design office use.

Without any doubt, the book fills an important knowledge gap in our energy conservation challenges!

Athens, Greece
February 2009

John A. Tegopoulos
Life Fellow IEEE
Professor Emeritus
National Technical University of Athens

Preface

Many of the standard books on transformers are now over ten years old and some much older. Much has changed in the transformer industry since these books were written. Newer and better materials are now available for core and winding construction. Powerful computers now make it possible to produce more detailed models of the electrical, mechanical and thermal behavior of transformers than previously possible. The ever-increasing competition in the global market has put tremendous responsibilities on the transformer industry to increase transformer reliability while reducing cost, since high quality, low cost products have become the key to survival. However, it is difficult, if at all possible, to meet today's transformer design demands via conventional design techniques.

Today, artificial intelligence is widely used in modeling nonlinear and large-scale systems, especially when explicit mathematical models are difficult to obtain or are completely lacking. Moreover, artificial intelligence is computationally efficient in solving hard optimization problems.

The limitations of the analytical techniques as well as the progress of computers facilitated the development of numerical techniques for the solution of electromagnetic field problems. Among the numerical techniques, the most popular method for the solution of electromagnetic field problems is the finite element method. A very real advantage of the finite element method is its ability to deal with complex geometries. Another advantage is that it yields stable and accurate solutions.

The subject of the book is *Modern Transformer Design*. This book introduces a novel approach to transformer design using artificial intelligence and numerical techniques.

The author worked in the transformer industry for 10 years before joining academia. He has vast experience in the design, development and manufacturing of transformers. The author has developed the bulk of the results presented in the book during the last 10 years, while some of the results appear for the first time.

There is no other book including shell type transformer design by means of magnetic field analysis and artificial intelligence techniques. Most of the material in the book is an expanded and detailed version of the author's original work in the field of transformer design. The basic philosophy of the book is that we learn by applying. That is why the book has many numerical examples that illustrate the use of the techniques for a variety of real-world transformer designs.

The book will be particularly useful to graduate and postgraduate students in electric power engineering devices, researchers in the design and implementation of power transformers, transformer designers and power engineering professionals. More specifically:

1. Graduate and postgraduate students as well as researchers will learn new methodologies for transformer design optimization (TDO). Moreover, they will be able to apply and extend the methodologies of the book to the optimization of different types of transformers or to the optimization of other electrical machines and devices. They will also find real and accurate data since all transformer design examples are from actual constructed and tested transformers.

2. Transformer designers will be helped to apply artificial intelligence to optimizing their transformer designs. In order to assist them, the book presents the basic principles of artificial intelligence methods in separate chapters and in stand-alone form, i.e., the transformer designers will find the majority of the information they need within the book. Moreover, transformer designers can extend the methodologies of the book to optimize the designs of specific transformer types and technologies they use at their transformer manufacturing plant.

3. Power engineering professionals working in electric utilities, industries, public authorities and design offices will find information to improve transformer specifications. They will find methodologies in the book that will help them in their transformer purchasing decisions. In particular, they will save money by purchasing the most cost-effective and energy-efficient transformers.

The material of the book is organized in three parts and eight chapters. Part I, which includes Chaps. 1 and 2, is devoted to the presentation of conventional transformer design. Part II, which includes Chapts. 3 to 5, presents the evaluation and optimization techniques that will be used in the third part of the book for the solution of a number of transformer design problems. Part III, which includes Chaps. 6 to 8, is dedicated to modern transformer design and it illustrates clearly how artificial intelligence and numerical techniques successfully solve a number of hard transformer design evaluation and optimization problems.

Chapter 1 is an introduction to transformer fundamentals. It describes the basic principles for the analysis of magnetic circuits, the correspondence between electric and magnetic circuits, and the modeling of magnetic materials used in the construction of the transformer magnetic circuit. It presents a transformer equivalent circuit, a method to determine the parameters of the equivalent circuit, and formulas to compute voltage regulation and efficiency. It defines the electrical characteristics of a transformer, e.g., rated power, rated voltages, frequency, no-load losses, load losses, and impedance voltage. It describes two interesting transformer operating modes, i.e., overloading and parallel operation. It gives a list of standards that are typically used for transformer manufacturing. It presents the type, routine, and special tests that are performed on transformers. It classifies transformers according to their use, cooling medium, insulating medium, and core construction.

Preface   xi

Finally, Chap. 1 describes the type and characteristics of transformers studied in this book.

Chapter 2 deals with the conventional design of wound core type transformers. It formulates the TDO problem and solves it using a multiple design method that is commonly referred to as the conventional TDO method. A design example of an actual commercial transformer is worked out throughout this chapter showing all the calculations that are needed to design a transformer. The example-driven presentation of the conventional TDO method makes this chapter unique in the transformer design literature.

Transformers involve magnetostatic problems. These problems can be solved by analytical and numerical techniques. The limitations of the analytical techniques as well as the progress of computers has facilitated the development of numerical techniques. Among the numerical techniques, the most popular method in the solution of magnetostatic problems is the finite element method. A very strong advantage of the finite element method is its ability to deal with complex geometries. Another advantage is that it yields stable and accurate solutions. Chapter 3 presents the finite element method for the solution of linear and nonlinear magnetostatic problems, the latter being very common in transformer design. Carefully selected arithmetic examples make clear the application of the finite element method in the solution of linear and nonlinear magnetostatic problems.

Classification aims at predicting the future class, and forecasting aims at predicting the future value of a system that is intrinsically uncertain. Chapter 4 briefly presents two artificial intelligence methods, namely decision trees and artificial neural networks. The decision tree methodology is a nonparametric inductive learning technique, able to produce classifiers for a given problem that can assess new, unseen situations and/or uncover the mechanisms driving this problem. The artificial neural network is a computer information processing system that is capable of adequately representing nonlinear functions. The decision tree technique is appropriate for the solution of classification problems. The artificial neural network method is suitable for the solution of both classification and forecasting problems.

Chapter 5 is devoted to optimization and is organized into five sections. Section 5.1 is an introduction to optimization. Section 5.2 presents an active set method that effectively solves quadratic programming problems. Section 5.3 describes the sequential quadratic programming method, which is one of the best methods for solving nonlinearly constrained optimization problems. The sequential quadratic programming method iteratively solves a sequence of quadratic programming subproblems. Section 5.4 presents the branch-and-bound method, which, in conjunction with sequential quadratic programming, effectively solves mixed-integer nonlinear programming problems (such as the TDO problem of Chap. 7). Section 5.5 is devoted to the genetic algorithm method, which successfully solves complex optimization problems (such as the transformer no-load loss minimization problem of Chap. 7). The four optimization methods that are presented in this chapter are accompanied by carefully selected and analytically solved arithmetic examples that make clear the application of the methods to the solution of a variety of optimization problems.

Chapter 6 is devoted to the evaluation of transformer technical characteristics. Decision trees and artificial neural networks solve the no-load loss classification problem. Artificial neural networks solve the no-load loss prediction problem. Impedance voltage evaluation is implemented using a particular finite element model with detailed representation of winding geometry.

Chapter 7 deals with modern design optimization of wound core type transformers. Four methods are presented that solve important transformer design problems. First, genetic algorithms are combined with artificial neural networks to optimally group $4 \cdot N$ available individual cores into N transformers so as to minimize the total no-load loss of N transformers. This method significantly reduces the no-load loss design margin as well as the cost of transformer main materials. Second, decision trees and artificial neural networks successfully solve the winding material selection problem, thus avoiding the need to optimize the transformer twice, once with copper and once with aluminum windings. Third, a mixed integer programming–finite element method is developed for solution of the TDO problem. Finally, a recursive genetic algorithm–finite element method is developed to solve the TDO problem and is compared with the mixed integer programming–finite element method. The recursive genetic algorithm approach can also be very useful for the solution of other optimization problems in electric machines and power systems.

Chapter 8 deals with transformer selection by electric utilities and industrial transformer users. It reviews the classical total owning cost formula and it also introduces the external environmental cost due to transformer losses. Using the methodologies of this chapter, transformer users will save money by purchasing the most cost-effective and energy-efficient transformers.

Much of the material presented in this book was obtained through teamwork with colleagues at the National Technical University of Athens, the Technical University of Crete and Schneider Electric AE.

I would like to express my most sincere thanks to Professor Nikos Hatziargyriou, supervisor of my PhD dissertation, for his continuous guidance, encouragement and support throughout my PhD and for introducing me to artificial intelligence based transformer design. Special thanks go to Professor Antonios Kladas for excellent and fruitful research collaboration in the area of numerical techniques for analysis of the transformer magnetic field.

I sincerely acknowledge the rich and ample experience gained while working in Schneider Electric AE and I am grateful to all my erstwhile senior colleagues. I would particularly like to express my sincere gratitude to Mr Athanasios Souflaris, Mr Yiannis Bakopoulos, Mr Spiros Elefsiniotis, Mr Dimitrios Paparigas, and Mr Dionissios Spiliopoulos for their support and guidance.

It was a great pleasure for me to collaborate with three PhD students in the area of transformer design. I would like to express my sincere thanks to Dr Marina Tsili, Dr Eleftherios Amoiralis and Dr Themistoklis Kefalas for our fruitful collaboration.

I would like to thank Professor Nikola Rajakovic, Professor Vlastimir Glamocanin, Professor Suad Halilcevic, Professor Antonios Kladas and three more anonymous reviewers for the time invested to review this book and for their constructive comments that helped me to improve the quality, presentation and organization of the book.

Thanks are also due to Mr Anthony Doyle from Springer for his invitation to write this book and for believing in the project from the beginning as well as to Mr Simon Rees and Ms Claire Protherough from Springer who gave very good editorial input.

This book would not have been possible without the understanding and patience of my wife Liza.

Chania, Greece   Pavlos S. Georgilakis
January 2009

Contents

Part I Conventional Transformer Design 1

1 Transformers 3
   1.1 Introduction 3
   1.2 Magnetic Circuits 4
       1.2.1 General 4
       1.2.2 Analysis of Magnetic Circuits 7
       1.2.3 Flux Linkage 9
       1.2.4 Magnetic Materials 10
   1.3 Transformer Fundamentals 12
       1.3.1 Equivalent Circuit 12
       1.3.2 Derivation of Equivalent Circuit Parameters 14
       1.3.3 Voltage Regulation 18
       1.3.4 Efficiency 23
   1.4 Transformer Electrical Characteristics 27
       1.4.1 Rated Power 27
       1.4.2 Temperature Rise 28
       1.4.3 Ambient Temperature 28
       1.4.4 Altitude of Installation 29
       1.4.5 Impedance Voltage 29
       1.4.6 No-Load Losses 29
       1.4.7 Load Losses 30
       1.4.8 Rated Voltages 31
       1.4.9 Vector Group 31
       1.4.10 Frequency 32
       1.4.11 Noise 32
       1.4.12 Short-Circuit Current 32
       1.4.13 No-Load Current 32
   1.5 Transformer Operation 33
       1.5.1 Overloading 33
       1.5.2 Parallel Operation 33
       1.5.3 Load Distribution to Transformers in Parallel Operation 34
   1.6 Transformer Standards and Tolerances 35
       1.6.1 Transformer Standards 35
       1.6.2 Tolerances 36

1.7		Transformer Tests 37
	1.7.1	Type Tests 37
	1.7.2	Routine Tests 37
	1.7.3	Special Tests 39
1.8		Transformer Types 39
	1.8.1	Classification According to Transformer Use 40
	1.8.2	Classification According to Transformer Cooling Method 40
	1.8.3	Classification According to Transformer Insulating Medium 41
	1.8.4	Classification According to Transformer Core Construction 41
1.9		Transformers Studied in this Book 42
References 43

2 Conventional Transformer Design 45

2.1	Nomenclature 45
2.2	Introduction 49
2.3	Problem Formulation 49
	2.3.1 Objective Function 50
	2.3.2 Constraints 52
	2.3.3 Mathematical Formulation of the TDO Problem 57
	2.3.4 Characteristics of the TDO Problem 58
2.4	Conventional Transformer Design Optimization Method 59
	2.4.1 Methodology 59
	2.4.2 Case Study 62
	2.4.3 Repetitive Transformer Design Process 66
2.5	Example of Transformer Design Data 68
	2.5.1 Values of Description Variables 70
	2.5.2 Values of Special Variables 70
	2.5.3 Values of Default Variables 70
	2.5.4 Values of Cost Variables 70
	2.5.5 Values of Various Variables 71
	2.5.6 Values of Conductor Cross-Section Calculation Variables 71
	2.5.7 Values of Design Variables 71
2.6	Calculation of Volts per Turn and Thickness of Core Leg 74
	2.6.1 Calculation of Volts per Turn 74
	2.6.2 Calculation of Thickness of Core Leg 74
	2.6.3 Example 2.1 76
2.7	Calculation of Layer Insulation 77
	2.7.1 Layer Insulation of LV Winding 78
	2.7.2 Layer Insulation of HV Winding 78
	2.7.3 Example 2.2 78
2.8	Calculation of Winding and Core Dimensions 79
	2.8.1 Example 2.3 79
2.9	Calculation of Core Weight and No-Load Loss 84
	2.9.1 Example 2.4 86

2.10 Calculation of Inductive Part of Impedance Voltage 87
  2.10.1 Example 2.5 89
2.11 Calculation of Load Loss 94
  2.11.1 Example 2.6 94
2.12 Calculation of Impedance Voltage 99
  2.12.1 Example 2.7 100
2.13 Calculation of Coil Length 100
  2.13.1 Example 2.8 101
2.14 Calculation of Tank Dimensions 102
  2.14.1 Example 2.9 102
2.15 Calculation of Winding Gradient and Oil Gradient 103
  2.15.1 Example 2.10 103
2.16 Calculation of Heat Transfer 106
  2.16.1 Example 2.11 108
2.17 Calculation of the Weight of Insulating Materials 110
  2.17.1 Example 2.12 110
2.18 Calculation of the Weight of Ducts 114
  2.18.1 Example 2.13 114
2.19 Calculation of the Weight of Oil 115
  2.19.1 Example 2.14 115
2.20 Calculation of the Weight of Sheet Steel 116
  2.20.1 Example 2.15 117
2.21 Calculation of the Weight of Corrugated Panels 117
  2.21.1 Example 2.16 117
2.22 Calculation of the Cost of Transformer Main Materials 117
  2.22.1 Example 2.17 118
2.23 Calculation of Transformer Manufacturing Cost 119
  2.23.1 Example 2.18 120
References 122

Part II Evaluation and Optimization Methods 123

3 Numerical Analysis 125
3.1 Introduction 125
  3.1.1 Magnetostatic Problems 125
  3.1.2 Methods for the Solution of Magnetostatic Problems 127
3.2 Finite Element Method 128
  3.2.1 Introduction 128
  3.2.2 Applications to Power Engineering 129
  3.2.3 Solution of Linear Magnetostatic Problems 130
  3.2.4 Solution of Nonlinear Magnetostatic Problems 146
References 153

4 Classification and Forecasting 157
4.1 Introduction 157
4.2 Automatic Learning 158
4.3 Data Mining 158
4.3.1 Representation 159
4.3.2 Attribute Selection 159
4.3.3 Model Selection 159
4.3.4 Interpretation and Validation 159
4.3.5 Model Use 160
4.4 Learning Set and Test Set 160
4.4.1 Classification 160
4.4.2 Forecasting 161
4.5 Decision Trees 162
4.5.1 Introduction 162
4.5.2 Applications to Power Systems 163
4.5.3 General Characteristics 164
4.5.4 Top Down Induction 165
4.5.5 Optimal Splitting Rule 167
4.5.6 Stop Splitting Rule 170
4.5.7 Overview of Decision Tree Building Algorithm 173
4.5.8 Example 4.1 174
4.5.9 Example 4.2 179
4.6 Artificial Neural Networks 185
4.6.1 Introduction 185
4.6.2 Applications to Power Systems 186
4.6.3 ANN Types 187
4.6.4 Neuron Mathematical Model 188
4.6.5 ANN Architectures 189
4.6.6 ANN Training 191
4.6.7 ANN Configuration 205
4.6.8 Example 4.5 207
4.7 Hybrid Decision Tree–Neural Network Classifier 210
4.7.1 Example 4.6 211
References 212

5 Optimization 219
5.1 Introduction 219
5.2 Quadratic Programming 222
5.2.1 Methodology 222
5.2.2 Applications to Power Systems 225
5.2.3 Example 5.1 225
5.3 Sequential Quadratic Programming 231
5.3.1 Methodology 231
5.3.2 Applications to Power Systems 233

		5.3.3	Example 5.2 233
	5.4	Branch-and-Bound 239
		5.4.1	Methodology 239
		5.4.2	Applications to Power Systems 241
		5.4.3	Example 5.3 241
	5.5	Genetic Algorithms 244
		5.5.1	Methodology 244
		5.5.2	Applications to Power Systems 248
		5.5.3	Example 5.4 249
	References 256

Part III Modern Transformer Design **263**

6	**Evaluation of Transformer Technical Characteristics** **265**
	6.1	Introduction 265
	6.2	No-Load Loss Classification with Decision Trees and Artificial Neural Networks 266
		6.2.1	Introduction 266
		6.2.2	Individual Core 267
		6.2.3	Transformer 281
	6.3	No-Load Loss Forecasting with Artificial Neural Networks 292
		6.3.1	Introduction 292
		6.3.2	Forecasting Accuracy 294
		6.3.3	Individual Core 294
		6.3.4	Transformer 298
	6.4	Impedance Voltage Evaluation with Numerical Models 301
		6.4.1	Introduction 301
		6.4.2	Finite Element Model 302
		6.4.3	Results and Discussion 317
	References 325

7	**Transformer Design Optimization** **331**
	7.1	Introduction 331
	7.2	No-Load Loss Reduction with Genetic Algorithms 332
		7.2.1	Introduction 332
		7.2.2	Conventional Core Grouping Process 332
		7.2.3	Genetic Algorithm Solution to the TNLLR Problem 334
		7.2.4	Results 341
	7.3	Winding Material Selection with Decision Trees and Artificial Neural Networks 343
		7.3.1	Introduction 343
		7.3.2	Creation of Knowledge Base 344

		7.3.3	Decision Trees	346
		7.3.4	Adaptive Trained Neural Networks	349
		7.3.5	Synthesis	359
	7.4	Transformer Design Optimization with Branch-and-Bound		359
		7.4.1	Introduction	359
		7.4.2	MIP-FEM Methodology	360
		7.4.3	Results and Discussion	364
	7.5	Transformer Design Optimization with Genetic Algorithms		368
		7.5.1	Introduction	368
		7.5.2	Recursive GA-FEM Methodology	368
		7.5.3	Results and Discussion	372
	References			374

8 Transformer Selection ... 377
8.1 Introduction ... 377
8.2 Total Owning Cost for Industrial and Commercial Users ... 378
8.2.1 Cost Evaluation Method ... 378
8.2.2 Example 8.1 ... 382
8.2.3 Example 8.2 ... 385
8.2.4 Example 8.3 ... 385
8.3 Total Owning Cost for Electric Utilities ... 391
8.3.1 Cost Evaluation Method ... 391
8.3.2 Example 8.4 ... 394
8.3.3 Example 8.5 ... 396
8.4 Proposed TOC Incorporating Environmental Cost ... 400
8.4.1 Introduction ... 400
8.4.2 Cost Evaluation Method ... 402
8.4.3 Example 8.6 ... 407
8.4.4 Example 8.7 ... 408
8.4.5 Example 8.8 ... 409
8.4.6 Example 8.9 ... 411
8.4.7 Example 8.10 ... 417
References ... 419

Index ... 423

Part I
Conventional Transformer Design

1 Transformers

Abstract This chapter is an introduction to transformer fundamentals. It describes the basic principles for the analysis of magnetic circuits, the correspondence between electric and magnetic circuits, and the modeling of magnetic materials used in the construction of a transformer magnetic circuit. It presents a transformer equivalent circuit, a method to determine the parameters of an equivalent circuit, and formulas to compute voltage regulation and efficiency. It defines the electrical characteristics of a transformer, e.g., rated power, rated voltages, frequency, no-load losses, load losses, and impedance voltage. It describes two interesting transformer operating modes, i.e., overloading and parallel operation. It gives a list of standards that are typically used for transformer manufacturing. It presents the type, routine, and special tests that are performed on transformers. It classifies transformers according to their use, cooling medium, insulating medium, and core construction. Finally, this chapter describes the type and characteristics of transformers studied in this book.

1.1 Introduction

A power transformer is a static device that, by electromagnetic induction, transmits electrical power from one alternating voltage level to another without changing the frequency. It has two or more windings of wire wrapped around a ferromagnetic core. These windings are not electrically connected, but they are magnetically coupled, i.e., the only connection between the windings is the magnetic flux present within the core.

One of the transformer windings, the primary winding, is connected to an alternating current (ac) electric power source. The second transformer winding, the secondary winding, supplies electric power to loads. If the transformer has three windings, then the third winding, the tertiary winding, also supplies electric power to loads.

The electrical energy received by the primary winding is first converted into magnetic energy that is reconverted back into a useful electrical energy in the secondary winding (and tertiary winding, if it exists).

A transformer is called a step-up transformer if its secondary winding voltage is higher than its primary winding voltage. In a step-up transformer, the primary winding is also called the low voltage winding and the secondary winding is also called the high voltage winding. On the other hand, if the transformer secondary winding voltage is lower than its primary winding voltage, the transformer is called a step-down transformer. In a step-down transformer, the primary winding is also called the high voltage winding and the secondary winding is also called the low voltage winding.

The transformer is an electrical machine that allows the transmission and distribution of electrical energy simply and inexpensively, since its efficiency is from 95% to 99%, i.e., the transformer operates more efficiently than most electrical devices (Kennedy 1998). This means that the transformer changes one ac voltage level to another while keeping the input power, i.e., the power at the first voltage level, practically equal to the output power, i.e., the power supplied to the loads. In a step-up transformer the secondary voltage is higher than the primary voltage, which means that the secondary current has to be lower than the primary current to keep the input power equal to the output power. Since the power losses in the transmission lines are proportional to the square of the current in the transmission lines, raising the transmission voltage and reducing the resulting transmission currents by a factor of 10 with step-up transformers reduces transmission line losses by a factor of 100 (Chapman 2005). That is why step-up transformers are used in power generating stations so that more power can be transmitted efficiently long distances. Step-down transformers are used in power distribution networks, factories, commercial buildings, and residences to reduce the voltage to a level at which the equipment and appliances can operate. Transformers play also a key role in the interconnection of power systems at different voltage levels. Without the transformer, it would simply not be possible to use electric power in many of the ways it is used today. Consequently, transformers occupy prominent positions in the electric power system, being the vital links between power generating stations and points of electric power utilization.

1.2 Magnetic Circuits

1.2.1 General

A simple magnetic circuit is shown in Fig. 1.1. The core is composed of ferromagnetic material with permeability μ (H/m) that is much greater than the permeability μ_0 (H/m) of the surrounding air. The core has a uniform cross-section area A_c (m^2). The core is excited by a winding of N turns carrying current i (A). This winding produces a magnetic field in the core. The source of this magnetic field is the ampere-turn ($\text{A} \cdot \text{t}$) product $N \cdot i$, which is called *magnetomotive force* $\mathcal{F}$ ($\text{A} \cdot \text{t}$):

$$\mathcal{F} = N \cdot i. \tag{1.1}$$

1.2 Magnetic Circuits

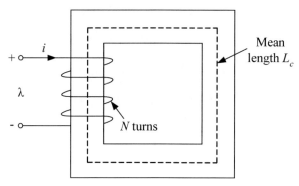

Fig. 1.1 Magnetic circuit

Ampere's law states that the line integral of the *magnetic field intensity* **H** around a closed path is equal to the net current enclosed by that path (the magnetomotive force for the magnetic circuit of Fig. 1.1):

$$\oint_C \mathbf{H} \cdot d\mathbf{L} = \mathcal{F} . \tag{1.2}$$

Assuming that for the magnetic circuit of Fig. 1.1 the *magnetic flux density* **B** is uniform across the core cross-section area, the line integral of **H** is equal to the scalar product $H_c \cdot L_c$, where H_c is the magnitude of **H** along the mean flux path whose length is L_c (m):

$$\oint_C \mathbf{H} \cdot d\mathbf{L} = H_c \cdot L_c . \tag{1.3}$$

If we substitute (1.1) and (1.3) into (1.2), we obtain:

$$H_c \cdot L_c = N \cdot i = \mathcal{F} . \tag{1.4}$$

For the ferromagnetic material of the core, the following relationship holds between the magnetic flux density **B** and the magnetic field intensity **H**:

$$\mathbf{B} = \mu \cdot \mathbf{H} , \tag{1.5}$$

where μ (H/m) is the *permeability* of the ferromagnetic material of the core. The permeability μ can be expressed in terms of the relative permeability μ_r of the ferromagnetic material and the permeability μ_0 of free space as follows:

$$\mu = \mu_0 \cdot \mu_r . \tag{1.6}$$

The permeability of free space is $\mu_0 = 4 \cdot \pi \cdot 10^{-7}$ H/m. It should be noted that the relative permeability μ_r varies with magnetic flux density.

The *magnetic flux* ϕ (Wb) crossing an area is the surface integral of the magnetic flux density **B**:

$$\phi = \oint_S \mathbf{B} \cdot d\mathbf{s} . \tag{1.7}$$

According to field theory for the continuity of flux, all the flux that enters the surface enclosing a volume must leave that volume over some other portion of that surface because magnetic flux lines form closed loops. If the magnetic flux outside the core of Fig. 1.1 is neglected, then (1.7) reduces to the following scalar form:

$$\phi_c = B_c \cdot A_c , \tag{1.8}$$

where ϕ_c (Wb) is the magnetic flux in the core, B_c (Wb/m^2 or T) is the magnetic flux density in the core, and A_c (m^2) is the cross-section area of the core.

Supposing that the permeability μ of the ferromagnetic material is constant, and since the magnetic flux density is uniform, from (1.5) the following expression is obtained for the calculation of B_c for the magnetic circuit of Fig. 1.1:

$$B_c = \mu \cdot H_c . \tag{1.9}$$

Solving (1.9) for H_c gives:

$$H_c = \frac{B_c}{\mu} . \tag{1.10}$$

Solving (1.8) for B_c gives:

$$B_c = \frac{\phi_c}{A_c} . \tag{1.11}$$

If we substitute (1.10) and (1.11) into (1.4), we obtain:

1.2 Magnetic Circuits

$$\mathcal{F} = H_c \cdot L_c \Rightarrow \mathcal{F} = \frac{B_c}{\mu} \cdot L_c \Rightarrow \mathcal{F} = \frac{\frac{\phi_c}{A_c}}{\mu} \cdot L_c \Rightarrow$$

$$\mathcal{F} = \phi_c \cdot \frac{L_c}{\mu \cdot A_c}. \tag{1.12}$$

The *reluctance* $\mathcal{R}_c$ of the magnetic core is defined from the following formula:

$$\mathcal{R}_c = \frac{L_c}{\mu \cdot A_c}. \tag{1.13}$$

The reluctance $\mathcal{R}_c$ is expressed in ampere-turns per weber (A · t / Wb). Substituting (1.13) into (1.12), we obtain:

$$\mathcal{F} = \phi_c \cdot \mathcal{R}_c. \tag{1.14}$$

1.2.2 Analysis of Magnetic Circuits

There is an analogy in the analysis of magnetic circuits with the analysis of electric circuits. This approximation derives from the mathematical similarity of electric and magnetic laws.

Table 1.1 Correspondence between electric and magnetic circuits

Magnetic circuits			Electric circuits		
Symbol	Name	Unit	Symbol	Name	Unit
ϕ	Magnetic flux	Wb	i	Current	A
B	Magnetic flux density	Wb/m²	J	Current density	A/m²
H	Magnetic field intensity	A·t/m	E	Electric field intensity	V/m
$\mathcal{F}$	Magnetomotive force	A·t	V	Voltage	V
μ	Permeability	H/m	γ	Conductivity	$(\Omega \cdot m)^{-1}$
$\mathcal{R}$	Reluctance	A·t/Wb	R	Resistance	Ω

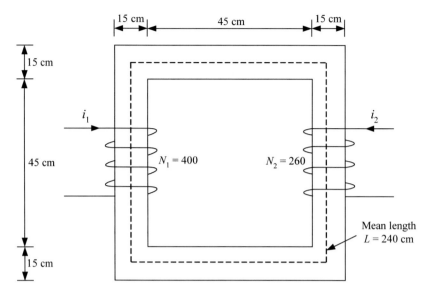

Fig. 1.2 Magnetic circuit for the Example 1.1

In electric circuits, Kirchhoff's current law states that the algebraic sum of the currents entering any node is zero, or equivalently, the sum of the currents entering a node is equal to the sum of the currents leaving the node. The approximate magnetic counterpart states that the algebraic sum of the magnetic fluxes entering any node is zero.

In electric circuits, Kirchhoff's voltage law states that the algebraic sum of the voltages around any loop is zero. The approximate magnetic counterpart states that the algebraic sum of the magnetomotive forces around any loop is zero.

Table 1.1 shows the correspondence between electric and magnetic circuits.

1.2.2.1 Example 1.1

A two-legged core is shown in Fig. 1.2. The winding on the left leg of the core has $N_1 = 400$ turns and the winding on the right has $N_2 = 260$ turns. The core depth is 15 cm. Calculate the flux that will be produced by currents $i_1 = 0.5$ A and $i_2 = 0.75$ A. Assume $\mu_r = 1000$ and is constant.

Solution

The magnetic circuit of Fig. 1.2 can be equivalently represented as shown in Fig. 1.3, using the correspondence with the electric circuit. The polarities of the magnetomotive sources $N_1 \cdot i_1$ and $N_2 \cdot i_2$ are determined by using the right-hand rule, according to which if the winding is grasped in the right hand with the fingers pointing in the direction of the current, the thumb will point to the positive

1.2 Magnetic Circuits

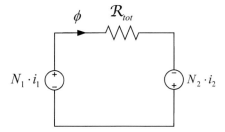

Fig. 1.3 Equivalent magnetic circuit for the Example 1.1

terminal of the magnetomotive force.

As can be seen from Fig. 1.3, the magnetic flux in the core is given by the following equation:

$$\phi = \frac{N_1 \cdot i_1 + N_2 \cdot i_2}{\mathcal{R}_{tot}},$$

where the total reluctance is calculated as follows:

$$\mathcal{R}_{tot} = \frac{L}{\mu_r \cdot \mu_0 \cdot A} = \frac{2.4 \text{ m}}{1000 \cdot (4 \cdot \pi \cdot 10^{-7} \text{ H/m}) \cdot [(0.15 \text{ m}) \cdot (0.15 \text{ m})]} = 84883 \, \frac{\text{A} \cdot \text{t}}{\text{Wb}}.$$

The magnetic flux in the core is:

$$\phi = \frac{N_1 \cdot i_1 + N_2 \cdot i_2}{\mathcal{R}_{tot}} = \frac{(400 \text{ t}) \cdot (0.5 \text{ A}) + (260 \text{ t}) \cdot (0.75 \text{ A})}{84883 \text{ A} \cdot \text{t}/\text{Wb}} = 0.00465 \text{ Wb}.$$

1.2.3 Flux Linkage

In magnetic circuits with windings, such as Fig. 1.1, when the magnetic field in the core varies with time, an induced voltage e is produced at the terminals, which is calculated by Faraday's law:

$$e = N \cdot \frac{d\phi}{dt} = \frac{d\lambda}{dt}, \quad (1.15)$$

where N is the number of turns, ϕ is the time-varying magnetic flux, and λ is the flux linkage of the winding (coil) in weber-turns (Wb · t).

For a magnetic circuit with a linear relationship between B and H, the $\lambda - i$ relationship is defined by:

$$L = \frac{\lambda}{i}, \qquad (1.16)$$

where L is the inductance in Henry (H).

1.2.4 Magnetic Materials

In the context of transformer manufacturing, the importance of magnetic materials is twofold. First, through their use it is possible to obtain large magnetic flux densities with relatively low levels of magnetizing force, which plays an important role in the performance of a transformer. Second, since magnetic materials can be used to constrain and direct magnetic fields in well-defined paths, in transformers the magnetic materials are used to maximize the coupling between the windings as well as to lower the excitation current required for transformer operation.

The relationship between B and H for a magnetic material is both nonlinear and multivalued. In general, the characteristics of the material cannot be described analytically. They are commonly presented in graphical form as a set of empirically determined curves based upon test samples of the material. The most common curve used to describe a magnetic material is the $B-H$ curve or hysteresis loop. For many engineering applications it is sufficient to describe the material using the dc or normal *magnetization curve*, which is a curve drawn through the maximum values of B and H at the tips of the hysteresis loops (Fitzgerald et al. 1990).

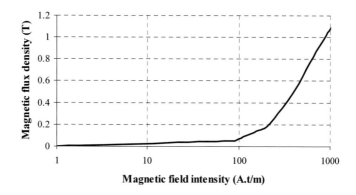

Fig. 1.4 Magnetization curve for the Example 1.2

1.2 Magnetic Circuits

1.2.4.1 Example 1.2

Solve again Example 1.1 and calculate the magnetic flux, considering that the core is made of a magnetic material whose magnetization curve is shown in Fig. 1.4.

Solution

The total magnetomotive force is:

$$\mathcal{F}_{tot} = N_1 \cdot i_1 + N_2 \cdot i_2 = (400 \text{ t}) \cdot (0.5 \text{ A}) + (260 \text{ t}) \cdot (0.75 \text{ A}) = 395 \text{ A} \cdot \text{t}.$$

The magnetic field intensity is:

$$H = \frac{\mathcal{F}_{tot}}{L} = \frac{395 \text{ A} \cdot \text{t}}{2.4 \text{ m}} = 165 \frac{\text{A} \cdot \text{t}}{\text{m}}.$$

From the magnetization curve of Fig. 1.4, for $H = 165 \text{ A} \cdot \text{t/m}$, we find that $B = 0.15 \text{ T}$.

The magnetic flux in the core is:

$$\phi_{new} = B \cdot A = (0.15 \text{ T}) \cdot [(0.15 \text{ m}) \cdot (0.15 \text{ m})] = 0.003375 \text{ Wb}.$$

The relative permeability of the core is calculated as follows:

$$\mathcal{R} = \frac{\mathcal{F}_{tot}}{\phi_{new}} = \frac{L}{\mu_{r,new} \cdot \mu_0 \cdot A} \Rightarrow \mu_{r,new} = \frac{L \cdot \phi_{new}}{\mathcal{F}_{tot} \cdot \mu_0 \cdot A} \Rightarrow$$

$$\mu_{r,new} = \frac{(2.4 \text{ m}) \cdot (0.003375 \text{ Wb})}{(395 \text{ A} \cdot \text{t}) \cdot (4 \cdot \pi \cdot 10^{-7} \text{ H/m}) \cdot [(0.15 \text{ m}) \cdot (0.15 \text{ m})]} = 725.$$

In Example 1.1, we found that $\phi = 0.00465 \text{ Wb}$, having assumed that $\mu_r = 1000$ and is constant. In Example 1.2, we found that $\phi_{new} = 0.003375 \text{ Wb}$, having used the magnetization curve of Fig. 1.4, which results in $\mu_{r,new} = 725$. It is clear that the assumption that $\mu_r = 1000$ and is constant is not very good in Example 1.1. In general, it is preferable to use the magnetization curve instead of assuming a constant value for the relative permeability of the magnetic material.

1.3 Transformer Fundamentals

1.3.1 Equivalent Circuit

The elementary transformer magnetic circuit is shown in Fig. 1.5, where for simplicity the primary and secondary windings are shown on opposite legs of the core. The primary winding has N_1 turns and the secondary winding has N_2 turns.

The leakage flux Φ_{l1} is generated by current i_1 flowing in winding 1 (primary) and it links only the turns of winding 1. The leakage flux Φ_{l2} is produced by current i_2 flowing in winding 2 (secondary) and it links only the turns of winding 2.

The magnetizing flux Φ_{m1} is generated by current i_1 flowing in winding 1 and it links all the turns of windings 1 and 2. The magnetizing flux Φ_{m2} is generated by current i_2 flowing in winding 2 and it links all the turns of windings 1 and 2.

It can be proved (Krause et al. 2002; Chapman 2005) that the transformer T equivalent circuit is as shown in Fig. 1.6, where $\mathbf{i}_\varphi$ is the phasor of the *excitation current* (also called no-load current), $\mathbf{i}_C$ is the core-loss current, and $\mathbf{i}_M$ is the magnetizing current. The parameter α denotes the transformer turns ratio or voltage ratio, i.e.:

$$\alpha = \frac{N_1}{N_2}. \tag{1.17}$$

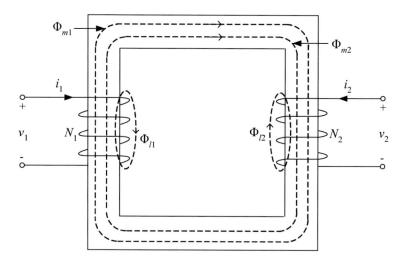

Fig. 1.5 Transformer magnetic circuit

1.3 Transformer Fundamentals

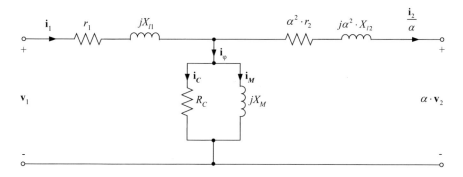

Fig. 1.6 Transformer T equivalent circuit with winding 1 being the reference winding

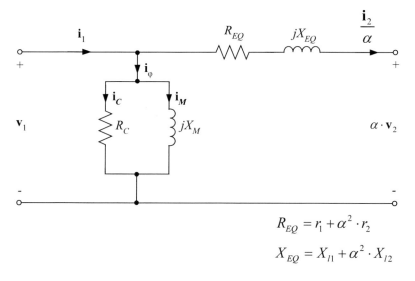

$$R_{EQ} = r_1 + \alpha^2 \cdot r_2$$
$$X_{EQ} = X_{l1} + \alpha^2 \cdot X_{l2}$$

Fig. 1.7 Transformer approximate equivalent circuit referred to the primary winding of the transformer

In Fig. 1.6, r_1 and r_2 denote the resistance of the primary and secondary winding, respectively, while X_{l1} and X_{l2} denote the reactance of the primary and secondary winding, respectively. The resistance R_C and the reactance X_M, shown in Fig. 1.6, model the core excitation effects. The resistances r_1 and r_2, shown in Fig. 1.6, model the transformer copper losses of the primary and secondary winding, respectively.

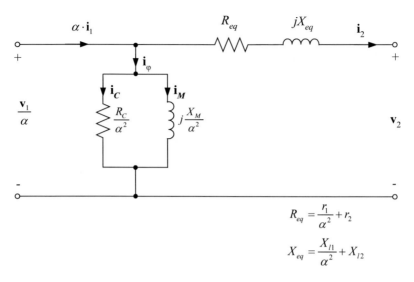

Fig. 1.8 Transformer approximate equivalent circuit referred to the secondary winding of the transformer

Moving the excitation branch (shunt branch) representing the excitation current out from the middle of the T circuit of Fig. 1.6 to either the primary or the secondary windings, as in Fig. 1.7 and Fig. 1.8, respectively, often can appreciably reduce the computational effort involved. Error is introduced by neglecting the voltage drop in the primary or the secondary leakage impedance caused by the excitation current, but this error is insignificant in most problems involving power transformers (Fitzgerald et al. 1990).

1.3.2 Derivation of Equivalent Circuit Parameters

In order to determine the parameters of the transformer approximate equivalent circuit of Figs. 1.7 and 1.8, the following two tests are used:

1. *Open-circuit test* measured from secondary side. During this test, the transformer primary side is open-circuited and rated voltage V_{OC} is applied on the secondary winding, while the current I_{OC} and the power P_{OC} on the secondary side are measured.

2. *Short-circuit test* measured from primary side. During this test, the transformer secondary side is short-circuited and appropriate voltage V_{SC} is applied on the

1.3 Transformer Fundamentals

primary winding so as to obtain full-load primary current I_{SC}, while the input power P_{SC} is measured.

The open-circuit test yields the values for the excitation branch R_C and X_M (referred to the secondary side).

The magnitude of the excitation admittance (referred to the secondary side) is:

$$Y_{EX} = \frac{I_{OC}}{V_{OC}}, \quad (1.18)$$

and the angle of the excitation admittance is:

$$\theta_{EX} = -\cos^{-1}\left(\frac{P_{OC}}{V_{OC} \cdot I_{OC}}\right), \quad (1.19)$$

so the excitation admittance is calculated as follows:

$$\mathbf{Y}_{EX} = Y_{EX} \angle \theta_{EX} = G_c - jB_M, \quad (1.20)$$

where:

$$R_C = \frac{1}{G_C} \text{ and } X_M = \frac{1}{B_M}. \quad (1.21)$$

The short-circuit test yields the values for the equivalent series impedance $\mathbf{Z}_{EQ} = R_{EQ} + jX_{EQ}$ (referred to the primary side).

The magnitude of the equivalent series impedance (referred to the primary side) is calculated as follows:

$$Z_{EQ} = \frac{V_{SC}}{I_{SC}}, \quad (1.22)$$

and the angle of the equivalent series impedance is computed as follows:

$$\theta_{EQ} = \cos^{-1}\left(\frac{P_{SC}}{V_{SC} \cdot I_{SC}}\right), \quad (1.23)$$

and the equivalent series impedance is calculated as follows:

$$\mathbf{Z}_{EQ} = Z_{EQ} \angle \theta_{EQ} = R_{EQ} + jX_{EQ}. \quad (1.24)$$

1.3.2.1 Example 1.3

A 20-kVA 20/0.48 kV 60-Hz single-phase distribution transformer is tested with the following results:

- Open-circuit test (measured from secondary side): $V_{OC} = 480$ V, $I_{OC} = 1.6$ A, $P_{OC} = 80$ W.

- Short-circuit test (measured from primary side): $V_{SC} = 1130$ V, $I_{SC} = 1$ A, $P_{SC} = 200$ W.

1. Find the equivalent circuit referred to the secondary side for this transformer at 60 Hz.

2. What would the rating of this transformer be, if it was operated at 50 Hz?

3. Draw the equivalent circuit referred to the secondary side, if the transformer was operated at 50 Hz.

Solution

1. The open-circuit test yields values for the excitation branch (referred to the secondary side):

$$Y_{EX} = \frac{I_{OC}}{V_{OC}} = \frac{1.6 \text{ A}}{480 \text{ V}} = 0.00333 \ \Omega^{-1},$$

$$\theta_{EX} = -\cos^{-1}\left(\frac{P_{OC}}{V_{OC} \cdot I_{OC}}\right) = -\cos^{-1}\left(\frac{80 \text{ W}}{(480 \text{ V}) \cdot (1.6 \text{ A})}\right) = -84.02°,$$

$$\mathbf{Y}_{EX} = Y_{EX} \angle \theta_{EX} = (0.00333 \angle -84.02°) \ \Omega^{-1} \Rightarrow$$

$$\mathbf{Y}_{EX} = (0.0003472 - j0.0033152) \ \Omega^{-1} = G_c - jB_M,$$

$$R_C = \frac{1}{G_C} = \frac{1}{0.0003472 \ \Omega^{-1}} = 2880 \ \Omega,$$

$$X_M = \frac{1}{B_M} = \frac{1}{0.0033152 \ \Omega^{-1}} = 301.64 \ \Omega.$$

1.3 Transformer Fundamentals

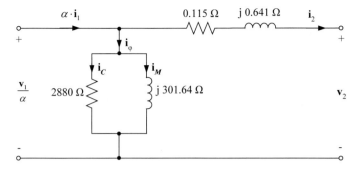

Fig. 1.9 Transformer equivalent circuit at 60 Hz referred to the secondary side

The short-circuit test yields values for the series impedances (referred to the primary side):

$$Z_{EQ} = \frac{V_{SC}}{I_{SC}} = \frac{1130\ V}{1\ A} = 1130\ \Omega,$$

$$\theta_{EQ} = \cos^{-1}\left(\frac{P_{SC}}{V_{SC} \cdot I_{SC}}\right) = \cos^{-1}\left(\frac{200\ W}{(1130\ V)\cdot(1\ A)}\right) = 79.81°,$$

$$Z_{EQ} = Z_{EQ} \angle \theta_{EQ} = (1130 \angle 79.81°)\ \Omega \Rightarrow$$

$$Z_{EQ} = (200 + j1112.16)\ \Omega = R_{EQ} + jX_{EQ}.$$

The voltage ratio is $\alpha = 20000/480 = 41.667$. The equivalent series transformer impedance referred to the secondary side is:

$$z_{eq} = R_{eq(s)} + jX_{eq(s)} \Rightarrow z_{eq} = \frac{R_{EQ}}{\alpha^2} + j\frac{X_{EQ}}{\alpha^2} \Rightarrow$$

$$z_{eq} = \frac{200}{41.667^2} + j\frac{1112.16}{41.667^2} \Rightarrow z_{eq} = (0.115 + j0.641)\ \Omega.$$

The transformer equivalent circuit at 60 Hz is shown in Fig. 1.9.

2. If the transformer was operated at 50 Hz, both the voltage and apparent power would have to be derated by a factor of 50/60, so the transformer ratings would be 16.67 kVA, 16667/400 V, and 50 Hz.

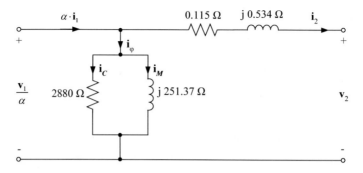

Fig. 1.10 Transformer equivalent circuit at 50 Hz referred to the secondary side

3. At 50 Hz, the resistance will be unaffected but the reactances are reduced in direct proportion to the decrease in frequency. At 50 Hz, the reactances are:

$$X_M = \frac{50 \text{ Hz}}{60 \text{ Hz}} \cdot (301.64 \, \Omega) = 251.37 \, \Omega,$$

$$X_{eq(s)} = \frac{50 \text{ Hz}}{60 \text{ Hz}} \cdot (0.641 \, \Omega) = 0.534 \, \Omega.$$

The equivalent circuit at 50 Hz is shown in Fig. 1.10.

1.3.3 Voltage Regulation

The voltage regulation, ΔV, of a transformer is defined as the difference in the magnitude of the secondary voltage at no-load, $v_{2,nl}$, and its value when loaded, v_2, divided by v_2 with the primary voltage held constant:

$$\Delta V = \frac{v_{2,nl} - v_2}{v_2}. \tag{1.25}$$

Let us consider the simplified transformer equivalent circuit of Fig. 1.11, where the effects of the excitation branch on voltage regulation are ignored, so the equivalent transformer impedance is:

$$\mathbf{z}_{eq} = R_{eq} + jX_{eq}. \tag{1.26}$$

1.3 Transformer Fundamentals

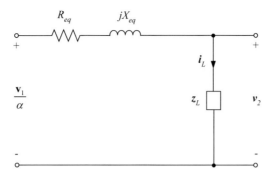

Fig. 1.11 Determination of transformer voltage regulation

In Fig. 1.11, it is assumed that v_2 is a reference phasor (zero phase angle), i.e., $v_2 = v_2 \angle 0^\circ$, and the load z_L has power factor $\cos\theta_L$ lagging, thus $i_L = i_L \angle -\theta_L$.

Using basic circuit analysis, it can be concluded from Fig. 1.11 that:

$$\Delta V = \frac{v_{2,nl} - v_2}{v_2} = \frac{\left|\frac{v_1}{\alpha} - v_2\right|}{v_2} = \frac{\left|v_2 \cdot \left|1 + \frac{\mathbf{z}_{eq}}{\mathbf{z}_L}\right| - v_2\right|}{v_2} \Rightarrow$$

$$\Delta V = \left|1 + \frac{\mathbf{z}_{eq}}{\mathbf{z}_L}\right| - 1, \qquad (1.27)$$

where in (1.27), the symbol $|\mathbf{w}|$ denotes the magnitude of the complex number $\mathbf{w}$.

Equation 1.27 can be further simplified and the following expression for the calculation of voltage regulation is obtained (MIT 1962; Del Vecchio et al. 2002):

$$\Delta V = \frac{i_L}{v_2} \cdot (R_{eq} \cdot \cos\theta_L + X_{eq} \cdot \sin\theta_L) + \frac{1}{2} \cdot \left[\frac{i_L}{v_2}\right]^2 \cdot (X_{eq} \cdot \cos\theta_L - R_{eq} \cdot \sin\theta_L)^2. \quad (1.28)$$

Equivalently, the voltage regulation can also be calculated by the following equation:

$$\Delta V = \frac{S}{S_n} \cdot (e_r \cdot \cos\theta_L + e_x \cdot \sin\theta_L) + \frac{1}{2} \cdot \left(\frac{S}{S_n}\right)^2 \cdot (e_r \cdot \sin\theta_L - e_x \cdot \cos\theta_L)^2, \qquad (1.29)$$

where:

$$e_r = \frac{LL}{S_n}, \tag{1.30}$$

and:

$$e_x = \sqrt{U_k^2 - e_r^2}, \tag{1.31}$$

where S (VA) is the transformer load, S_n (VA) is the transformer rated power, $\cos\theta_L$ is the power factor (θ_L is positive for lagging load and negative for leading load), LL (W) are the transformer load losses, and U_k (%) is the short-circuit impedance of the transformer.

1.3.3.1 Example 1.4

For the Example 1.3, calculate the full-load voltage regulation at 0.85 lagging power factor, 1.0 power factor, and at 0.85 leading power factor. The frequency is 60 Hz.

Solution
In Example 1.3, it was found that the equivalent series transformer impedance referred to the secondary side is:

$$\mathbf{z}_{eq} = (0.115 + j0.641)\,\Omega.$$

The magnitude of the full-load current on the transformer secondary side is:

$$i_L = \frac{S}{v_2} = \frac{20000\text{ VA}}{480\text{ V}} = 41.667\text{ A}.$$

Case 1 Power factor 0.85 lagging
We have $\cos\theta_L = 0.85 \Rightarrow \theta_L = 31.79°$. Assuming that $\mathbf{v}_2 = 480\angle 0°$ V, and since the load is lagging, the load current is $\mathbf{i}_L = i_L\angle -\theta_L = 41.667\angle -31.79°$ A.
From the equivalent circuit of Fig. 1.11, it can be obtained that:

$$\frac{\mathbf{v}_1}{\alpha} = \mathbf{i}_L \cdot \mathbf{z}_{eq} + \mathbf{v}_2 \Rightarrow$$

1.3 Transformer Fundamentals

$$\frac{\mathbf{v}_1}{\alpha} = (41.667\angle -31.79^0 \text{ A}) \cdot \left[(0.115 + j0.641)\ \Omega\right] + 480\angle 0^0\ \text{V} \Rightarrow$$

$$\frac{\mathbf{v}_1}{\alpha} = 498.55\angle 2.3^0\ \text{V} \Rightarrow \frac{v_1}{\alpha} = 498.55\ \text{V}.$$

The voltage regulation is:

$$\Delta V = \frac{498.55 - 480}{480} = 0.0386 \Rightarrow \Delta V = 3.86\%.$$

Case 2 Power factor 1.0
We have $\cos\theta_L = 1.0 \Rightarrow \theta_L = 0^0$. Assuming that $\mathbf{v}_2 = 480\angle 0^0$ V, the load current is $\mathbf{i}_L = i_L\angle\theta_L = 41.667\angle 0^0$ A.
We have:

$$\frac{\mathbf{v}_1}{\alpha} = \mathbf{i}_L \cdot \mathbf{z}_{eq} + \mathbf{v}_2 = (41.667\angle 0^0 \text{ A}) \cdot \left[(0.115 + j0.641)\ \Omega\right] + 480\angle 0^0\ \text{V} \Rightarrow$$

$$\frac{\mathbf{v}_1}{\alpha} = 485.53\angle 3.2^0\ \text{V} \Rightarrow \frac{v_1}{\alpha} = 485.53\ \text{V}.$$

The voltage regulation is:

$$\Delta V = \frac{485.53 - 480}{480} = 0.0115 \Rightarrow \Delta V = 1.15\%.$$

Case 3 Power factor 0.85 leading
We have $\cos\theta_L = 0.85 \Rightarrow \theta_L = 31.79^0$. Assuming that $\mathbf{v}_2 = 480\angle 0^0$ V, and since the load is leading, the load current is $\mathbf{i}_L = i_L\angle\theta_L = 41.667\angle 31.79^0$ A.
We have:

$$\frac{\mathbf{v}_1}{\alpha} = \mathbf{i}_L \cdot \mathbf{z}_{eq} + \mathbf{v}_2 = (41.667\angle 31.79^0 \text{ A}) \cdot \left[(0.115 + j0.641)\ \Omega\right] + 480\angle 0^0\ \text{V} \Rightarrow$$

$$\frac{\mathbf{v}_1}{\alpha} = 470.70\angle 3.1^0\ \text{V} \Rightarrow \frac{v_1}{\alpha} = 470.70\ \text{V}.$$

The voltage regulation is:

$$\Delta V = \frac{470.70 - 480}{480} = -0.0194 \Rightarrow \Delta V = -1.94\%.$$

1.3.3.2 Example 1.5

Let us assume that a three-phase transformer with rated power 630 kVA, rated primary voltage 20 kV and rated secondary voltage 0.4 kV, has 1200 W no-load losses, 9300 W load losses, and 6% short-circuit impedance. Determine the voltage regulation at full load and at 75% load for power factor 0.8 lagging.

Solution

The values of e_r and e_x are calculated as follows:

$$e_r = \frac{LL}{S_n} = \frac{9300}{630000} = 0.014762,$$

$$e_x = \sqrt{U_k^2 - e_r^2} = \sqrt{0.06^2 - 0.014762^2} = 0.05816.$$

Full load means that $S/S_n = 1$, while 75% load means that $S/S_n = 0.75$. The voltage regulation at the two different cases is calculated below.

Case 1 Full load and power factor 0.8 lagging

We have $\sin\theta_L = \sqrt{1-\cos^2\theta_L} = \sqrt{1-0.8^2} \Rightarrow \sin\theta_L = 0.6$. The voltage regulation is computed using (1.29):

$$\Delta V = \frac{S}{S_n} \cdot (e_r \cdot \cos\theta_L + e_x \cdot \sin\theta_L) + \frac{1}{2} \cdot \left(\frac{S}{S_n}\right)^2 \cdot (e_r \cdot \sin\theta_L - e_x \cdot \cos\theta_L)^2 \Rightarrow$$

$$\Delta V = 1.0 \cdot (0.014762 \cdot 0.8 + 0.05816 \cdot 0.6) +$$

$$+ \frac{1}{2} \cdot (1.0)^2 \cdot (0.014762 \cdot 0.6 - 0.05816 \cdot 0.8)^2 \Rightarrow \Delta V = 0.047 \Rightarrow \Delta V = 4.7\%.$$

Case 2 Per-unit load 0.75 and power factor 0.8 lagging
The voltage regulation is:

$$\Delta V = 0.75 \cdot (0.014762 \cdot 0.8 + 0.05816 \cdot 0.6) +$$

$$+ \frac{1}{2} \cdot (0.75)^2 \cdot (0.014762 \cdot 0.6 - 0.05816 \cdot 0.8)^2 \Rightarrow \Delta V = 0.035 \Rightarrow \Delta V = 3.5\%.$$

1.3.4 Efficiency

The power efficiency of any electrical machine is defined as the ratio of the useful power output, P_{out} (W), to the total power input, P_{in} (W). The efficiency can be defined by simultaneously measuring the output and the input power. However, this measurement is expensive and difficult, especially for large machines. Moreover, in the case of high efficiency machines (e.g., transformer), higher precision can be achieved, if the efficiency is expressed through the losses. Consequently, the transformer efficiency, n, is calculated using the following formula:

$$n = \frac{P_{out}}{P_{in}} = \frac{P_{out}}{P_{out} + losses} = \frac{S \cdot \cos\theta_L}{S \cdot \cos\theta_L + losses}, \quad (1.32)$$

where S is the transformer load (VA), $losses$ are the transformer losses (W) and $\cos\theta_L$ is the power factor.

The transformer efficiency is increased with a decrease of transformer losses.

The transformer losses are divided into no-load losses and load losses. The no-load losses are constant, while the load losses are proportional to transformer load. Consequently, the efficiency of transformer is calculated using the following formula:

$$n = \frac{S \cdot \cos\theta_L}{S \cdot \cos\theta_L + NLL + LL \cdot \left(\frac{S}{S_n}\right)^2}, \quad (1.33)$$

where NLL are the no-load losses (W), LL are the load losses (W), and S_n is the rated power of the transformer (VA).

If L is the per-unit load:

$$L = \frac{S}{S_n}, \quad (1.34)$$

then by substituting (1.34) into (1.33), we obtain the following expression for transformer efficiency:

$$n = \frac{L \cdot S_n \cdot \cos\theta_L}{L \cdot S_n \cdot \cos\theta_L + NLL + LL \cdot L^2}. \quad (1.35)$$

Taking L as an independent variable, the value of L that maximizes efficiency is calculated as follows:

$$\frac{dn}{dL} = 0 \Rightarrow \frac{d}{dL}\left[\frac{L \cdot S_n \cdot \cos\theta_L}{L \cdot S_n \cdot \cos\theta_L + NLL + LL \cdot L^2}\right] = 0 \Rightarrow$$

$$\frac{(L \cdot S_n \cdot \cos\theta_L + NLL + LL \cdot L^2) \cdot S_n \cdot \cos\theta_L}{(L \cdot S_n \cdot \cos\theta_L + NLL + LL \cdot L^2)^2} -$$

$$\frac{L \cdot S_n \cdot \cos\theta_L \cdot (S_n \cdot \cos\theta_L + 2 \cdot L \cdot LL)}{(L \cdot S_n \cdot \cos\theta_L + NLL + LL \cdot L^2)^2} = 0 \Rightarrow$$

$$L \cdot S_n^2 \cdot \cos^2\theta_L + NLL \cdot S_n \cdot \cos\theta_L + L^2 \cdot LL \cdot S_n \cdot \cos\theta_L - L \cdot S_n^2 \cdot \cos^2\theta_L -$$
$$-2 \cdot L^2 \cdot LL \cdot S_n \cdot \cos\theta_L = 0 \Rightarrow$$

$$NLL \cdot S_n \cdot \cos\theta_L - L^2 \cdot LL \cdot S_n \cdot \cos\theta_L = 0 \Rightarrow NLL - L^2 \cdot LL = 0 \Rightarrow$$

$$L_{opt} = \sqrt{\frac{NLL}{LL}}. \tag{1.36}$$

As can be seen from (1.36), the optimum per-unit load, L_{opt}, i.e., the per-unit load that maximizes transformer efficiency is independent of the power factor of the load.

Substituting (1.36) into (1.35), we obtain the following expression for the *maximum efficiency*:

$$n_{max} = \frac{\sqrt{\frac{NLL}{LL}} \cdot S_n \cdot \cos\theta_L}{\sqrt{\frac{NLL}{LL}} \cdot S_n \cdot \cos\theta_L + NLL + LL \cdot \frac{NLL}{LL}} \Rightarrow$$

$$n_{max} = \frac{\sqrt{NLL} \cdot S_n \cdot \cos\theta_L}{\sqrt{NLL} \cdot S_n \cdot \cos\theta_L + 2 \cdot NLL \cdot \sqrt{LL}}. \tag{1.37}$$

1.3.4.1 Example 1.6

For the transformer of Example 1.3, calculate the full-load efficiency at 0.85 lagging power factor.

Solution
The load losses are:

1.3 Transformer Fundamentals

$$LL = i_L^2 \cdot R_{eq(s)} = (41.667 \text{ A})^2 \cdot (0.115 \text{ Ω}) = 200 \text{ W}.$$

The above result confirms that $LL = P_{SC} = 200$ W, which means that the load losses at full-load are equal to P_{SC} measured during the short-circuit test, since the short-circuit test was done with full-load primary current, as can be seen from the data of Example 1.3.

It can be seen from the data of Example 1.3 that the open-circuit test was done at rated secondary voltage, so the rated no-load losses are:

$$NLL = P_{OC} = 80 \text{ W}.$$

Full load means that $S/S_n = 1$, i.e., the transformer load is $S = S_n = 20000$ VA.

The transformer efficiency is:

$$n = \frac{S \cdot \cos\theta_L}{S \cdot \cos\theta_L + NLL + LL \cdot \left(\dfrac{S}{S_n}\right)^2} \Rightarrow$$

$$n = \frac{20000 \cdot 0.85}{20000 \cdot 0.85 + 80 + 200 \cdot 1^2} = 0.9838 \Rightarrow n = 98.38\%.$$

1.3.4.2 Example 1.7

The three-phase transformer of Example 1.5 with rated power 630 kVA, rated primary voltage 20 kV and rated secondary voltage 0.4 kV, has 1200 W no-load losses and 9300 W load losses.

1. Determine the transformer efficiency at full load and at 75% load for power factor 1.0 and 0.8.

2. Calculate the maximum efficiency for power factor 1.0 and 0.8.

3. Draw the efficiency curves versus per-unit load for power factor 1.0 and 0.8.

Solution

1. Full load means that $S/S_n = 1$, i.e., the transformer load is $S = S_n = 630000$ VA. On the other hand, 75% load means that $S/S_n = 0.75$,

i.e., the transformer load is $S = 0.75 \cdot S_n = 0.75 \cdot 630000 \Rightarrow S = 472500$ VA.
The efficiency at the four different cases is calculated below.

Case 1 Full load and power factor 1.0
The efficiency at full load and power factor equal to 1.0 ($\cos\theta_L = 1.0$) is calculated as follows:

$$n = \frac{S \cdot \cos\theta_L}{S \cdot \cos\theta_L + NLL + LL \cdot \left(\dfrac{S}{S_n}\right)^2} = \frac{630000 \cdot 1.0}{630000 \cdot 1.0 + 1200 + 9300 \cdot (1.0)^2} \Rightarrow$$

$n = 0.9836 \Rightarrow n = 98.36\%$.

Case 2 Full load and power factor 0.8
The efficiency at full load and power factor equal to 0.8 is:

$$n = \frac{630000 \cdot 0.8}{630000 \cdot 0.8 + 1200 + 9300 \cdot (1.0)^2} \Rightarrow n = 97.96\%.$$

Case 3 Per-unit load 0.75 and power factor 1.0
The efficiency at 75% load and power factor equal to 1.0 is:

$$n = \frac{472500 \cdot 1.0}{472500 \cdot 1.0 + 1200 + 9300 \cdot (0.75)^2} \Rightarrow n = 98.66\%.$$

Case 4 Per-unit load 0.75 and power factor 0.8
The efficiency at 75% load and power factor equal to 0.8 is:

$$n = \frac{472500 \cdot 0.8}{472500 \cdot 0.8 + 1200 + 9300 \cdot (0.75)^2} \Rightarrow n = 98.33\%.$$

2. The maximum efficiency corresponds to the following per-unit load:

$$L_{opt} = \sqrt{\frac{NLL}{LL}} = \sqrt{\frac{1200}{9300}} = 0.36.$$

The transformer load is:

$$S = L_{opt} \cdot S_n = 0.36 \cdot 630000 \Rightarrow S = 226800 \text{ VA}.$$

1.4 Transformer Electrical Characteristics

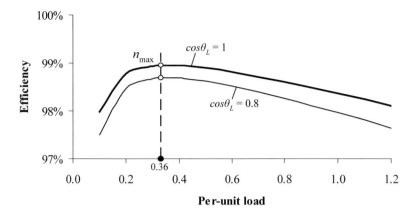

Fig. 1.12 Efficiency curves versus per-unit load

Case 1 Power factor 1.0
The maximum efficiency is:

$$n_{max} = \frac{226800 \cdot 1}{226800 \cdot 1 + 1200 + 9300 \cdot 0.36^2} \Rightarrow n_{max} = 98.95\%.$$

Case 2 Power factor 0.8
The maximum efficiency is:

$$n_{max} = \frac{226800 \cdot 0.8}{226800 \cdot 0.8 + 1200 + 9300 \cdot 0.36^2} \Rightarrow n_{max} = 98.69\%.$$

3. The efficiency curves versus per-unit load are shown in Fig. 1.12.

1.4 Transformer Electrical Characteristics

1.4.1 Rated Power

The rated power (kVA) of a transformer is the output that can be delivered at rated secondary voltage and rated frequency without exceeding the specified temperature rise limitations.

The rated power, S_n, of the three-phase transformer is calculated using the following formula:

$$S_n = \sqrt{3} \cdot U_n \cdot I_n, \tag{1.38}$$

where U_n is the rated voltage and I_n is the rated current of the transformer.

Similarly, the rated power, S_n, of the single-phase transformer is calculated by the following formula:

$$S_n = U_n \cdot I_n. \tag{1.39}$$

1.4.2 Temperature Rise

The temperature rise is the difference between the temperature of the part under consideration (usually the average winding rise or the hottest-spot winding rise) and the ambient temperature.

The *average winding temperature rise* of a transformer is the arithmetic difference between the average winding temperature of the hottest winding and the ambient temperature. The *top-oil temperature rise* is the arithmetic difference between the top-oil temperature (the temperature of the top layer of the insulating liquid in a transformer) and the ambient temperature (IEEE 2002).

Typical characteristics for oil-immersed transformers:

- The average temperature rise of the winding is 65 K, i.e., 65°C above the ambient temperature.

- The top-oil temperature rise is 60 K, i.e., 60°C above the ambient temperature.

1.4.3 Ambient Temperature

The ambient temperature is the temperature of the air into which the heat of the transformer is dissipated.

The rated power of the transformer is typically calculated for the following conditions:

- Maximum ambient temperature of 40°C.

- Average daily ambient temperature of 30°C.

- Average annual ambient temperature of 20°C.

1.4.4 Altitude of Installation

The rated power of a transformer is valid for installation altitudes up to 1000 m. If the transformer is going to be installed at an altitude higher than 1000 m, this should be mentioned in the transformer specification.

1.4.5 Impedance Voltage

The impedance voltage or short-circuit impedance or short-circuit voltage is the percentage of the rated primary voltage that has to be applied at the transformer primary winding, when the secondary winding is short-circuited, in order to have the rated current at the primary winding.

The impedance voltage is very important because it represents the transformer's impedance. The higher the short-circuit impedance, the higher the voltage regulation. The lower the short-circuit impedance, the higher the short-circuit current, in the case of short-circuit. Based on short-circuit impedance, the following are determined: the voltage regulation due to transformer loading, the distribution of loads in the case of parallel operation of transformers, and short-circuit current.

1.4.6 No-Load Losses

Core loss is the power dissipated in the magnetic core subjected to a time-varying magnetizing force. Core loss includes hysteresis and eddy current losses of the core.

No-load losses or excitation losses are incident to the excitation of the transformer. No-load losses include core loss, dielectric loss, conductor loss in the winding due to excitation current, and conductor loss due to circulating current in parallel windings (IEEE 2002).

Table 1.2 presents the three lists (A', B', C') of no-load losses for transformers from 50 to 2500 kVA (CENELEC 1992).

Table 1.2 Lists of no-load losses (CENELEC 1992)

Rated power (kVA)	No-load losses (W)			Short-circuit impedance (%)
	List A'	List B'	List C'	
50	190	145	125	4
100	320	260	210	4
160	460	375	300	4
250	650	530	425	4
400	930	750	610	4
630	1300	1030	860	4
630	1200	940	800	6
1000	1700	1400	1100	6
1600	2600	2200	1700	6
2500	3800	3200	2500	6

Table 1.3 Lists of load losses (CENELEC 1992)

Rated power (kVA)	Load losses (W)			Short-circuit impedance (%)
	List A	List B	List C	
50	1100	1350	875	4
100	1750	2150	1475	4
160	2350	3100	2000	4
250	3250	4200	2750	4
400	4600	6000	3850	4
630	6500	8400	5400	4
630	6750	8700	5600	6
1000	10500	13000	9500	6
1600	17000	20000	14000	6
2500	26500	32000	22000	6

1.4.7 Load Losses

Load losses are incident to the carrying of a specified load. Load losses include $I^2 \cdot R$ loss in the current carrying parts (windings, leads, busbars, bushings), eddy losses in conductors due to eddy currents, and stray loss induced by leakage flux in the tank, core clamps, or other parts (IEEE 2002).

1.4 Transformer Electrical Characteristics

Table 1.3 presents the three lists (A, B, C) of load losses for transformers from 50 to 2500 kVA (CENELEC 1992).

For example, it is said that a transformer has a combination of losses of AC', if its load losses belong to list A and its no-load losses belong to list C'. More specifically, one transformer with rated power of 1000 kVA and combination of losses AC', has load losses equal to 10500 W (Table 1.3) and no-load losses equal to 1100 W (Table 1.2).

1.4.8 Rated Voltages

The rated primary voltage (input voltage) is the voltage at which the transformer is designed to operate. The rated primary voltage determines the *basic insulation level* (BIL) of the transformer, according to international standards (IEC 60076). The BIL is a basic transformer characteristic, since it indicates the ability of the transformer to withstand the overvoltages that can appear in the network. The calculation of the winding insulation is based on the BIL.

The rated secondary voltage (output voltage) is the voltage at the terminals of the secondary winding at no-load, under rated primary voltage and rated frequency.

1.4.9 Vector Group

The vector group determines the phase displacement between the primary and the secondary winding.

The primary or secondary windings can be connected in different ways in order to have a three-phase transformer. These connections are the following:

- **D (d)**: delta connection for primary (secondary) winding

- **Y (y)**: star connection for primary (secondary) winding

- **Z (z)**: zigzag connection for primary (secondary) winding

- **N (n)**: the neutral exists in primary (secondary) winding for connection outside the transformer

1.4.10 Frequency

The frequency at which the transformer is designed to operate is 50 Hz or 60 Hz in accordance with the network frequency.

1.4.11 Noise

The transformer noise is mainly due to the magnetostriction of the sheets of the magnetic circuit (Valkovic 1994). In general, a transformer operating at low magnetic induction has low noise level. Other sources of transformer noise are the windings and the cooling equipment. Transformers located in residential areas should have sound level as low as possible.

1.4.12 Short-Circuit Current

The short-circuit current is composed of the asymmetrical and the symmetrical short-circuit current. The asymmetrical short-circuit current stresses the transformer mechanically, while the symmetrical short-circuit current stresses the transformer thermally.

In the case of sudden short-circuits, mechanical forces increase many times and can be dangerous for the transformer. In some cases, steady state short-circuit current reaches as high as 10 to 15 times the transformer rated current (Mittle and Mittal 1996). Since the mechanical forces are proportional to the square of the current, they increase to as much as 100 to 225 times the mechanical forces at rated current. Such large mechanical forces can cause appreciable damage to the transformer. Hence, the transformer windings must be designed and constructed to withstand the mechanical forces during short-circuits.

1.4.13 No-Load Current

The no-load current or excitation current represents the current that the transformer absorbs, when rated voltage is applied to the primary winding and the secondary winding is open-circuited. The no-load current is expressed as a percentage of the value of the rated primary current.

1.5 Transformer Operation

1.5.1 Overloading

The rated overload of a transformer depends on the transformer's previous load or the corresponding oil temperature at the beginning of the overloading. The permissible duration and the respective levels of the acceptable overload of commercial oil-immersed distribution transformers are shown in Table 1.4 (Schneider Electric AE 2002). For example, if the transformer is loaded to 50% of its rated power continuously, then the transformer can be overloaded to 150% of its rated power for 15 minutes or to 120% of its rated power for 90 minutes, as shown in Table 1.4.

It should be noted that the oil temperature is not a safe measure for the winding temperature, since the time constant of the oil is 2 to 4 hours, while the time constant of the winding is 2 to 6 minutes (Schneider Electric AE 2002). Therefore, the determination of the permissible duration of the overload must be done very carefully, since there is a danger that the winding temperature exceeds the critical temperature of 105°C, without this being indicated by the oil temperature.

1.5.2 Parallel Operation

The parallel operation of two or more transformers is feasible, when the following requirements are met:

- The ratio of their rated power should be less than 3:1.

- Their voltage ratio should be the same (the permitted tolerance is according to IEC 60076-1, Table 1.6).

- Their short-circuit impedance should be the same (the permitted tolerance is according to IEC 60076-1, Table 1.6).

Table 1.4 Permissible duration of overload and level of acceptable overload

Previous continuous load (% of rated power)	Oil temperature (°C)	Level of overload (% of rated power)				
		10%	20%	30%	40%	50%
		Permissible duration of overload (min)				
50	55	180	90	60	30	15
75	68	120	60	30	15	8
90	78	60	30	15	8	4

- Their vector groups should be the same and the connection should be implemented with the corresponding terminals U-u, V-v, W-w. In other words, the transformers must have the same inherent phase angle difference between primary and secondary terminals, the same polarity and the same phase sequence.

1.5.3 Load Distribution to Transformers in Parallel Operation

If parallel operated transformers have the same voltage ratio but different short-circuit impedance, then the load is distributed among them in such a way that each transformer accepts a specific level of load for which the short-circuit impedance becomes the same for all the parallel operated transformers.

When none of the parallel operated transformers is permitted to be overloaded, the transformer with the minimum short-circuit impedance must operate at maximum under its rated power. Consequently, the load distribution is given by the following equation:

$$S_i = S_{n,i} \cdot \frac{U_{k,\min}}{U_{k,i}}, \tag{1.40}$$

where S_i is the load that is distributed to transformer i, $S_{n,i}$ is the rated power of transformer i, $U_{k,i}$ is the rated short-circuit impedance of transformer i, and $U_{k,\min}$ is the minimum rated short-circuit impedance of the n parallel operated transformers.

Finally, the total rated power, S_{tot}, of the n parallel operated transformers is:

$$S_{tot} = \sum_{i=1}^{n} S_i = \sum_{i=1}^{n} S_{n,i} \cdot \frac{U_{k,\min}}{U_{k,i}} \Rightarrow S_{tot} < \sum_{i=1}^{n} S_{n,i}. \tag{1.41}$$

1.5.3.1 Example 1.8

Let us assume that three transformers operate in parallel. The first transformer has 800 kVA rated power and 4.4% short-circuit impedance. The rated power and the short-circuit impedance of the other two transformers is 500 kVA and 4.8%, and 315 kVA and 4.0%, respectively. Calculate the maximum total load of the three transformers.

Solution

Among the three transformers, the third transformer has the minimum short-circuit impedance, i.e., $U_{k,\min} = 4.0\%$.

The load of transformer 1 is:

$$S_1 = S_{n,1} \cdot \frac{U_{k,\min}}{U_{k,1}} = 800 \cdot \frac{4}{4.4} \Rightarrow S_1 = 728 \text{ kVA}.$$

The load of transformer 2 is:

$$S_2 = S_{n,2} \cdot \frac{U_{k,\min}}{U_{k,2}} = 500 \cdot \frac{4}{4.8} \Rightarrow S_2 = 417 \text{ kVA}.$$

The load of transformer 3 is:

$$S_3 = S_{n,3} \cdot \frac{U_{k,\min}}{U_{k,3}} = 315 \cdot \frac{4}{4} \Rightarrow S_3 = 315 \text{ kVA}.$$

The maximum total load of the three transformers is:

$$S_{tot} = S_1 + S_2 + S_3 = 728 + 417 + 315 \Rightarrow S_{tot} = 1460 \text{ kVA}.$$

The three transformers have total installed power:

$$S_{inst} = S_{n,1} + S_{n,2} + S_{n,3} = 800 + 500 + 315 \Rightarrow S_{inst} = 1615 \text{ kVA}.$$

From the above, it is concluded that the maximum total load (1460 kVA) represents 90.4% of the total installed power (1615 kVA).

It should be noted that, in order for the maximum total load to be equal to the total installed power, the transformers must have the same short-circuit impedance.

1.6 Transformer Standards and Tolerances

1.6.1 Transformer Standards

Transformer manufacturing is based on international standards as well as on specific customer needs. From time to time, some of the standards may be modified and in that case they are republished.

Table 1.5 IEC transformer standards

Standard	Description
IEC 60076-1	Power transformers: general
IEC 60076-2	Power transformers: temperature rise
IEC 60076-3	Power transformers: insulation levels and dielectric tests
IEC 60076-5	Power transformers: ability to withstand short circuit
IEC 60137	Bushings for alternating voltages above 1000 V
IEC 60354	Loading guide for oil-immersed power transformers
IEC 60726	Dry-type power transformers
IEC 60905	Loading guide for dry-type power transformers

A list of IEC transformer standards is shown in Table 1.5.

1.6.2 Tolerances

Constructional requirements result in deviations between the measured parameters and the values that are defined in the transformer "offer" (i.e., the guaranteed values). Table 1.6 presents tolerances that are applied to certain items, according to IEC 60076-1.

Table 1.6 Tolerances on certain transformer items according to IEC 60076-1

Item	Tolerance
Voltage ratio	The lower of the following values: a) ±0.5% of guaranteed voltage ratio b) ±1/10 of the measured short-circuit impedance on the principal tapping
Short-circuit impedance (U_k)	a) ±7.5% of the guaranteed U_k, when $U_k \geq 10\%$ b) ±10% of the guaranteed U_k, when $U_k < 10\%$
No-load losses	+15% of the guaranteed no-load losses
Load losses	+15% of the guaranteed load losses
Total losses (load and no-load)	+10% of the guaranteed total losses (load and no-load)
No-load current	+30% of the guaranteed no-load current

1.7 Transformer Tests

Transformer tests are classified, in accordance with IEC 60076-1 standard, as follows:

- Type tests
- Routine tests
- Special tests

1.7.1 Type Tests

Type tests, which are made on one transformer from every transformer type, are the following:

1. **Temperature Rise Test**   The temperature rise test procedure is typically performed according to IEC 60076-2. The objective of this test is to verify guaranteed temperature rises for oil and windings.

2. **Lightning Impulse Test**   The lightning impulse test procedure is typically performed according to IEC 60076-3. This specific test checks if the transformer can withstand overvoltages. These overvoltages are caused by (a) traveling waves (caused by lightning) in transmission lines, (b) sudden on/off switching of breakers, and (c) short-circuits. It should be noted that the lightning impulse test is a routine test for transformers with higher voltage for equipment, U_m, greater than 72.5 kV and a type test for $U_m \leq 72.5$ kV.

1.7.2 Routine Tests

Routine tests are performed on every transformer separately, and include:

1. **Winding Resistance**   The winding resistance is defined as the direct current (dc) resistance of a winding. The procedure for the measurement of windings resistance is typically performed according to IEC 60076-1. During this test the resistance of each winding is measured and the temperature is recorded. The test is performed with direct current.

2. **Voltage Ratio and Check of Phase Displacement**   Measurement of the voltage ratio is typically performed according to IEC 60076-1. The objective of

the test is to compare the measured values of the transformer ratio with the respective guaranteed values. For the transformer, the voltage ratio is equal to the turns ratio, namely:

$$\frac{V_1}{V_2} = \frac{N_1}{N_2}, \qquad (1.42)$$

where V_1 and V_2 is the phase voltage of the primary and secondary winding, respectively, and N_1 and N_2 is the number of turns of the primary and secondary winding, respectively.

3. **Impedance Voltage** Measurement of impedance voltage is typically performed according to IEC 60076-1. The impedance voltage, which is expressed as a percentage of the rated voltage, represents the transformer's impedance. The IEC standard requires the impedance voltage to be calculated at the reference temperature of 75°C. The transformer impedance voltage is guaranteed by the manufacturer and is verified for the customer during the impedance voltage routine test.

4. **Load Loss** The transformer load loss is guaranteed by the manufacturer and is verified for the customer during the load loss routine test. The measurement of load loss is implemented with the secondary winding short-circuited and by increasing the voltage of the primary winding until the current of the primary winding reaches its nominal value. The load losses are calculated at the reference temperature of 75°C according to the IEC standard.

5. **No-Load Current and No-Load Losses** The measurement is typically performed according to IEC 60076-1. The no-load current represents the real value of current that is required to magnetize the magnetic core. The no-load losses represent the power that is absorbed by the transformer core when rated voltage and rated frequency are applied to one winding (e.g., secondary) and the other winding (e.g., primary) is open-circuited.

6. **Dielectric Routine Tests** The dielectric routine tests are the following:

    – **Applied Voltage Dielectric Test** The duration of the test, according to IEC 60076-3, is 1 min. With this specific test, the following are checked: (a) the insulation between primary and secondary windings, (b) the insulation between the tested winding and the tank, and (c) the insulation between the tested winding and the magnetic circuit.

    – **Induced Voltage Dielectric Test** A three-phase voltage, twice the rated voltage, is applied to the transformer for 1 minute. However, the doubling of the voltage will double the magnetic induction resulting in transformer saturation and, consequently, there is a danger of the transformer being destroyed. In order to avoid saturation, the frequency is also doubled, so the magnetic induction remains constant. Consequently, during this test, the volts per turn and therefore the volts per layer are doubled. This test verifies the dielectric strength between turns and layers.

1.7.3 Special Tests

Special tests are not included in the category of type or routine tests and are executed after agreement between customer and manufacturer. The special tests are the following:

1. Dielectric special tests.

2. Determination of capacitances of windings-to-earth and between windings.

3. **Short-Circuit Withstand Test** According to this test, the transformer is subjected to successive short-circuits of 0.5 second duration and the transformer must withstand these short-circuits. Since this test requires high power, it is executed in special test centers.

4. **Determination of Sound Levels** The transformer is energized at no-load and at rated voltage and rated frequency, so the noise peripheral to the transformer can be measured.

5. Measurement of the harmonics of the no-load current.

6. Measurement of insulation resistance and/or measurement of dissipation factor of the insulation system capacitances.

7. Radio interference voltage.

8. Measurement of zero-sequence impedance.

1.8 Transformer Types

The transformers are classified into various categories, according to their:

- Use

- Cooling method

- Insulating medium

- Core construction

1.8.1 Classification According to Transformer Use

Transformers are classified according to their use into the following categories:

1. **Distribution Transformers** They are used in distribution networks in order to transmit energy from the medium voltage (MV) network to the low voltage (LV) network of the consumers. Their rated power usually ranges from 50 to 1600 kVA.

2. **Power Transformers** They are used in high-power generating stations for voltage step-up and in transmission substations for voltage step-up or step-down. Usually they are of power greater than 2 MVA.

3. **Autotransformers** They are used for voltage transformation within relativity small limits, for connection of electric energy systems of various voltages, for starting alternating current motors, etc.

4. **Test Transformers** They are used for the execution of performance tests with high or ultra-high voltage.

5. **Special Power Transformers** They are used for special applications, e.g., in furnaces and in welding.

6. **Instrument Transformers** They are used for the accurate measurement of voltage or current.

7. **Telecommunication Transformers** They are used in telecommunication applications aiming at the reliable reproduction of a signal over a wide range of frequency and voltage.

1.8.2 Classification According to Transformer Cooling Method

The identification of oil-immersed transformers according to the cooling method is expressed by a four-letter code. The first letter expresses the internal cooling medium in contact with the windings. The second letter identifies the circulation mechanism for the internal cooling medium. The third letter identifies the external cooling medium. The fourth letter identifies the circulation mechanism for external cooling medium. For example, if the internal cooling medium is mineral oil, which is circulated by natural flow, and the external cooling medium is air, which is circulated by natural convection, then this cooling method is coded as ONAN (Oil Natural Air Natural).

In power transformers, various cooling methods are used including oil circulation by pumps, or forced air circulation by fans, or both of the above. As a result, the following cooling methods exist:

1. **ONAF**   Oil Natural Air Forced

2. **OFAN**   Oil Forced Air Natural

3. **OFAF**   Oil Forced Air Forced

4. **OFWF**   Oil Forced Water Forced

1.8.3 Classification According to Transformer Insulating Medium

Transformers are classified according to their insulating medium into the following categories:

1. **Oil-Immersed Transformers**   The insulating medium is mineral oil or synthetic (silicon) oil.

2. **Dry Type Transformers**   The cooling is implemented with natural air circulation and the windings are usually insulated with materials of H or F class. The materials of H class are designed to operate, in normal conditions, at temperatures up to 180°C and the materials of F class at temperatures up to 155°C.

3. **Resin Type Transformers**   The resin type transformer is a dry type transformer insulated with epoxy resin cast under vacuum.

1.8.4 Classification According to Transformer Core Construction

Construction of the magnetic circuit of three-phase transformers can be implemented, alternatively, as follows:

1. **With three legs (vertical limbs)**   The magnetic flux of one leg must flow through the other two legs and the flux also flows through the windings of the other phases, i.e., the transformer has no free return of the flux.

2. **With five legs (vertical limbs)**   Free return of the flux through the external legs.

There are two different technologies for stacking the sheets of the magnetic material of the core:

1. **Stack Core**   The layers of the sheets of the magnetic material are placed one over the other and vertical and horizontal layers are overlapped.

2. **Wound Core**   The magnetic circuit is of shell type and the sheets are wound.

Two different materials are used for core construction:

1. **Silicon Steel Sheet**   The silicon steel sheet that is used for core construction is an alloy consisting of 97% iron and 3% silicon. This material is crystalline. The silicon steel sheets have thickness from 0.18 to 0.5 mm. There are also silicon steel sheets for operation at high magnetic induction (Hi-B).

2. **Amorphous Metal Sheet**   The amorphous metal sheet that is used for core construction is an alloy consisting of 92% iron, 5% silicon and 3% boron. This material is not crystalline. It has 70% lower no-load loss than silicon steel. The thickness of the amorphous metal sheet is 0.025 mm, i.e., it is about 10 times thinner than the typical thickness of silicon steel sheet.

1.9 Transformers Studied in this Book

The rest of this book studies three-phase, wound core type distribution transformers, mineral oil-immersed with ONAN cooling method, with voltages up to 33 kV and rated power up to 2000 kVA.

The methods that are presented in this book can also be applied to other transformer types, e.g., power transformers, single-phase transformers, dry type transformers, stack core transformers, etc. In order to enable readers to apply the methods discussed in this book to other transformer types, the rest of the book is structured as follows:

1. Chapter 2 presents, step-by-step, the conventional design methodology of three-phase wound core type transformers.

2. The second part of the book presents the basic principles of the methods that will be used in the third part of the book to solve transformer design problems. In particular, the second part of the book briefly presents the following methods:

    –   The finite element method (Chap. 3).

    –   Artificial intelligence methods for classification and forecasting (Chap. 4).

- Deterministic and stochastic optimization methods (Chap. 5).

3. The third part of the book (Chaps. 6 to 8) solves challenging design problems for three-phase wound core type transformers using the methods of the second part of the book.

References

CENELEC (1992) Three phase oil-immersed distribution transformers 50 Hz, from 50 to 2500 kVA with highest voltage for equipment not exceeding 36 kV. CENELEC Harmonization Document 428.1 S1, CENELEC, Brussels, Belgium

Chapman SJ (2005) Electric machinery fundamentals, 4th edn. McGraw-Hill, New York

Del Vecchio RM, Poulin B, Feghali PT, Shah DM, Ahuja R (2002) Transformer design principles with applications to core-form power transformers. CRC Press, Boca Raton, Florida

Fitzgerald AE, Kingsley C Jr, Umans SD (1990) Electric machinery, 5th edn. McGraw-Hill, New York

IEEE (2002) IEEE standard terminology for power and distribution transformers. IEEE Standard C57.12.80, IEEE, New York

Kennedy BW (1998) Energy efficient transformers. McGraw-Hill, New York

Krause PC, Wasynczuk O, Sudhoff SD (2002) Analysis of electric machinery and drive systems, 2nd edn. IEEE Press, Piscataway, NJ

MIT (1962) Magnetic circuits and transformers, 14th edn. John Wiley and Sons, New York

Mittle VN, Mittal A (1996) Design of electrical machines, 4th edn. Standard Publishers Distributors, Nai Sarak, Delhi

Schneider Electric AE (2002) Use and maintenance of ELVIM oil-immersed distribution transformers. Schneider Electric AE, Athens, Greece

Valkovic Z (1994) Effect of electrical steel grade on transformer core audible noise. Journal of Magnetism and Magnetic Materials 133:607–609

2 Conventional Transformer Design

Abstract This chapter deals with conventional design of wound core type transformers. It formulates the transformer design optimization (TDO) problem and solves it using a multiple design method that is commonly referred to as the conventional TDO method. A design example of an actual commercial transformer is worked out throughout this chapter showing all the calculations that are needed to design a transformer. The example-driven presentation of the conventional TDO method makes this chapter unique in transformer design texts.

2.1 Nomenclature

A	No-load loss cost rate throughout the transformer lifetime ($/W)
$area_{HV}$	Cross-section area of high voltage (HV) conductor (mm^2). See Fig. 2.6
$area_{LV}$	Cross-section area of low voltage (LV) conductor (mm^2). See Fig. 2.6
B	Load loss cost rate throughout the transformer lifetime ($/W)
BIL_{HV}	Basic insulation level of HV winding (kV)
BIL_{LV}	Basic insulation level of LV winding (kV)
BLD_{HV}	Thickness of HV winding (mm). See Fig. 2.6
BLD_{LV}	Thickness of LV winding (mm). See Fig. 2.6
BP	Transformer bid price ($)
$CCEE$	Core to coil each end (mm). See Fig. 2.6
C_{Lab}	Labor cost to manufacture the transformer ($)
CM	Cost of transformer materials ($)
CMM	Cost of transformer main materials ($)
CPA	Area of corrugated panels (m^2)
CRM	Cost of the remaining materials (i.e., not the main materials) of transformer ($)
CSA	Cross-section area (mm^2)
CSF	Core stacking factor
CTM	Transformer manufacturing cost ($)
D	Width of core leg (mm). See Fig. 2.3
$D13$	Coil equivalent external diameter immediately after the HV winding, including the HV cooling ducts (mm). See Fig. 2.9
$D3$	Coil equivalent external diameter immediately after the tube paper (mm). See Fig. 2.9

$D7$	Coil equivalent external diameter immediately after the LV winding, including the LV cooling ducts (mm). See Fig. 2.9
$D9$	Coil equivalent external diameter immediately after the I_{HV-LV} insulation (mm). See Fig. 2.9
d_{HV}	Diameter of HV conductor (mm). See Fig. 2.5
D_{HV-C}	Distance between HV winding and core (mm). See Fig. 2.6
D_{LV-C}	Distance between LV winding and core (mm). See Fig. 2.6
D_{Panel}	Width of corrugated panel (mm). See Fig. 2.15
$Ducts_{HV}$	Number of ducts of HV winding
$Ducts_{LV}$	Number of ducts of LV winding
D_w	Width of cooling duct (mm)
$DWPG_{HV}$	Width of HV duct strip plus gap (mm)
$DWPG_{LV}$	Width of LV duct strip plus gap (mm)
EdL_{HV}	Eddy current loss of HV winding (W)
EdL_{LV}	Eddy current loss of LV winding (W)
E_u	Thickness of core leg (mm). See Fig. 2.3
f	Frequency (Hz)
$F1$	Window width of small individual core (mm). See Fig. 2.3
$F2$	Window width of large individual core (mm). See Fig. 2.3
FD_{max}	Maximum flux density (Gauss)
G	Height of core window (mm). See Fig. 2.3
g_{CP}	Weight per unit area of corrugated panels (kg/m^2)
g_{DS}	Mass density of duct strips (kg/m^3)
g_{HV}	Mass density of HV conductor (kg/m^3)
g_{LV}	Mass density of LV conductor (kg/m^3)
g_{MM}	Mass density of magnetic material (kg/m^3)
g_O	Mass density of mineral oil (kg/m^3)
HCP	Height of corrugated panel (mm). See Fig. 2.15
HV	High voltage
$HVCC$	Connection of external (HV) winding
I_{HV-HV}	Insulation outside HV winding (mm). See Fig. 2.6
I_{HVL}	Insulation between layers of HV winding (mm). See Fig. 2.5
I_{HV-LV}	Insulation between LV and HV winding (mm). See Fig. 2.6
I_{LV-C}	Insulation between LV winding and core (mm). See Fig. 2.6
I_{LVL}	Insulation between layers of LV winding (mm). See Fig. 2.4
$Impulse_{max}$	Maximum impulse voltage that an insulating paper can withstand (kV)
$Induced_{max}$	Maximum induced voltage that an insulating paper can withstand (kV)
I^p_{LV}	Phase current of LV winding (A)
IR	Ohmic (resistive) part of impedance voltage (%)
IX	Inductive part of impedance voltage (%)
K	Distance between two adjacent cores (mm). See Fig. 2.3
$Layers_{HV}$	Number of layers of HV winding

2.1 Nomenclature

$Layers_{LV}$	Number of layers of LV winding
$LDSP_{HV}$	Layer direction space factor of HV winding
$LDSP_{LV}$	Layer direction space factor of LV winding
LG_{HV}	Dimension of cooling ducts of HV winding (mm). A practical computation formula is given in Example 2.5
LL_1	Transformer load loss (W) at voltage $V_{HV,1}^l$
LL_2	Transformer load loss (W) at minimum voltage of HV winding
LL_g	Guaranteed load loss (W)
$LL_{HV,1}$	Load loss (W) of HV winding at voltage $V_{HV,1}^l$
$LL_{HV,2}$	Load loss (W) of HV winding at minimum voltage of HV winding
LL_{LV}	Load loss (W) of LV winding
LV	Low voltage
LVCC	Connection of internal (LV) winding
ML	Length of the mould of the coil. See Fig. 2.10
MS	Margin ($) in the sale of transformer
MT_{HV}	Mean turn length of HV winding (mm). See Fig. 2.10
MT_{LV}	Mean turn length of LV winding (mm). See Fig. 2.10
MW	Width of the mould of the coil (mm). See Fig. 2.10
NCP	Total number of corrugated panels
NLL	Transformer no-load loss (W)
NLL_g	Guaranteed no-load loss (W)
OH	Height of mineral oil (mm)
Pitch	Distance between two adjacent corrugated panels (mm). See Fig. 2.15
S_n	Transformer rated power (kVA)
SM	Transformer sales margin (%)
$SNLL_{TF}$	Transformer specific no-load loss (W/kg). See Fig. 2.8
$t_{a,max}$	Maximum ambient temperature (°C)
TAOR	Transformer average oil rise (°C)
$Taps_{HV,max}$	Upper limit of taps of HV winding (%)
$Taps_{HV,min}$	Lower limit of taps of HV winding (%)
TD_{HV}	Width of HV layer (mm). See Fig. 2.5
TD_{LV}	Width of LV layer (mm). See Fig. 2.4
TDO	Transformer design optimization
T_{DS}	Thickness of duct strips without insulation (mm)
$TDSP_{HV}$	Turn direction space factor of HV winding
$TDSP_{LV}$	Turn direction space factor of LV winding
TE	Tolerances and elongation of coil (mm)
TH	Tank height (mm). See Fig. 2.11
TL	Tank length (mm). See Fig. 2.11
TLC	Total length of the coil (mm)
TLT_{HV}	Total thickness of the HV leads (mm)

TLT_{LV}	Total thickness of the LV leads (mm)
t_{LV}	Thickness of LV conductor (mm). See Fig. 2.4
$t_{o,\max}$	Maximum oil temperature (°C)
TOC	Transformer total owning cost ($) throughout transformer lifetime
$Turns_{HV,\max}$	Maximum number of turns of HV winding
$turns_{LV}$	Number of turns of LV winding
$TurnsMain_{HV}$	Number of turns of HV winding at voltage $V_{HV,1}^l$
$TurnWidth_{HV}$	Width of HV conductor with insulation (mm)
$TurnWidth_{LV}$	Width of LV conductor with insulation (mm)
TW	Tank width (mm). See Fig. 2.11
$t_{w,\max}$	Maximum winding temperature (°C)
TI_{HV}	Insulation of taps of HV winding (mm)
uc_1	Unit cost of LV winding ($/kg)
uc_2	Unit cost of HV winding ($/kg)
uc_3	Unit cost of magnetic material ($/kg)
uc_4	Unit cost of insulating paper ($/kg)
uc_5	Unit cost of duct strips ($/kg)
uc_6	Unit cost of mineral oil ($/kg)
uc_7	Unit cost of sheet steel ($/kg)
uc_8	Unit cost of corrugated panels ($/kg)
U_k	Impedance voltage (%)
$U_{k,g}$	Guaranteed impedance voltage (%)
V_{CT}	Volume of oil conservator (L)
$V_{HV,1}^l$	First rated line voltage (V) of HV winding
$V_{HV,2}^l$	Second rated line voltage (V) of HV winding
V_{LV}^l	Rated line voltage (V) of LV winding
V_{LV}^p	Rated phase voltage (V) of LV winding
VPT	Volts per turn (V/turn)
uc_1	Total weight of LV winding (kg)
uc_2	Total weight of HV winding (kg)
uc_3	Total weight of magnetic material (kg)
uc_4	Total weight of insulating paper (kg)
uc_5	Total weight of duct strips (kg)
uc_6	Total weight of mineral oil (kg)
uc_7	Total weight of sheet steel (kg)
uc_8	Total weight of corrugated panels (kg)
Δd_{HV}	Insulation of HV conductor (mm). See Fig. 2.5
ρ_{HV}	Resistivity of HV conductor ($\Omega \cdot \text{mm}^2 / \text{m}$)
ρ_{LV}	Resistivity of LV conductor ($\Omega \cdot \text{mm}^2 / \text{m}$)

2.2 Introduction

The objective of *transformer design optimization* (TDO) is to design the transformer so as to minimize the transformer manufacturing cost, i.e., the sum of materials cost plus labor cost, subject to constraints imposed by international standards and transformer user specification.

The aim of transformer design is to obtain the dimensions of all parts of the transformer in order to supply these data to the manufacturer. The transformer design should be carried out based on the specification given, using available materials economically in order to achieve low cost, low weight, small size and good operating performance.

The transformer design is worked out using various methods based on accumulated experience realized in different formulas, equations, tables and charts. Transformer design methods vary among transformer manufacturers (Mittle and Mittal 1996).

While designing a transformer, much emphasis should be placed on lowering its cost by saving materials and reducing to a minimum labor-consuming operations in its manufacture. The design should be satisfactory with respect to dielectric strength and mechanical endurance, and windings must withstand dynamic and thermal stresses in the event of short-circuit.

In order to meet the above requirements, the transformer designer should be familiar with the prices of basic materials used in the transformer. He should also be familiar with the amount of labor consumed in the production of transformer parts and assemblies.

This chapter presents a conventional transformer design methodology based on a multiple design technique (Georgilakis et al. 2007) for solution of the TDO problem. This conventional transformer design method is a heuristic technique that assigns many alternative values to the design variables so as to generate a large number of alternative designs and finally to select the design that satisfies all the problem constraints with minimum manufacturing cost.

2.3 Problem Formulation

The TDO problem is formulated as follows: minimize an objective function subject to several constraints.

Among the various objective functions of the TDO problem that are defined in Sect. 2.3.1, the most commonly used objective functions are:

1. The minimization of transformer manufacturing cost. This is mainly used when designing transformers for industrial and commercial users, since these users usually do not evaluate the cost of losses when purchasing transformers.

2. The minimization of transformer total owning cost. This is mainly used when designing transformers for electric utilities, since these users usually evaluate the cost of losses when purchasing transformers.

The constraints of the TDO problem are related to transformer operation, manufacturing capabilities, and transformer user special needs. These constraints are presented in Sect. 2.3.2.

2.3.1 Objective Function

In the bibliography of transformer design, several objective functions are optimized:

1. Minimization of active part mass (Jabr 2005)

2. Minimization of active part cost (Rubaai 1994, Amoiralis et al. 2008)

3. Minimization of main materials cost (Amoiralis et al. 2009)

4. Minimization of manufacturing cost (Odessey 1974; Georgilakis et al. 2007; Georgilakis 2008; Georgilakis 2009)

5. Minimization of total owning cost (Andersen 1991; Del Vecchio et al. 2002)

6. Maximization of transformer rated power (Judd and Kressler 1977, Jabr 2005)

These objective functions are analyzed in the following paragraphs.

2.3.1.1 Active Part Mass

The objective is to minimize the active part mass, APM:

$$\min APM = \min \sum_{i=1}^{3} w_i , \qquad (2.1)$$

where w_1 (kg) is the total weight of the low voltage (LV) winding, w_2 is the total weight of the high voltage (HV) winding, and w_3 is the total weight of the magnetic material.

2.3 Problem Formulation

2.3.1.2 Active Part Cost

The objective is to minimize the active part cost, APC:

$$\min APC = \min \sum_{i=1}^{3} uc_i \cdot w_i ,\qquad(2.2)$$

where uc_1 ($/kg) is the unit cost of the LV winding, uc_2 is the unit cost of the HV winding, uc_3 is the unit cost of magnetic material, w_1 (kg) is the total weight of the LV winding, w_2 is the total weight of the HV winding, and w_3 is the total weight of magnetic material.

2.3.1.3 Main Materials Cost

The objective is to minimize the cost of transformer main materials, CMM:

$$\min CMM = \min \sum_{i=1}^{8} uc_i \cdot w_i ,\qquad(2.3)$$

where uc_1 ($/kg) is the unit cost of the LV winding, uc_2 is the unit cost of the HV winding, uc_3 is the unit cost of magnetic material, uc_4 is the unit cost of insulating paper, uc_5 is the unit cost of duct strips, uc_6 is the unit cost of mineral oil, uc_7 is the unit cost of sheet steel, uc_8 is the unit cost of corrugated panels, w_1 (kg) is the total weight of the LV winding, w_2 is the total weight of the HV winding, w_3 is the total weight of magnetic material, w_4 is the total weight of insulating paper, w_5 is the total weight of duct strips, w_6 is the total weight of mineral oil, w_7 is the total weight of sheet steel, and w_8 is the total weight of corrugated panels.

2.3.1.4 Manufacturing Cost

The objective is to minimize the cost of transformer manufacturing, CTM:

$$\min CTM = \min \left[CMM + CRM + C_{Lab} \right],\qquad(2.4)$$

where CMM ($) is the cost of transformer main materials as computed by (2.3), CRM is the cost of the remaining materials (not included in CMM) of the transformer, and C_{Lab} is the labor cost to manufacture the transformer.

2.3.1.5 Total Owning Cost

The objective is to minimize the transformer total owning cost, TOC, which includes the cost to purchase the transformer and the cost of losses throughout the transformer lifetime:

$$\min TOC = \min\left[BP + A \cdot NLL + B \cdot LL\right], \qquad (2.5)$$

where A ($/W) is the no-load loss cost, B ($/W) is the load loss cost, NLL (W) is the no-load loss, LL (W) is the load loss, and BP ($) is the transformer purchase cost (also called sales price or bid price) that is computed as follows:

$$BP = \frac{CTM}{1-SM} = \frac{CMM + CRM + C_{Lab}}{1-SM}, \qquad (2.6)$$

where CTM ($) is the transformer manufacturing cost, SM (%) is the transformer sales margin, CMM ($) is the cost of transformer main materials as computed by (2.3), CRM ($) is the cost of the remaining materials (not included in CMM) of the transformer, and C_{Lab} ($) is the labor cost to manufacture the transformer.

The A and B coefficients of (2.5) are computed according to the methodologies of Chap. 8.

2.3.1.6 Rated Power

The objective is to maximize transformer rated power, S_n:

$$\max S_n. \qquad (2.7)$$

2.3.2 Constraints

The constraints of the TDO problem are the following:

1. Induced voltage constraint

2. Turns ratio constraint

3. No-load loss constraint

2.3 Problem Formulation

4. Load loss constraint

5. Total loss constraint

6. Impedance voltage constraint

7. Magnetic induction constraint

8. Heat transfer constraint

9. Temperature rise constraint

10. Efficiency constraint

11. No-load current constraint

12. Voltage regulation constraint

13. Induced voltage constraints

14. Impulse voltage constraints

15. Tank dimensions constraints

These constraints are analyzed in the following paragraphs.

2.3.2.1 Induced Voltage Constraint

This expresses the relation between the induced voltage in the secondary winding and the maximum flux density:

$$V_2 = 4.44 \cdot f \cdot N_2 \cdot FD_{max} \cdot CSF \cdot D \cdot 2 \cdot E_u, \qquad (2.8)$$

where V_2 is the effective value of the induced phase voltage in the secondary winding, f is the frequency, N_2 is the number of turns of the secondary winding, FD_{max} is the maximum flux density, CSF is the core stacking factor, D is the width of the core leg, and E_u is the thickness of the core leg. In (2.8), the quantity $CSF \cdot D \cdot 2 \cdot E_u$ expresses the effective core cross-section area of the magnetic flux in the shell-type transformer. Section 2.6.2 explains how (2.8) is obtained.

2.3.2.2 Turns Ratio Constraint

The voltage ratio is equal to the turns ratio:

$$\frac{V_1}{V_2} = \frac{N_1}{N_2}, \qquad (2.9)$$

where V_1 and V_2 are the phase voltage of the primary and secondary winding, respectively, and N_1 and N_2 are the number of turns of the primary and secondary winding, respectively.

2.3.2.3 No-Load Loss Constraint

The designed no-load loss, NLL, must be smaller than a maximum no-load loss, NLL_{max}:

$$NLL < NLL_{max}. \qquad (2.10)$$

2.3.2.4 Load Loss Constraint

The designed load loss, LL, must be smaller than a maximum load loss, LL_{max}:

$$LL < LL_{max}. \qquad (2.11)$$

2.3.2.5 Total Loss Constraint

The total loss, TTL, of the transformer must be smaller than a maximum total loss, TTL_{max}:

$$TTL < TTL_{max}. \qquad (2.12)$$

It should be noted that the total loss of the transformer is equal to the sum of its no-load loss and load loss, i.e.:

$$TTL = NLL + LL. \qquad (2.13)$$

2.3.2.6 Impedance Voltage Constraint

The transformer impedance voltage, U_k, must be between a minimum impedance voltage, $U_{k,\min}$, and a maximum impedance voltage, $U_{k,\max}$:

$$U_{k,\min} < U_k < U_{k,\max}. \tag{2.14}$$

2.3.2.7 Flux Density Constraint

The maximum flux density, $FD_{\max}$, is required to be smaller than a saturation flux density, FD_{sat} (Judd and Kressler 1977):

$$FD_{\max} < FD_{sat}. \tag{2.15}$$

2.3.2.8 Heat Transfer Constraint

The total heat (W) produced by the total loss, TTL, of the transformer must be smaller than the total heat (W), TH_{CCR}, that can be carried away by the combined effects of conduction, convection, and radiation (MIT 1962):

$$TTL < TH_{CCR}. \tag{2.16}$$

2.3.2.9 Temperature Rise Constraint

The transformer temperature rise, ΔT, that is due to the heat generated by the total loss of the transformer must be smaller than a maximum temperature rise, $\Delta T_{\max}$ (Odessey 1974):

$$\Delta T < \Delta T_{\max}. \tag{2.17}$$

2.3.2.10 Efficiency Constraint

The transformer efficiency, n, is sometimes required to be greater than a minimum efficiency, $n_{\min}$:

$$n > n_{\min}. \tag{2.18}$$

2.3.2.11 No-Load Current Constraint

The transformer no-load current, i_φ, is sometimes required to be smaller than a maximum no-load current, $i_{\varphi,\max}$:

$$i_\varphi < i_{\varphi,\max}. \tag{2.19}$$

2.3.2.12 Voltage Regulation Constraint

The transformer voltage regulation, ΔV, is sometimes required to be smaller than a maximum voltage regulation, $\Delta V_{\max}$ (Odessey 1974):

$$\Delta V < \Delta V_{\max}. \tag{2.20}$$

2.3.2.13 Induced Voltage Constraints

The thickness of layer insulation must withstand the induced voltage test. In particular, the induced voltage in the internal winding, $Induced_{LV}$, is required to be smaller than the maximum induced voltage, $Induced_{LV,\max}$, that the insulating paper between the layers of the internal winding can withstand:

$$Induced_{LV} < Induced_{LV,\max}. \tag{2.21}$$

The induced voltage in the external winding, $Induced_{HV}$, is required to be smaller than the maximum induced voltage, $Induced_{HV,\max}$, that the insulating paper between the layers of the external winding can withstand:

$$Induced_{HV} < Induced_{HV,\max}. \tag{2.22}$$

2.3 Problem Formulation

2.3.2.14 Impulse Voltage Constraints

The thickness of layer insulation must withstand the impulse voltage test. In particular, the impulse voltage in the internal winding, $Impulse_{LV}$, is required to be smaller than the maximum impulse voltage, $Impulse_{LV,max}$, that the insulating paper between the layers of the internal winding can withstand:

$$Impulse_{LV} < Impulse_{LV,max}. \qquad (2.23)$$

The impulse voltage in the external winding, $Impulse_{HV}$, is required to be smaller than the maximum impulse voltage, $Impulse_{HV,max}$, that the insulating paper between the layers of the external winding can withstand:

$$Induced_{HV} < Induced_{HV,max}. \qquad (2.24)$$

2.3.2.15 Tank Dimensions Constraints

The tank length, TL, is sometimes required to be smaller than a maximum tank length, TL_{max}:

$$TL < TL_{max}, \qquad (2.25)$$

the tank width, TW, is sometimes required to be smaller than a maximum tank width, TW_{max}:

$$TW < TW_{max}, \qquad (2.26)$$

and the tank height, TH, is sometimes required to be smaller than a maximum tank height, TH_{max}:

$$TH < TH_{max}. \qquad (2.27)$$

2.3.3 Mathematical Formulation of the TDO Problem

One common formulation of the TDO problem is the minimization of transformer manufacturing cost (Georgilakis et al. 2007):

$$\min CTM = \min\left[CRM + C_{Lab} + \sum_{i=1}^{8} uc_i \cdot w_i\right], \quad (2.28)$$

subject to the following constraints:

$$V_2 = 4.44 \cdot f \cdot N_2 \cdot FD_{max} \cdot CSF \cdot D \cdot 2 \cdot E_u, \quad (2.29)$$

$$\frac{V_1}{V_2} = \frac{N_1}{N_2}, \quad (2.30)$$

$$NLL < NLL_{max}, \quad (2.31)$$

$$LL < LL_{max}, \quad (2.32)$$

$$TTL < TTL_{max}, \quad (2.33)$$

$$U_{k,min} < U_k < U_{k,max}, \quad (2.34)$$

$$FD_{max} < FD_{sat}, \quad (2.35)$$

$$TTL < TH_{CCR}, \quad (2.36)$$

$$\Delta T < \Delta T_{max}, \quad (2.37)$$

$$Induced_{LV} < Induced_{LV,max}, \quad (2.38)$$

$$Induced_{HV} < Induced_{HV,max}, \quad (2.39)$$

$$Impulse_{LV} < Impulse_{LV,max}, \quad (2.40)$$

$$Induced_{HV} < Induced_{HV,max}, \quad (2.41)$$

$$TL < TL_{max}, \quad (2.42)$$

$$TW < TW_{max}, \quad (2.43)$$

$$TH < TH_{max}. \quad (2.44)$$

2.3.4 Characteristics of the TDO Problem

The TDO problem is a complex constrained mixed-integer nonlinear programming problem. The TDO problem is further complicated by the fact that the objective function is discontinuous (Andersen 1991).

2.4 Conventional Transformer Design Optimization Method

2.4.1 Methodology

This section describes a conventional heuristic methodology for solution of the TDO problem of Sect. 2.3.3. This methodology, also known as multiple design method, is a heuristic technique that assigns many alternative values to the design variables so as to generate a large number of alternative designs and finally to select the design that satisfies all the problem constraints with minimum manufacturing cost.

The methodology concerns the optimization of transformers with the following technical characteristics:

1. Three-phase, oil-immersed distribution transformers

2. Magnetic circuit of shell type and wound cores

3. Foil, round wire, or rectangular wire technology for both low voltage (LV) and high voltage (HV) conductors

The process of finding the optimum transformer is implemented with the help of a suitable computer program, which uses at maximum 134 input parameters in order to make the transformer program as parametric as possible. These 134 input parameters are split into the following seven types (Georgilakis et al. 2007):

1. *Description variables* (e.g., rated power, rated low voltage (LV) and high voltage (HV), frequency, LV and HV connection). Table 2.6 shows an example of description variables.

2. *Special variables* (e.g., core stacking factor, turns direction space factor, mass density of materials used). Table 2.7 shows an example of special variables.

3. *Default variables* (e.g., LV and HV taps, guaranteed no-load loss, load loss, and impedance voltage). Table 2.8 shows an example of default variables.

4. *Cost variables* (e.g., unit cost for LV and HV conductor, magnetic steel, mineral oil, insulating paper, duct strips, corrugated panels). Table 2.9 shows an example of cost variables.

5. *Various variables* (e.g., number of LV and HV ducts). Table 2.10 shows an example of various variables.

6. *Conductor cross-section calculation variables* (LV and HV conductor cross-sections can be defined by the user or can be calculated using current density,

or thermal short-circuit test). Table 2.11 shows an example of conductor cross-section calculation variables.

7. *Design variables* (i.e., number of LV turns, width of core leg, height of core window, magnetic induction, LV and HV cross-section area). It should be noted that the magnetic material properties (e.g., type, grade, thickness, specific no-load loss) are given as input data when defining the values of magnetic induction within the design variables. Table 2.12 shows an example of design variables.

The computer program allows many variations in design variables. These variations permit the investigation of sufficient candidate solutions. For each one of the candidate solutions, it is checked if all the specifications (constraints) are satisfied, and if they are satisfied, the manufacturing cost is estimated and the solution is characterized as *acceptable*. On the other hand, the candidate solutions that violate the specification are characterized as *non-acceptable* solutions. Finally, among the acceptable solutions, the transformer with the minimum manufacturing cost is selected, which is the optimum transformer.

There are six design variables:

1. The number of turns of low voltage coil, $turns_{LV}$

2. The width of core leg, D

3. The magnetic induction, FD_{max}

4. The height of core window, G

5. The cross-section area of the LV conductor, $area_{LV}$

6. The cross-section area of the HV voltage conductor, $area_{HV}$

Giving n_{LV} different values for the number of turns of low voltage coil, n_D values for the width of core leg, n_{FD} tries for the magnetic induction, n_G different values for the height of core window, cs_{LV} different values for the calculation of cross-section area of low voltage coil and cs_{HV} different values for the calculation of cross-section of high voltage coil, the total candidate solutions (loops of the computer program), n_{loops}, are calculated from the following equation:

$$n_{loops} = n_{LV} \cdot n_D \cdot n_{FD} \cdot n_G \cdot cs_{LV} \cdot cs_{HV} . \tag{2.45}$$

The search algorithm of the optimum transformer is presented in Table 2.1 (Georgilakis et al. 2007). This methodology is already applied in the transformer manufacturing industry.

2.4 Conventional Transformer Design Optimization Method

Table 2.1 Conventional TDO method

For $i = 1$ to n_{LV}
   For $j = 1$ to n_D
      For $k = 1$ to n_{FD}
         For $l = 1$ to n_G
            Calculate the volts per turn and the thickness of core leg based on Sect. 2.6.
            For $m = 1$ to cs_{LV}
               For $n = 1$ to cs_{HV}
                  Calculate the layer insulations based on Sect. 2.7.
                  If one or more of the constraints of (2.38), (2.39), (2.40), and (2.41) is violated, then the solution is rejected and the next loop is executed.
                  Calculate winding and core dimensions based on Sect. 2.8.
                  Calculate core weight and no-load loss based on Sect. 2.9.
                  If the no-load loss constraint of (2.31) is violated, then the solution is rejected and the next loop is executed.
                  Calculate the inductive part of impedance voltage based on Sect. 2.10.
                  Calculate the load loss based on Sect. 2.11.
                  If the load loss constraint of (2.32) is violated, then the solution is rejected and the next loop is executed.
                  If the total loss constraint of (2.33) is violated, then the solution is rejected and the next loop is executed.
                  Calculate the impedance voltage based on Sect. 2.12.
                  If the impedance voltage constraint of (2.34) is violated, then the solution is rejected and the next loop is executed.
                  Calculate the coil length based on Sect. 2.13.
                  Calculate the tank dimensions based on Sect. 2.14.
                  If one or more of tank constraints of (2.42), (2.43), and (2.44) is violated, then the solution is rejected and the next loop is executed.
                  Calculate the winding gradient (temperature rise) and the oil gradient based on Sect. 2.15.
                  If the temperature rise constraint of (2.37) is violated, then the solution is rejected and the next loop is executed.
                  Calculate the heat that can be dissipated based on Sect. 2.16.
                  If the heat transfer constraint of (2.36) is violated, then the solution is rejected and the next loop is executed.
                  Calculate the weight of insulating materials based on Sect. 2.17.
                  Calculate the weight of duct strips based on Sect. 2.18.
                  Calculate the weight of mineral oil based on Sect. 2.19.
                  Calculate the weight of sheet steel based on Sect. 2.20.
                  Calculate the weight of corrugated panels based on Sect. 2.21.
                  Calculate the cost of transformer main materials based on Sect. 2.22.
                  Calculate the transformer manufacturing cost based on Sect. 2.23.
The optimum transformer is the one with the minimum manufacturing cost.

2.4.2 Case Study

2.4.2.1 Transformer Design Data

The efficiency of the TDO methodology of Sect. 2.4.1 is presented through an actual design example of a three-phase, 160 kVA, 20/0.4 kV (i.e., primary voltage 20 kV and secondary voltage 0.4 kV), 50 Hz, transformer. The internal (LV) winding is star-connected (Y) and the external (HV) winding is delta-connected (Δ).

The maximum load loss is 2350 W and the maximum no-load loss is 425 W. The impedance voltage is 4% with tolerance ±10%.

Copper sheet is used for the low voltage conductor and copper wire is used for the high voltage conductor.

The calculation of the cross-section area of the conductors is implemented using the current density. More specifically, current density of 3.2 A/mm² is chosen for the internal coil and current density of 3.7 A/mm² is chosen for the external coil.

2.4.2.2 Selection of the Values of Design Variables

For the number of turns of the low voltage conductor, eight alternative values (i.e., $n_{LV} = 8$) are considered: 28, 29, 30, 31, 32, 33, 34, and 35 turns.

For the width of core leg (*D*), two alternative values (i.e., $n_D = 2$) are selected: 170 and 190 mm.

For the magnetic induction, seven alternative values (i.e., $n_{FD} = 7$) are used: 14000, 14500, 15000, 15500, 16000, 16500, and 17000 Gauss. For each one of these seven different values of the magnetic induction, the respective specific no-load losses (W/kg) of the transformer are given.

For the height of core window (*G*), 10 alternative values (i.e., $n_G = 10$) are considered: 190, 195, 200, 205, 210, 215, 220, 225, 230, and 235 mm.

For each value of *G*, only one cross-section area of the LV conductor is used, i.e., $cs_{LV} = 1$.

Four values for the cross-section area of the external conductor are considered, i.e., $cs_{HV} = 4$.

The total number of candidate solutions is computed using (2.45):

$$n_{loops} = n_{LV} \cdot n_D \cdot n_{FD} \cdot n_G \cdot cs_{LV} \cdot cs_{HV} \Rightarrow$$

$$n_{loops} = 8 \cdot 2 \cdot 7 \cdot 10 \cdot 1 \cdot 4 \Rightarrow n_{loops} = 4480.$$

2.4 Conventional Transformer Design Optimization Method

Table 2.2 Acceptable solutions sorted by manufacturing cost (*CTM*)

Number	$turns_{LV}$	D (mm)	FD_{max} (Gauss)	G (mm)	$area_{LV}$ (mm^2)	$area_{HV}$ (mm^2)	CTM ($)
1	29	190	16500	210	76.80	0.8825	6144.55
2	29	170	16500	205	74.80	0.8825	6148.28
3	29	190	16500	215	78.80	0.8825	6171.58
4	30	170	16500	215	78.80	0.8825	6175.65
5	28	190	16000	200	72.80	0.8825	6186.10
197	28	170	14000	215	78.80	0.9503	6805.78
198	29	170	14000	215	78.80	0.9852	6810.43
199	29	170	14000	220	80.80	0.9852	6812.55
200	28	170	14000	210	76.80	0.9852	6832.53
201	28	170	14000	215	78.80	0.9852	6834.68

Since the maximum load loss is 2350 W, then, among the 4480 candidate solutions, those that have load loss over 2350 W will be rejected.

Since the maximum no-load loss is 425 W, then, among the 4480 candidate solutions, those that have no-load loss over 425 W will be rejected.

Since the impedance voltage is 4% with tolerance ±10%, then, among the 4480 candidate solutions, those that have impedance voltage less than 3.6% or greater than 4.4% will be rejected.

2.4.2.3 Results

A computer program calculates which of the 4480 candidate solutions are acceptable (all the constraints are satisfied) and which are rejected.

For all the acceptable solutions, their technical characteristics are calculated and their manufacturing cost is estimated. The manufacturing cost is equal to the sum of transformer materials cost plus labor cost, given by (2.4).

For all the non-acceptable solutions, the reasons for rejection are recorded to a computer file.

The computer program finds that among the 4480 candidate solutions, 201 are acceptable solutions and the remaining 4279 are rejected. More specifically, 383 are rejected due to violation of the no-load loss (NLL) specification, 3453 are rejected due to violation of the load loss (LL) specification, and 443 are rejected due to violation of the impedance voltage (U_k) specification.

Table 2.2 presents the first five (cheapest) and the last five (most expensive) solutions from the total 201 accepted solutions. It can be seen from Table 2.2 that the manufacturing cost of the cheapest solution (optimum transformer) is $6144.55 and the most expensive solution costs $6834.68. Namely, the optimum solution is 10.1% cheaper in comparison with the most expensive solution. The optimum transformer is the transformer number 1 of Table 2.2, which has the following technical characteristics: $NLL = 415$ W , $LL = 2325$ W , and $U_k = 3.90\%$. As can be seen from Table 2.2, the values of the design variables for the optimum transformer are as follows: $turns_{LV} = 29$, $D = 190$ mm, $FD_{max} = 16500$ Gauss, $G = 210$ mm, $area_{LV} = 76.80$ mm^2, and $area_{HV} = 0.8825$ mm^2. The LV conductor of the optimum transformer is made of copper sheet with 192 mm width and 0.4 mm thickness, so the cross-section area of the LV conductor is 76.80 mm^2. The HV conductor of the optimum transformer is made of copper wire with 1.06 mm diameter, so the cross-section area of the HV conductor is 0.8825 mm^2.

2.4.2.4 Sensitivity Analysis

2.4.2.4.1 Variation of Magnetic Induction

Table 2.3 presents the values of four output variables, namely, (1) no-load loss (*NLL*), (2) load loss (*LL*), (3) impedance voltage (U_k), and (4) manufacturing cost (*CTM*), when only one of the design variables is varied, and more specifically the magnetic induction FD_{max}, which takes 31 values from 14000 to 17000 Gauss with a step of 100 Gauss. All the remaining design variables ($turns_{LV}$, D, G, $area_{LV}$, and $area_{HV}$) remain constant and equal to the respective values of the optimum transformer (transformer number 1 of Table 2.2). Table 2.3 shows that among the 31 candidate solutions, only seven are accepted. It can be seen from Table 2.3 that the new optimum solution has a manufacturing cost of $6129.68. This means that with specific variation of the magnetic induction, a cheaper solution was found (since the previous optimum solution of Table 2.2 has a manufacturing cost of $6144.55). In Fig. 2.1, the no-load loss versus the magnetic induction is plotted, while Fig. 2.2 plots the load loss versus the magnetic induction. From Figs. 2.1 and 2.2 it can be concluded that, in general, the no-load loss is increased and the load loss is decreased with an increase of magnetic induction FD_{max}.

Table 2.3 Variation of magnetic induction FD_{max} (accepted solutions sorted by manufacturing cost, *CTM*)

Input variables						Output variables			
$turns_{LV}$	D (mm)	FD_{max} (Gauss)	G (mm)	$area_{LV}$ (mm^2)	$area_{HV}$ (mm^2)	NLL (W)	LL (W)	U_k (%)	CTM ($)
29	190	16600	210	76.80	0.8825	420	2321	3.90	6129.68
29	190	16500	210	76.80	0.8825	415	2325	3.90	6144.55
29	190	16400	210	76.80	0.8825	409	2329	3.91	6159.20
29	190	16300	210	76.80	0.8825	402	2334	3.92	6174.90
29	190	16200	210	76.80	0.8825	395	2338	3.93	6190.33
29	190	16100	210	76.80	0.8825	388	2342	3.94	6205.53
29	190	16000	210	76.80	0.8825	381	2347	3.94	6212.68

2.4 Conventional Transformer Design Optimization Method

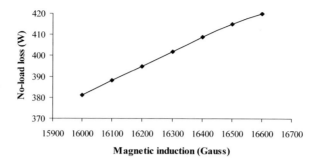

Fig. 2.1 No-load loss versus magnetic induction

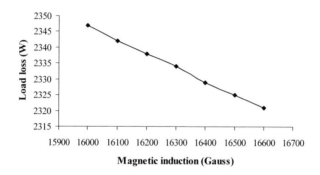

Fig. 2.2 Load loss versus magnetic induction

2.4.2.4.2 Variation of Cross-Section Area of Low Voltage Coil

Table 2.4 presents the values of four output variables, namely, (1) no-load loss (*NLL*), (2) load loss (*LL*), (3) impedance voltage (U_k), and (4) manufacturing cost (*CTM*), when only one of the design variables is varied, and more specifically the cross-section area of the low voltage conductor, which takes nine values: 192×0.36, 192×0.37, 192×0.38, 192×0.39, 192×0.40, 192×0.41, 192×0.42, 192×0.43, and 192×0.44 mm². All the remaining design variables remain constant and equal to the respective values of the optimum transformer (transformer number 1 of Table 2.2). Table 2.4 shows that among the nine candidate solutions, only four are accepted. It can be seen from Table 2.4 that the new optimum solution has a manufacturing cost of $6129.58. This means that with the specific variation of the cross-section area of the LV conductor, a marginally cheaper solution was found (since the previous optimum solution of Table 2.3 has a manufacturing cost of $6129.68).

It is concluded that the optimum transformer is the first transformer of Table 2.4, which has the following technical characteristics: $NLL = 415\,W$, $LL = 2346\,W$, and $U_k = 3.89\,\%$, while $CTM = \$\,6129.58$. As can be seen from Table 2.4, the values of the design variables for the optimum transformer are as follows: $turns_{LV} = 29$, $D = 190\,mm$, $FD_{max} = 16500\,Gauss$, $G = 210\,mm$, $area_{LV} = 74.88\,mm^2$, and $area_{HV} = 0.8825\,mm^2$. The LV conductor of the optimum transformer is copper sheet with 192 mm width and 0.39 mm thickness, so $area_{LV} = 192 \times 0.39 = 74.88\,mm^2$. The HV conductor of the optimum transformer is copper wire with 1.06 mm diameter, so the cross-section area of the HV conductor is $area_{HV} = (\pi/4) \cdot 1.06^2 = 0.8825\,mm^2$.

2.4.3 Repetitive Transformer Design Process

Table 2.5 shows how changing core and conductor design can reduce no-load and load losses but also affects the cost of the transformer, when we try to further improve the optimum design.

The optimum design is implemented through the following steps:

1. Initially the input variables are entered in a computer program. Many different values of the design variables are given, so many candidate solutions are considered.

2. A computer program determines which candidate solutions are acceptable and which are rejected (they violate one or more of the constraints).

3. The acceptable solutions are sorted according to their manufacturing cost. The optimum transformer corresponds to the least-cost solution.

Table 2.4 Variation of cross-section area of low voltage conductor (solutions sorted by manufacturing cost, *CTM*)

Input variables						Output variables			
$turns_{LV}$	D (mm)	FD_{max} (Gauss)	G (mm)	$area_{LV}$ (mm²)	$area_{HV}$ (mm²)	NLL (W)	LL (W)	U_k (%)	CTM ($)
29	190	16500	210	74.88	0.8825	415	2346	3.89	6129.58
29	190	16500	210	76.80	0.8825	415	2325	3.90	6144.55
29	190	16500	210	78.72	0.8825	415	2303	3.92	6159.60
29	190	16500	210	80.64	0.8825	415	2283	3.94	6175.48

2.4 Conventional Transformer Design Optimization Method

Table 2.5 Loss reduction alternatives

	No-load loss	Load loss	Cost
To decrease no-load loss			
A. Use lower-loss core material	Lower	No change	Higher
B. Decrease flux density by:			
1. Increasing core cross-section area (CSA)	Lower	Higher	Higher
2. Decreasing volts per turn	Lower	Higher	Higher
C. Decrease flux path length by decreasing conductor CSA	Lower	Higher	Lower
To decrease load loss			
A. Decrease current density by increasing conductor CSA	Higher	Lower	Higher
B. Decrease current path length by:			
1. Decreasing core CSA	Higher	Lower	Lower
2. Increasing volts per turn	Higher	Lower	Lower

It is possible that all the candidate solutions are rejected. Then the computer file of non-acceptable solutions should be studied and the reasons for rejection understood.

Generally, the following cases may appear:

1. Necessity to decrease or increase no-load loss

2. Necessity to decrease or increase load loss

3. Necessity to decrease or increase impedance voltage

The no-load loss is decreased by one of the following methods (linked to design variables):

1. Increasing the number of turns of LV coil

2. Decreasing the magnetic induction, FD_{max}

3. Decreasing the height of core window, G

The no-load loss is increased by one of the following methods (linked to design variables):

1. Decreasing the number of turns of LV coil

2. Increasing the magnetic induction

3. Increasing the height of core window

The load loss is decreased with the following ways (related to design variables):

1. Decreasing the number of turns of LV coil
2. Increasing the magnetic induction
3. Increasing the cross-section area of HV coil
4. Increasing the cross-section area of LV coil
5. Increasing the height of core window

The impedance voltage is decreased as follows:

1. Decreasing the number of turns of LV coil
2. Increasing the height of core window

Generally, the cost of transformer is decreased as follows:

1. Increasing the no-load loss
2. Increasing the load loss

From the above it is seen that there is an interaction between the design and output variables. For example, the no-load loss is decreased with the decrease of magnetic induction (with the remaining design parameters kept constant), but unfortunately the load loss is increased. The optimum solution is derived by selecting values for the design variables such that the transformer satisfies the constraints with minimum manufacturing cost. The selection is implemented through many tries (many different values for the design parameters) and is executed with the help of a suitable computer program.

2.5 Example of Transformer Design Data

This section gives the input data that are necessary for the design of a 630 kVA, three-phase, oil-immersed, distribution transformer. The transformer windings are made of copper. The input data are split into seven types, as already described in Sect. 2.4.1. Sections 2.6 to 2.23 present the design of this transformer.

2.5 Example of Transformer Design Data

Table 2.6 Values of description variables for the 630 kVA transformer designed in Example 2.1 to Example 2.18

#	Symbol	Value	Unit	Description
1	f	50	Hz	Frequency
2	$HVCC$	Δ	-	Connection of HV winding
3	$LVCC$	Y	-	Connection of LV winding
4	S_n	630	kVA	Rated power
5	$V^l_{HV,1}$	20000	V	First rated line voltage of HV winding
6	$V^l_{HV,2}$	6600	V	Second rated line voltage of HV winding
7	V^l_{LV}	400	V	Rated line voltage of LV winding

Table 2.7 Values of special variables for the 630 kVA transformer designed in Example 2.1 to Example 2.18

#	Symbol	Value	Unit	Description
1	BIL_{HV}	125	kV	Basic insulation level of HV winding
2	BIL_{LV}	10	kV	Basic insulation level of LV winding
3	CSF	0.965	-	Core stacking factor
4	g_{CP}	9.87	kg/m²	Weight per unit area of corrugated panels
5	g_{DS}	1.25	kg/m³	Mass density of duct strips
6	g_{HV}	8856	kg/m³	Mass density of HV winding
7	g_{LV}	8856	kg/m³	Mass density of LV winding
8	g_{MM}	7650	kg/m³	Mass density of magnetic steel
9	g_O	870	kg/m³	Mass density of mineral oil
10	$LDSP_{HV}$	1	-	Layer direction space factor of HV winding
11	$LDSP_{LV}$	0.909	-	Layer direction space factor of LV winding
12	$t_{a,max}$	45	°C	Maximum ambient temperature
13	$TDSP_{HV}$	0.98	-	Turns direction space factor of HV winding
14	$TDSP_{LV}$	1	-	Turns direction space factor of LV winding
15	$t_{o,max}$	100	°C	Maximum oil temperature
16	$t_{w,max}$	105	°C	Maximum winding temperature
17	ρ_{HV}	0.020968	Ω·mm²/m	Resistivity of HV winding
18	ρ_{LV}	0.020968	Ω·mm²/m	Resistivity of LV winding

2.5.1 Values of Description Variables

The values of description variables are given in Table 2.6. Internal winding is the low voltage winding, while external winding is the high voltage winding.

2.5.2 Values of Special Variables

The values of special variables are given in Table 2.7.

2.5.3 Values of Default Variables

The values of default variables are given in Table 2.8. The tolerance for no-load loss, load-loss and impedance voltage are according to IEC 60076-1 (Table 1.6). There are ±2.5% and ±5% voltage taps at the HV winding. There are no voltage taps at the LV winding.

2.5.4 Values of Cost Variables

The values of cost variables are given in Table 2.9.

Table 2.8 Values of default variables for the 630 kVA transformer designed in Example 2.1 to Example 2.18

#	Symbol	Value	Unit	Description
1	EdL_{HV}	266	W	Eddy current loss of HV winding
2	EdL_{LV}	399	W	Eddy current loss of LV winding
3	LL_g	8900	W	Guaranteed load loss
4	NLL_g	1100	W	Guaranteed no-load loss
5	$Taps_{HV,max}$	5	%	Upper limit of taps of HV winding
6	$Taps_{HV,min}$	5	%	Lower limit of taps of HV winding
7	$U_{k,g}$	6	%	Guaranteed impedance voltage

2.5 Example of Transformer Design Data

Table 2.9 Values of cost variables for the 630 kVA transformer designed in Example 2.1 to Example 2.18

#	Symbol	Value	Unit	Description
1	A	13.39	\$/W	No-load loss factor (cost rate)
2	B	2.09	\$/W	Load loss factor (cost rate)
3	C_{Lab}	4541	\$	Labor cost
4	CRM	1236	\$	Cost of the rest materials
5	SM	35	%	Sales margin
6	uc_1	12.01	\$/kg	Unit cost of LV winding
7	uc_2	12.01	\$/kg	Unit cost of HV winding
8	uc_3	6.01	\$/kg	Unit cost of magnetic steel
9	uc_4	7.72	\$/kg	Unit cost of insulating paper
10	uc_5	8.58	\$/kg	Unit cost of duct strips
11	uc_6	1.72	\$/kg	Unit cost of mineral oil
12	uc_7	1.03	\$/kg	Unit cost of sheet steel
13	uc_8	1.20	\$/kg	Unit cost of corrugated panels

2.5.5 Values of Various Variables

The values of various variables are given in Table 2.10.

2.5.6 Values of Conductor Cross-Section Calculation Variables

The values of conductor cross-section calculation variables are given in Table 2.11.

2.5.7 Values of Design Variables

The values of design variables are given in Table 2.12. Using (2.45), we can see that the number of loops to solve this design problem is:

$$n_{loops} = n_{LV} \cdot n_D \cdot n_{FD} \cdot n_G \cdot cs_{LV} \cdot cs_{HV} \Rightarrow$$

$$n_{loops} = 1 \cdot 1 \cdot 1 \cdot 1 \cdot 1 \cdot 1 \Rightarrow n_{loops} = 1.$$

This one iteration (loop) of calculations will be presented in Example 2.1 to Example 2.18.

Table 2.10 Values of various variables for the 630 kVA transformer designed in Example 2.1 to Example 2.18

#	Symbol	Value	Unit	Description
1	$CCEE$	3	mm	Core to coil each end
2	$D_{HV\text{-}C}$	39	mm	Distance between HV winding and core
3	$D_{LV\text{-}C}$	6.5	mm	Distance between LV winding and core
4	D_{Panel}	220	mm	Width of corrugated panel
5	$Ducts_{HV}$	12	-	Number of ducts of HV winding
6	$Ducts_{LV}$	10	-	Number of ducts of LV winding
7	Dw	15	mm	Width of cooling ducts
8	$DWPG_{HV}$	35	mm	Width of HV winding duct strip plus gap
9	$DWPG_{LV}$	25	mm	Width of LV winding duct strip plus gap
10	HCP	800	mm	Height of corrugated panel
11	$I_{HV\text{-}HV}$	6.64	mm	Insulation outside external winding
12	$I_{HV\text{-}LV}$	6.92	mm	Insulation between LV winding and HV winding
13	$I_{LV\text{-}C}$	1.5	mm	Insulation between LV winding and core
14	K	9	mm	Distance between two adjacent cores
15	$Pitch$	44	mm	Distance between two adjacent fins
16	T_{DS}	3	mm	Thickness of duct strips (without insulation)
17	TE	38.1	mm	Tolerances and elongation of coil
18	TI_{HV}	1.4	mm	Insulation of HV taps
19	TLT_{HV}	14.2	mm	Total thickness of the HV leads
20	TLT_{LV}	12.48	mm	Total thickness of the LV leads
21	V_{CT}	25	L	Volume of oil conservator

2.5 Example of Transformer Design Data

Table 2.11 Values of conductor cross-section calculation variables for the 630 kVA transformer designed in Example 2.1 to Example 2.18

#	Symbol	Value	Unit	Description
1	$area_{LV}$	191.18	mm²	Cross-section area of LV winding
2	d_{HV}	1.8	mm	Diameter of HV conductor (without insulation Δd)
3	$HVCM$	Copper	-	Material of HV conductor
4	$LVCM$	Copper	-	Material of LV conductor
5	t_{LV}	0.79	mm	Thickness of LV conductor
6	$Type_{HV}$	Round wire	-	Type of HV conductor
7	$Type_{LV}$	Foil	-	Type of LV conductor
8	w_{LV}	242	mm	Width of LV conductor
9	Δd_{HV}	0.111	mm	Insulation of HV conductor

Table 2.12 Values of design variables for the 630 kVA transformer designed in Example 2.1 to Example 2.18

#	Symbol	Value	Unit	Description
1	cs_{HV}	1	-	Number of iterations for calculation of the cross-section area of the HV conductor
2	cs_{LV}	1	-	Number of iterations for calculation of the cross-section area of the LV conductor
3	D	220	mm	Width of core leg
4	FD_{max}	17000	Gauss	Magnetic induction
5	G	261	mm	Height of core window
6	n_D	1	-	Number of iterations for the width of core leg
7	n_{FD}	1	-	Number of iterations for the magnetic induction
8	n_G	1	-	Number of iterations for the height of core window
9	n_{LV}	1	-	Number of iterations for the number of LV turns
10	$SNLL_{TF}$	1.87	W/kg	Specific no-load loss
11	$turns_{LV}$	15	-	Number of turns of LV winding

2.6 Calculation of Volts per Turn and Thickness of Core Leg

2.6.1 Calculation of Volts per Turn

The volts per turn, *VPT*, are calculated from the following equation:

$$VPT = \frac{V_{LV}^p}{turns_{LV}} \qquad (2.46)$$

where V_{LV}^p (V) is the rated phase voltage of the LV winding and $turns_{LV}$ is the number of turns of the LV winding.

2.6.2 Calculation of Thickness of Core Leg

Let us suppose that the *magnetic flux* ϕ is sinusoidal:

$$\phi = \Phi_{max} \cdot \sin(\omega \cdot t), \qquad (2.47)$$

where Φ_{max} is the maximum flux and ω (rad/s) is the angular frequency.
The induced voltage e is:

$$e = N \cdot \frac{d\phi}{dt}, \qquad (2.48)$$

where N is the number of turns of the winding.
By combining (2.47) and (2.48), the induced voltage e is computed from the following equation:

$$e = \omega \cdot N \cdot \Phi_{max} \cdot \cos(\omega \cdot t). \qquad (2.49)$$

The effective value E of the induced voltage e is:

$$E = \frac{\omega \cdot N \cdot \Phi_{max}}{\sqrt{2}} = \frac{2 \cdot \pi \cdot f \cdot N \cdot \Phi_{max}}{\sqrt{2}} \Rightarrow$$

$$E = 4.44 \cdot f \cdot N \cdot \Phi_{max}, \qquad (2.50)$$

where f (Hz) is the frequency.

2.6 Calculation of Volts per Turn and Thickness of Core Leg

Based on (2.50), the maximum flux Φ_{max} is computed as follows:

$$\Phi_{max} = \frac{E}{4.44 \cdot f \cdot N}. \tag{2.51}$$

The maximum flux Φ_{max} is also computed as follows:

$$\Phi_{max} = FD_{max} \cdot A_{eff}, \tag{2.52}$$

where FD_{max} is the maximum magnetic induction and A_{eff} is the effective core cross-section area of the magnetic flux.

Based on Fig. 2.3, the effective core cross-section area A_{eff} of the magnetic flux is computed as follows:

$$A_{eff} - CSF \cdot 2 \cdot A_C = CSF \cdot 2 \cdot (D \cdot E_u), \tag{2.53}$$

where CSF is the core stacking factor, D (mm) is the width of core leg and E_u (mm) is the thickness of the core leg. The core stacking factor expresses the net cross-section area of the magnetic flux, i.e., the insulation of the magnetic material is subtracted.

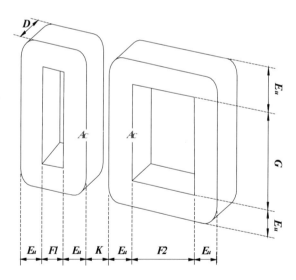

Fig. 2.3 Geometrical characteristics of the small and the large individual core. The magnetic flux passes throught the cross-section area A_c of the small and the large individual core

The volts per turn, *VPT*, can be computed as follows:

$$VPT = \frac{E}{N}. \tag{2.54}$$

By combining (2.51) to (2.54), an analytical expression for the calculation of the thickness of core leg, E_u, is derived as follows:

$$\Phi_{max} = \frac{VPT}{4.44 \cdot f} \Rightarrow FD_{max} \cdot A_{eff} = \frac{VPT}{4.44 \cdot f} \Rightarrow FD_{max} \cdot CSF \cdot 2 \cdot D \cdot E_u = \frac{VPT}{4.44 \cdot f} \Rightarrow$$

$$E_u = \frac{VPT}{8.88 \cdot CSF \cdot D \cdot FD_{max} \cdot f}. \tag{2.55}$$

2.6.3 Example 2.1

It is desired to design a 630 kVA distribution transformer having the input data shown in Tables 2.6 to 2.12. Compute the volts per turn and thickness of the core leg.

Solution

Since the LV winding is star-connected (Table 2.6), the rated phase voltage of the LV winding is:

$$V_{LV}^P = \frac{V_{LV}^l}{\sqrt{3}} = \frac{400 \text{ V}}{\sqrt{3}} \Rightarrow V_{LV}^P = 230.94 \text{ V}.$$

The volts per turn are computed using (2.46):

$$VPT = \frac{V_{LV}^P}{turns_{LV}} = \frac{230.94}{15} \Rightarrow VPT = 15.396 \, \frac{\text{V}}{\text{turn}}.$$

The thickness of the core leg is computed using (2.55):

$$E_u = \frac{VPT}{8.88 \cdot CSF \cdot D \cdot FD_{max} \cdot f} = \frac{15.396}{8.88 \cdot 0.965 \cdot (220 \cdot 10^{-3}) \cdot 1.7 \cdot 50} = 0.096 \text{ m} \Rightarrow$$

$$E_u = 96 \text{ mm}.$$

2.7 Calculation of Layer Insulation

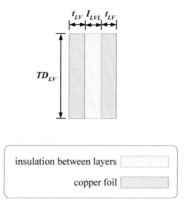

Fig. 2.4 Insulation between two layers of the LV winding. Each LV layer has one turn made of copper foil

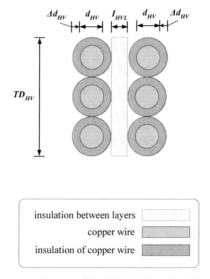

Fig. 2.5 Insulation between two layers of the HV winding. Each layer has three turns made of copper wire

2.7 Calculation of Layer Insulation

The thickness of the insulation between the layers of the LV winding must be sufficient to withstand the induced voltage as well as the impulse voltage of the LV winding. Similarly, the thickness of the insulation between the layers of the HV winding must be sufficient to withstand the induced voltage as well as the impulse voltage of the HV winding.

2.7.1 Layer Insulation of LV Winding

The thickness of the insulation between the layers of the LV winding, I_{LVL}, is computed using appropriate tables. Such a table contains, for example, the empirical rule that if the LV winding is made of copper foil with thickness between 0.4 and 1 mm, then $I_{LVL} = 0.28$ mm. Figure 2.4 shows the insulation between two layers of the LV winding, where each layer has one turn made of copper foil.

2.7.2 Layer Insulation of HV Winding

The thickness of the insulation between the layers of the HV winding, I_{HVL}, is typically computed using appropriate tables. Such a table contains, for example, the empirical rule that if the HV winding is made of copper wire with diameter less than 2 mm, then $I_{HVL} = 0.28$ mm. Figure 2.5 shows the insulation between two layers of the HV winding, where each layer has three turns made of copper wire.

2.7.3 Example 2.2

It is desired to design a 630 kVA distribution transformer having the input data shown in Tables 2.6 to 2.12. Compute the layer insulations.

Solution

As can be seen from Table 2.11, the LV winding is made of copper foil with 0.79 mm thickness, so, using the empirical rule of Sect. 2.7.1, $I_{LVL} = 0.28$ mm, since the thickness of the copper foil is between 0.4 and 1 mm.

The HV winding is made of copper wire with 1.8 mm diameter (Table 2.11), so, using the empirical rule of Sect. 2.7.2, $I_{HVL} = 0.28$ mm, since the diameter of the copper wire is less than 2 mm.

2.8 Calculation of Winding and Core Dimensions

During this step, the following technical characteristics are computed:

1. The width of the LV and the HV layer, TD_{LV} and TD_{HV}, respectively.

2. The width of the LV and the HV conductor with insulation, $TurnWidth_{LV}$ and $TurnWidth_{HV}$, respectively.

3. The thickness of the LV and the HV winding, BLD_{LV} and BLD_{HV}, respectively.

4. The window width of the small and the large individual core, $F1$ and $F2$, respectively.

5. The number of layers of the LV and the HV winding, $Layers_{LV}$ and $Layers_{HV}$, respectively.

6. The number of turns of the HV winding, $TurnsMain_{HV}$, at voltage $V^I_{HV,1}$.

7. The maximum number of turns of the HV winding, $Turns_{HV,\max}$.

Moreover, it is checked if the layer insulations I_{LVL} and I_{HVL} can withstand the induced and impulse voltage.

The sequence of calculations is illustrated in Example 2.3.

The above-mentioned dimensions (TD_{LV}, TD_{HV}, BLD_{LV}, BLD_{HV}, $F1$, $F2$) can be seen in Fig. 2.6.

2.8.1 Example 2.3

It is desired to design a 630 kVA distribution transformer having the input data shown in Tables 2.6 to 2.12. The layer insulation of 0.28 mm can withstand 6 kV maximum induced voltage and 23.5 kV maximum impulse voltage.

1. Compute the dimensions of the LV winding.

2. Compute the dimensions of the HV winding.

3. Calculate the window width of the small and large individual core.

4. Check if the layer insulations are correctly selected.

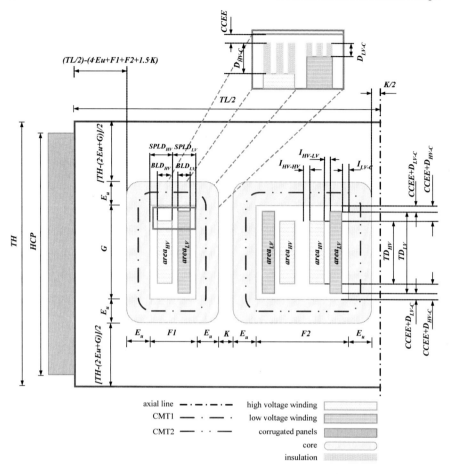

Fig. 2.6 Geometrical dimensions of transformer intersection

Solution

1. Since there are no taps at the LV winding, the number of turns of the LV winding is:

$$TurnsMain_{LV} = turns_{LV} \Rightarrow TurnsMain_{LV} = 15.$$

Since the LV winding is made of foil, each layer has one turn, so the number of layers of the LV winding is equal to the number of LV turns:

2.8 Calculation of Winding and Core Dimensions

$$Layers_{LV} = TurnsMain_{LV} \Rightarrow Layers_{LV} = 15.$$

It can be seen from Fig. 2.6 that the width of the LV layer is:

$$TD_{LV} = G - 2 \cdot CCEE - 2 \cdot D_{LV-C} = 261 - 2 \cdot 3 - 2 \cdot 6.5 \Rightarrow TD_{LV} = 242 \text{ mm}.$$

The width of the LV conductor with the insulation, $TurnWidth_{LV}$, is:

$$TurnWidth_{LV} = TD_{LV} \Rightarrow TurnWidth_{LV} = 242 \text{ mm}.$$

The thickness of the LV winding is:

$$BLD_{LV} = (t_{LV} + I_{LVL}) \cdot \frac{Layers_{LV}}{LDSP_{LV}} = (0.79 + 0.28) \cdot \frac{15}{0.909} \Rightarrow$$

$$BLD_{LV} = 17.66 \text{ mm}.$$

It can be seen from Fig. 2.6 that the total thickness of the LV winding is:

$$SPLD_{LV} = BLD_{LV} + I_{LV-C} + I_{HV-LV} = 17.66 + 1.5 + 6.92 \Rightarrow SPLD_{LV} = 26.08 \text{ mm}.$$

2. The HV winding is delta-connected, so the phase voltages are equal to the respective line voltages:

$$V^p_{HV,1} = V^l_{HV,1} = 20000 \text{ V and } V^p_{HV,2} = V^l_{HV,2} = 6600 \text{ V}.$$

Since the maximum voltage tap at the HV winding is +5% (Table 2.8), the maximum HV winding phase voltage at the maximum tap is:

$$V^p_{HV,\text{max tap}} = V^p_{HV,1} \cdot (1 + 0.05) = 20000 \cdot (1 + 0.05) \Rightarrow V^p_{HV,\text{max tap}} = 21000 \text{ V}.$$

At the maximum primary voltage of 21 kV, the maximum number of turns of the HV winding is:

$$Turns_{HV,\text{max}} = \frac{V^p_{HV,\text{max tap}}}{VPT} = \frac{21000}{15.396} \Rightarrow Turns_{HV,\text{max}} = 1364.$$

At the rated primary voltage of 20 kV, the rated number of turns of the HV winding is:

$$TurnsMain_{HV} = Turns_{HV,\max} \cdot \frac{V^p_{HV,1}}{V^p_{HV,\max tap}} = 1364 \cdot \frac{20000}{21000} \Rightarrow$$

$$TurnsMain_{HV} = 1299.$$

It can be seen from Fig. 2.6 that the width of the HV layer is:

$$TD_{HV} = G - 2 \cdot CCEE - 2 \cdot D_{HV-C} = 261 - 2 \cdot 3 - 2 \cdot 39 \Rightarrow TD_{HV} = 177 \text{ mm}.$$

The width of the HV conductor with the insulation, $TurnWidth_{HV}$, is:

$$TurnWidth_{HV} = d_{HV} + \Delta d_{HV} = 1.8 + 0.111 \Rightarrow TurnWidth_{HV} = 1.911 \text{ mm}.$$

The turns per layer of the HV winding are:

$$TurnsPerLayer_{HV} = \left[\frac{TD_{HV}}{TurnWidth_{HV}} - 1\right] \cdot TDSP_{HV} = \left[\frac{177}{1.911} - 1\right] \cdot 0.98 \Rightarrow$$

$$TurnsPerLayer_{HV} = 89.79.$$

The external winding is composed of three sub-coils, because the second primary voltage of 6.6 kV is obtained from the first primary voltage of 20 kV divided by three:

$$6.6 \text{ kV} \approx \frac{20 \text{ kV}}{3}.$$

In particular, if the three sub-coils are connected in series, a voltage of 20 kV is obtained. On the other hand, if the three sub-coils are connected in parallel, a voltage of 6.6 kV is obtained.

The number of turns per layer for each one of the three sub-coils is:

$$TurnsPerLayer_{HV,sub-coil} = \frac{TurnsPerLayer_{HV}}{3} = \frac{89.79}{3} \Rightarrow$$

$$TurnsPerLayer_{HV,sub-coil} = 29.93.$$

After rounding off:

$$TurnsPerLayer_{HV,sub-coil} = 29, \text{ and}$$

$$TurnsPerLayer_{HV} = 3 \cdot TurnsPerLayer_{HV,sub-coil} = 3 \cdot 29 \Rightarrow TurnsPerLayer_{HV} = 87.$$

2.8 Calculation of Winding and Core Dimensions

The number of layers of the HV winding is:

$$Layers_{HV} = \frac{Turns_{HV,max}}{TurnsPerLayer_{HV}} = \frac{1364}{87} \Rightarrow Layers_{HV} = 15.68.$$

After rounding off, $Layers_{HV} = 16$.

The thickness of the HV winding is:

$$BLD_{HV} = (TurnWidth_{HV} + I_{HVL}) \cdot \frac{Layers_{HV}}{LDSP_{HV}} + TI_{HV} = (1.911 + 0.28) \cdot \frac{16}{1} + 1.4 \Rightarrow$$

$$BLD_{HV} = 36.47 \text{ mm}.$$

It can be seen from Fig. 2.6 that the total thickness of the HV winding is:

$$SPLD_{HV} = BLD_{HV} + I_{HV-HV} = 36.47 + 6.64 \Rightarrow SPLD_{HV} = 43.11 \text{ mm}.$$

3. It can be seen from Fig. 2.6 that the window width of the small individual core is:

$$F1 = SPLD_{LV} + SPLD_{HV} = 26.08 + 43.11 \Rightarrow F1 = 69.19 \text{ mm}.$$

After rounding off, $F1 = 69$ mm.

The window width of the large individual core is:

$$F2 = 2 \cdot F1 = 2 \cdot 69 \Rightarrow F2 = 138 \text{ mm}.$$

4. The layer insulation of 0.28 mm can withstand 6 kV maximum induced voltage and 23.5 kV maximum impulse voltage, i.e.:

$$Induced_{max} = 6 \text{ kV} \text{ and } Impulse_{max} = 23.5 \text{ kV}.$$

The layer insulation $I_{LVL} = 0.28$ mm of the LV winding can withstand the induced voltage, since:

$$Induced_{LV} = VPT \cdot 2 \cdot 2 \cdot 10^{-3} = 15.396 \cdot 2 \cdot 2 \cdot 10^{-3} \Rightarrow$$

$$Induced_{LV} = 0.062 \text{ kV} < Induced_{max} = 6 \text{ kV}.$$

The layer insulation $I_{LVL} = 0.28$ mm of the LV winding can also withstand the impulse voltage, since:

$$Impulse_{LV} = \frac{2 \cdot BIL_{LV}}{turns_{LV}} = \frac{2 \cdot 10}{15} \Rightarrow$$

$$Impulse_{LV} = 1.33 \text{ kV} < Impulse_{max} = 23.5 \text{ kV}.$$

The layer insulation $I_{HVL} = 0.28$ mm of the HV winding can withstand the induced voltage, since:

$$Induced_{HV} = VPT \cdot 2 \cdot TurnsPerLayer_{HV} \cdot 2 \cdot 10^{-3} = 15.396 \cdot 2 \cdot 87 \cdot 2 \cdot 10^{-3} \Rightarrow$$

$$Induced_{HV} = 5.36 \text{ kV} < Induced_{max} = 6 \text{ kV}.$$

Since the minimum voltage tap at the HV winding is 5% (Table 2.8), the minimum HV winding phase voltage at the minimum tap is:

$$V^p_{HV, \min tap} = V^p_{HV,1} \cdot (1 - 0.05) = 20000 \cdot (1 - 0.05) \Rightarrow V^p_{HV, \min tap} = 19000 \text{ V}.$$

The layer insulation $I_{HVL} = 0.28$ mm of the HV winding can also withstand the impulse voltage, since:

$$Impulse_{HV} = \frac{2 \cdot BIL_{HV} \cdot TurnsPerLayer_{HV}}{TurnsMain_{HV} \cdot \frac{V^p_{HV, \min tap}}{V^p_{HV,1}}} = \frac{2 \cdot 125 \cdot 87}{1299 \cdot \frac{19000}{20000}} \Rightarrow$$

$$Impulse_{HV} = 17.62 \text{ kV} < Impulse_{max} = 23.5 \text{ kV}.$$

Consequently, the insulation between the layers of the LV winding can withstand the induced and the impulse voltage. Similarly, the insulation between the layers of the HV winding can withstand the induced and the impulse voltage.

2.9 Calculation of Core Weight and No-Load Loss

The core constructional parameters are shown in Figs. 2.3 and 2.7. The mean turn length of the small individual core is (Hatziargyriou et al. 1998):

$$CMT1 = 2 \cdot (F1 + G) + 2 \cdot \pi \cdot \left[\frac{E_u}{2} + 3.5 \right] - 8 \cdot 3.5. \tag{2.56}$$

2.9 Calculation of Core Weight and No-Load Loss

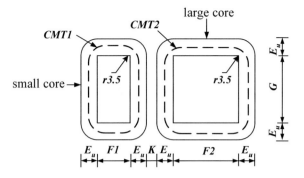

Fig. 2.7 Constructional parameters of the small and large individual core

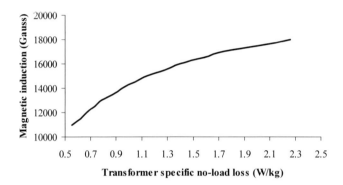

Fig. 2.8 Typical no-load loss curve

The weight of the small individual core is:

$$SCW = CMT1 \cdot D \cdot E_u \cdot CSF \cdot g_{MM}. \tag{2.57}$$

The mean turn length of the large individual core is (Hatziargyriou et al. 1998):

$$CMT2 = 2 \cdot (F2 + G) + 2 \cdot \pi \cdot \left[\frac{E_u}{2} + 3.5\right] - 8 \cdot 3.5. \tag{2.58}$$

The weight of the large individual core is:

$$LCW = CMT2 \cdot D \cdot E_u \cdot CSF \cdot g_{MM}. \tag{2.59}$$

The total weight of transformer magnetic material is equal to the sum of the weight of the two small and the two large individual cores:

$$w_3 = 2 \cdot (SCW + LCW). \tag{2.60}$$

The transformer no-load loss is (Hatziargyriou et al. 1998):

$$NLL = w_3 \cdot SNLL_{TF}, \tag{2.61}$$

where $SNLL_{TF}$ is the transformer specific no-load loss (W/kg) obtained from the no-load loss curve for a given magnetic induction FD_{max}. Different no-load loss curves correspond to different magnetic materials. A typical no-load loss curve is shown in Fig. 2.8 (Hatziargyriou et al. 1998).

2.9.1 Example 2.4

It is desired to design a 630 kVA distribution transformer having the input data shown in Tables 2.6 to 2.12.

1. Calculate the weight of the small individual core.

2. Compute the weight of the large individual core.

3. Calculate the total weight of transformer magnetic material.

4. Compute the transformer no-load loss. Check if the no-load loss tolerance of IEC 60076-1 is satisfied.

Solution

1. The mean turn length of the small individual core is computed from (2.56):

$$CMT1 = 2 \cdot (F1 + G) + 2 \cdot \pi \cdot \left[\frac{E_u}{2} + 3.5 \right] - 8 \cdot 3.5 \Rightarrow$$

$$CMT1 = 2 \cdot (69 + 261) + 2 \cdot \pi \cdot \left[\frac{96}{2} + 3.5 \right] - 8 \cdot 3.5 \Rightarrow CMT1 = 955.6 \text{ mm}.$$

The weight of the small individual core is calculated using (2.57):

$$SCW = CMT1 \cdot D \cdot E_u \cdot CSF \cdot g_{MM} \Rightarrow$$

$$SCW = (955.6 \cdot 10^{-3}) \cdot (220 \cdot 10^{-3}) \cdot (96 \cdot 10^{-3}) \cdot 0.965 \cdot 7650 \Rightarrow SCW = 149 \text{ kg}.$$

2. The mean turn length of the large individual core is:

$$CMT2 = 2 \cdot (F2 + G) + 2 \cdot \pi \cdot \left[\frac{E_u}{2} + 3.5\right] - 8 \cdot 3.5 \Rightarrow$$

$$CMT2 = 2 \cdot (138 + 261) + 2 \cdot \pi \cdot \left[\frac{96}{2} + 3.5\right] - 8 \cdot 3.5 \Rightarrow CMT2 = 1093.6 \text{ mm}.$$

The weight of the large individual core is:

$$LCW = CMT2 \cdot D \cdot E_u \cdot CSF \cdot g_{MM} \Rightarrow$$

$$LCW = (1093.6 \cdot 10^{-3}) \cdot (220 \cdot 10^{-3}) \cdot (96 \cdot 10^{-3}) \cdot 0.965 \cdot 7650 \Rightarrow LCW = 170.5 \text{ kg}.$$

3. The total weight of transformer magnetic material is computed from (2.60):

$$w_3 = 2 \cdot (SCW + LCW) = 2 \cdot (149 + 170.5) \Rightarrow w_3 = 639 \text{ kg}.$$

4. The transformer no-load loss is calculated using (2.61):

$$NLL = w_3 \cdot SNLL_{TF} = 639 \cdot 1.87 \Rightarrow NLL = 1195 \text{ W}.$$

In order for the no-load loss tolerance of IEC 60076-1 to be satisfied, the following constraint must be met:

$$NLL < 1.15 \cdot NLL_g \Rightarrow 1195 < 1.15 \cdot 1100 \Rightarrow 1195 < 1265,$$

which is fulfilled.

2.10 Calculation of Inductive Part of Impedance Voltage

During this step, the following technical characteristics are computed:

1. The equivalent diameters $D3$, $D7$, $D9$, and $D13$ of the coil. These diameters are shown in Fig. 2.9.

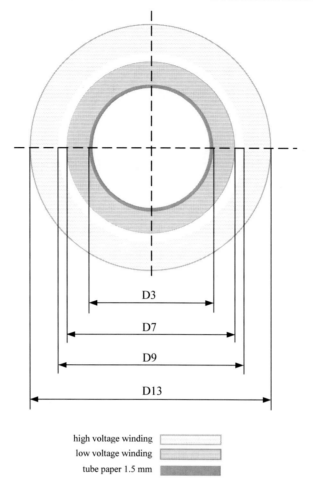

Fig. 2.9 Equivalent diameters of coil

2. The length *ML* and width *MW* of the coil former. These dimensions are shown in Fig. 2.10.

3. The inductive part of impedance voltage, *IX*.

   The necessary calculations are illustrated in Example 2.5.

2.10 Calculation of Inductive Part of Impedance Voltage

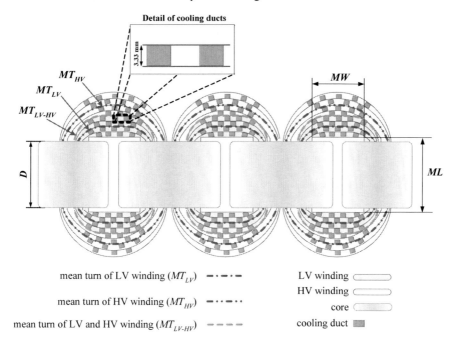

Fig. 2.10 Transformer active part

2.10.1 Example 2.5

It is desired to design a 630 kVA distribution transformer having the input data shown in Tables 2.6 to 2.12. Calculate the inductive part of the impedance voltage.

Solution

The length ML and width MW of the coil former are:

$$ML = D + 2 \cdot CCEE = 220 + 2 \cdot 3 \Rightarrow ML = 226 \text{ mm},$$

$$MW = 2 \cdot E_u + K = 2 \cdot 96 + 9 \Rightarrow MW = 201 \text{ mm}.$$

The equivalent diameter DMC of coil former is:

$$DMC = \frac{2 \cdot (ML + MW) - 10.992}{\pi} = \frac{2 \cdot (226 + 201) - 10.992}{\pi} \Rightarrow DMC = 268.34 \text{ mm}.$$

The equivalent external diameter $D3$ including the tube paper of thickness $I_{LV-C} = 1.5$ mm is:

$$D3 = DMC + 2 \cdot I_{LV-C} = 268.34 + 2 \cdot 1.5 \Rightarrow D3 = 271.34 \text{ mm}.$$

The area $A3$ corresponding to diameter $D3$ is:

$$A3 = \frac{\pi \cdot D3^2}{4} = \frac{\pi \cdot 271.34^2}{4} \Rightarrow A3 = 57825 \text{ mm}^2.$$

The equivalent external diameter $D5$ of the LV winding (without taking into account the cooling ducts of the LV winding) is:

$$D5 = D3 + 2 \cdot BLD_{LV} = 271.34 + 2 \cdot 17.66 \Rightarrow D5 = 306.66 \text{ mm}.$$

The area $A5$ corresponding to diameter $D5$ is:

$$A5 = \frac{\pi \cdot D5^2}{4} = \frac{\pi \cdot 306.66^2}{4} \Rightarrow A5 = 73859 \text{ mm}^2.$$

The dimension LG_{LV} of the cooling ducts of the LV winding is:

$$LG_{LV} = MW \Rightarrow LG_{LV} = 201 \text{ mm}.$$

The thickness of the LV cooling ducts is 3.33 mm, so the area $A6$ of the cooling ducts of the LV winding is:

$$A6 = Ducts_{LV} \cdot 3.33 \cdot LG_{LV} \cdot 2 = 10 \cdot 3.33 \cdot 201 \cdot 2 \Rightarrow A6 = 13387 \text{ mm}^2.$$

The area $A7$ of the LV winding, including the area of the cooling ducts of the LV winding, is:

$$A7 = A5 + A6 = 73859 + 13387 \Rightarrow A7 = 87246 \text{ mm}^2.$$

The diameter $D7$ corresponding to area $D7$ is:

$$D7 = \sqrt{\frac{4 \cdot A7}{\pi}} = \sqrt{\frac{4 \cdot 87246}{\pi}} \Rightarrow D7 = 333.29 \text{ mm}.$$

The diameter $D9$ over the insulation I_{HV-LV} is:

2.10 Calculation of Inductive Part of Impedance Voltage

$$D9 = D7 + 2 \cdot I_{HV-LV} = 333.29 + 2 \cdot 6.92 \Rightarrow D9 = 347.13 \text{ mm}.$$

The area $A9$ corresponding to diameter $D9$ is:

$$A9 = \frac{\pi \cdot D9^2}{4} = \frac{\pi \cdot 347.13^2}{4} \Rightarrow A9 = 94640 \text{ mm}^2.$$

The equivalent external diameter $D11$ of the HV winding (without taking into account the cooling ducts of the HV winding) is:

$$D11 = D9 + 2 \cdot BLD_{HV} = 347.13 + 2 \cdot 36.47 \Rightarrow D11 = 420.07 \text{ mm}.$$

The area $A11$ corresponding to diameter $D11$ is:

$$A11 = \frac{\pi \cdot D11^2}{4} = \frac{\pi \cdot 420.07^2}{4} \Rightarrow A11 = 138590 \text{ mm}^2.$$

The dimension LG_{HV} of the cooling ducts of the HV winding is:

$$LG_{HV} = A + 2 \cdot BLD_{LV} + 2 \cdot I_{HV-LV} = 201 + 2 \cdot 17.66 + 2 \cdot 6.92 \Rightarrow$$

$$LG_{HV} = 250.16 \text{ mm}.$$

The thickness of the HV cooling ducts is 3.33 mm, so the area $A12$ of the cooling ducts of the HV winding is:

$$A12 = Ducts_{HV} \cdot 3.33 \cdot LG_{HV} \cdot 2 = 12 \cdot 3.33 \cdot 250.16 \cdot 2 \Rightarrow A12 = 19993 \text{ mm}^2.$$

The area $A13$ of the HV winding, including the area of the cooling ducts of the HV winding, is:

$$A13 = A11 + A12 = 138590 + 19993 \Rightarrow A13 = 158583 \text{ mm}^2.$$

The diameter $D13$ corresponding to area $D13$ is:

$$D13 = \sqrt{\frac{4 \cdot A13}{\pi}} = \sqrt{\frac{4 \cdot 158583}{\pi}} \Rightarrow D13 = 449.35 \text{ mm}.$$

The factor k_L is computed as follows:

$$k_L = \frac{-2\cdot\sqrt{A3}+\sqrt{A7}+\sqrt{A9}}{3.54\cdot TD_{LV}} = \frac{-2\cdot\sqrt{57825}+\sqrt{87246}+\sqrt{94640}}{3.54\cdot 242} \Rightarrow k_L = 0.142.$$

The dimension $L17$ is computed as follows:

$$L17 = (k_L^2 + k_L + 1)\cdot TD_{LV} = (0.142^2 + 0.142 + 1)\cdot 242 \Rightarrow L17 = 281.24 \text{ mm}.$$

The factor k_P is computed as follows:

$$k_P = \frac{2\cdot\sqrt{A13}-\sqrt{A7}-\sqrt{A9}}{3.54\cdot TD_{HV}} = \frac{2\cdot\sqrt{158583}-\sqrt{87246}-\sqrt{94640}}{3.54\cdot 177} \Rightarrow k_P = 0.309.$$

The dimension $L21$ is computed as follows:

$$L21 = (k_P^2 + k_P + 1)\cdot TD_{HV} = (0.309^2 + 0.309 + 1)\cdot 177 \Rightarrow L21 = 248.59 \text{ mm}.$$

The dimension $LH23$ is computed as follows:

$$LH23 = L17 + L21 = 281.24 + 248.59 \Rightarrow LH23 = 529.83 \text{ mm}.$$

The inductance $L25$ is computed as follows:

$$L25 = \frac{(A5-A3)\cdot 0.396}{L17}\cdot \mu_0 \Rightarrow$$

$$L25 = \left[\frac{(73859-57825)\cdot 10^{-6}\text{ m}^2}{281.24\cdot 10^{-3}\text{ m}}\right]\cdot 0.396\cdot\left[4\cdot\pi\cdot 10^{-7}\,\frac{\text{H}}{\text{m}}\right] \Rightarrow$$

$$L25 = 2.84\cdot 10^{-8}\text{ H}.$$

The inductance $L26$ is computed as follows:

$$L26 = \frac{(A11-A9)\cdot 0.396}{L21}\cdot \mu_0 \Rightarrow$$

$$L26 = \left[\frac{(138590-94640)\cdot 10^{-6}\text{ m}^2}{248.59\cdot 10^{-3}\text{ m}}\right]\cdot 0.396\cdot\left[4\cdot\pi\cdot 10^{-7}\,\frac{\text{H}}{\text{m}}\right] \Rightarrow$$

$$L26 = 8.80\cdot 10^{-8}\text{ H}.$$

The inductance $L27$ is computed as follows:

2.10 Calculation of Inductive Part of Impedance Voltage

$$L27 = \frac{(A9-A7)\cdot 2}{LH23}\cdot \mu_0 \Rightarrow$$

$$L27 = \left[\frac{(94640-87246)\cdot 10^{-6}\ \text{m}^2}{529.83\cdot 10^{-3}\ \text{m}}\right]\cdot 2 \cdot \left[4\cdot \pi \cdot 10^{-7}\ \frac{\text{H}}{\text{m}}\right] \Rightarrow$$

$$L27 = 3.51\cdot 10^{-8}\ \text{H}.$$

The inductance $L28$ is computed as follows:

$$L28 = \frac{A6\cdot FN(Ducts_{LV})}{L17}\cdot \mu_0 \Rightarrow$$

$$L28 = \left[\frac{13387\cdot 10^{-6}\ \text{m}^2}{281.24\cdot 10^{-3}\ \text{m}}\right]\cdot 0.318\cdot \left[4\cdot \pi \cdot 10^{-7}\ \frac{\text{H}}{\text{m}}\right] \Rightarrow L28 = 1.90\cdot 10^{-8}\ \text{H},$$

where the value of parameter $FN(Ducts_{LV}) = FN(10) = 0.318$ is obtained from tables.

The inductance $L29$ is computed as follows:

$$L29 = \frac{A12\cdot FN(Ducts_{HV})}{L21}\cdot \mu_0 \Rightarrow$$

$$L29 = \left[\frac{19993\cdot 10^{-6}\ \text{m}^2}{248.59\cdot 10^{-3}\ \text{m}}\right]\cdot 0.320\cdot \left[4\cdot \pi \cdot 10^{-7}\ \frac{\text{H}}{\text{m}}\right] \Rightarrow L29 = 3.23\cdot 10^{-8}\ \text{H},$$

where the value of parameter $FN(Ducts_{HV}) = FN(12) = 0.320$ is obtained from tables.

The total inductance, L_{tot}, is computed as follows:

$$L_{tot} = L25 + L26 + L27 + L28 + L29 \Rightarrow$$

$$L_{tot} = (2.84 + 8.80 + 3.51 + 1.90 + 3.23)\cdot 10^{-8} \Rightarrow L_{tot} = 2.03\cdot 10^{-7}\ \text{H}.$$

The phase current I_{LV}^p that flows through the star-connected LV winding is:

$$I_{LV}^p = \frac{S}{3\cdot V_{LV}^p} = \frac{630000\ \text{VA}}{3\cdot (230.94\ \text{V})} \Rightarrow I_{LV}^p = 909.33\ \text{A}.$$

The inductive part IX of the impedance voltage is calculated as follows:

$$IX = \frac{I_{LV}^p \cdot 2 \cdot \pi \cdot f \cdot turns_{LV}^2 \cdot L_{tot}}{V_{LV}^p} \Rightarrow$$

$$IX = \frac{(909.33 \text{ A}) \cdot 2 \cdot \pi \cdot (50 \text{ Hz}) \cdot 15^2 \cdot (2.03 \cdot 10^{-7} \text{ H})}{230.94 \text{ V}} = 0.0565 \Rightarrow IX = 5.65\%.$$

2.11 Calculation of Load Loss

During this step, the following technical characteristics are computed:

1. The mean turn length of LV winding

2. The load loss of LV winding

3. The weight of LV winding

4. The mean turn length of HV winding

5. The load loss of HV winding

6. The weight of HV winding

7. The transformer load loss

8. The transformer total loss

The sequence of calculations is illustrated in Example 2.6.

2.11.1 Example 2.6

It is desired to design a 630 kVA distribution transformer having the input data shown in Tables 2.6 to 2.12.

1. Calculate the load loss of the LV winding.

2. Calculate the load loss of the HV winding at voltage $V_{HV,1}^I$.

3. Compute the transformer load loss at voltage $V_{HV,1}^I$. Check if the IEC 60076-1 tolerance for load loss at voltage $V_{HV,1}^I$ is satisfied.

2.11 Calculation of Load Loss

4. Check if the IEC 60076-1 tolerance for total loss at voltage $V_{HV,1}^{\prime l}$ is satisfied.

5. Calculate the load loss of the HV winding at the minimum high voltage.

6. Compute the transformer load loss at the minimum high voltage. Check if the IEC 60076-1 tolerance for load loss at the minimum high voltage is satisfied.

7. Check if the IEC 60076-1 tolerance for total loss at the minimum high voltage is satisfied.

8. Calculate the weight of the LV winding.

9. Calculate the weight of the HV winding.

Solution

1. The mean turn length of the LV winding is:

$$MT_{LV} = \left[\frac{D3+D7}{2}\right]\cdot\pi = \left[\frac{271.34+333.29}{2}\right]\cdot\pi \Rightarrow MT_{LV} = 949.75 \text{ mm}.$$

The length of the LV winding is:

$$CL_{LV} = MT_{LV} \cdot Layers_{LV} \cdot 3 = (949.75\cdot 10^{-3} \text{ m})\cdot 15\cdot 3 \Rightarrow CL_{LV} = 42.74 \text{ m}.$$

The cross-section area of the LV winding is:

$$area_{LV} = TurnWidth_{LV} \cdot t_{LV} = (242 \text{ mm})\cdot(0.79 \text{ mm}) \Rightarrow$$

$$area_{LV} = 191.18 \text{ mm}^2.$$

The resistance of the LV winding is:

$$R_{LV} = \frac{\rho_{LV}\cdot\dfrac{CL_{LV}}{3}}{area_{LV}} = \frac{\left[0.020968\dfrac{\Omega\cdot\text{mm}^2}{\text{m}}\right]\cdot\dfrac{(42.74 \text{ m})}{3}}{191.18 \text{ mm}^2} \Rightarrow R_{LV} = 1.56\cdot 10^{-3}\ \Omega.$$

The load loss of the LV winding is:

$$LL_{LV} = 3\cdot R_{LV} \cdot (I_{LV}^p)^2 \cdot 1.04 = 3\cdot(1.56\cdot 10^{-3}\ \Omega)\cdot(909.33 \text{ A})^2 \cdot 1.04 \Rightarrow$$

$$LL_{LV} = 4025 \text{ W}.$$

2. The mean turn length of the HV winding is:

$$MT_{HV} = \left[\frac{D9+D13}{2}\right] \cdot \pi = \left[\frac{347.13+449.35}{2}\right] \cdot \pi \Rightarrow MT_{HV} = 1251.11 \text{ mm}.$$

The length of the HV winding at the rated voltage $V^I_{HV,1} = 20 \text{ kV}$ is:

$$CL_{HV,1} = MT_{HV} \cdot TurnsMain_{HV} \cdot 3 = (1251.11 \cdot 10^{-3} \text{ m}) \cdot 1299 \cdot 3 \Rightarrow$$

$$CL_{HV,1} = 4875.6 \text{ m}.$$

The cross-section area of the HV winding is:

$$area_{HV} = \frac{\pi \cdot d^2_{HV}}{4} = \frac{\pi \cdot (1.8 \text{ mm})^2}{4} \Rightarrow area_{HV} = 2.54 \text{ mm}^2.$$

The phase current $I^p_{HV,1}$ that flows through the delta-connected HV winding at the rated voltage $V^p_{HV,1} = V^I_{HV,1} = 20 \text{ kV}$ is:

$$I^p_{HV,1} = \frac{S_n}{3 \cdot V^p_{HV,1}} = \frac{630000 \text{ VA}}{3 \cdot (20000 \text{ V})} \Rightarrow I^p_{HV,1} = 10.5 \text{ A}.$$

The resistance of the HV winding at the rated voltage $V^I_{HV,1} = 20 \text{ kV}$ is:

$$R_{HV,1} = \frac{\rho_{HV} \cdot \frac{CL_{HV,1}}{3}}{area_{HV}} = \frac{\left[0.020968 \frac{\Omega \cdot \text{mm}^2}{\text{m}}\right] \cdot \frac{(4875.6 \text{ m})}{3}}{2.54 \text{ mm}^2} \Rightarrow R_{HV,1} = 13.4 \text{ }\Omega.$$

The load loss of the HV winding at the rated voltage $V^I_{HV,1} = 20 \text{ kV}$ is:

$$LL_{HV,1} = 3 \cdot R_{HV,1} \cdot (I^p_{HV,1})^2 \cdot 1.06 = 3 \cdot (13.4 \text{ }\Omega) \cdot (10.5 \text{ A})^2 \cdot 1.06 \Rightarrow$$

$$LL_{HV,1} = 4698 \text{ W}.$$

3. The transformer load loss at the rated voltage $V^I_{HV,1} = 20 \text{ kV}$ is:

$$LL_1 = LL_{LV} + LL_{HV,1} + EdL_{LV} + EdL_{HV} = 4025 + 4698 + 399 + 266 \Rightarrow$$

2.11 Calculation of Load Loss

$$LL_1 = 9388 \text{ W}.$$

In order for the IEC 60076-1 tolerance on the load loss at the rated voltage $V'_{HV,1} = 20 \text{ kV}$ to be satisfied, the following constraint must be met:

$$LL_1 < 1.15 \cdot LL_g \Rightarrow 9388 < 1.15 \cdot 8900 \Rightarrow 9388 < 10235,$$

which is fulfilled.

4. In order for the IEC 60076-1 tolerance on the total loss at the rated voltage $V'_{HV,1} = 20 \text{ kV}$ to be satisfied, the following constraint must be met:

$$(NLL + LL_1) < 1.1 \cdot (NLL_g + LL_g) \Rightarrow$$

$$(1195 + 9388) < 1.1 \cdot (1100 + 8900) \Rightarrow 10583 < 11000,$$

which is fulfilled.

5. The transformer operates with 20 kV or with 6.6 kV primary voltages, while the minimum tap is 5% for both primary voltages. This means that the minimum high voltage is:

$$HV^P_{min} = V^P_{HV,2} \cdot (1 - Taps_{HV,min}) = 6600 \cdot (1 - 0.05) \Rightarrow HV^P_{min} = 6270 \text{ V}.$$

The number of turns that give the minimum high voltage of 6270 V is:

$$Turns_{HV,min} = TurnsMain_{HV} \cdot \frac{HV^P_{min}}{V^P_{HV,1}} = 1299 \cdot \left[\frac{6270 \text{ V}}{20000 \text{ V}}\right] \Rightarrow Turns_{HV,min} = 407.$$

The length of the external winding at the minimum high voltage of 6270 V is:

$$CL_{HV,2} = MT_{HV} \cdot Turns_{HV,min} \cdot 3 = (1251.11 \cdot 10^{-3} \text{ m}) \cdot 407 \cdot 3 \Rightarrow$$

$$CL_{HV,2} = 1527.6 \text{ m}.$$

The phase current $I^P_{HV,2}$ that flows through the delta-connected HV winding at the minimum high voltage of 6270 V is:

$$I^p_{HV,2} = \frac{S}{3 \cdot HV^p_{min}} = \frac{630000 \text{ VA}}{3 \cdot (6270 \text{ V})} \Rightarrow I^p_{HV,2} = 33.5 \text{ A}.$$

Since there are three sub-coils connected in parallel, the current that flows in each sub-coil is:

$$I^p_{HV,2,sub-coil} = \frac{I^p_{HV,2}}{3} = \frac{33.5 \text{ A}}{3} \Rightarrow I^p_{HV,2,sub-coil} = 11.17 \text{ A}.$$

The resistance of the HV winding at the minimum high voltage of 6270 V is:

$$R_{HV,2} = \frac{\rho_{HV} \cdot \dfrac{CL_{HV,2}}{3}}{area_{HV}} = \frac{\left[0.020968 \dfrac{\Omega \cdot \text{mm}^2}{\text{m}}\right] \cdot \dfrac{(1527.6 \text{ m})}{3}}{2.54 \text{ mm}^2} \Rightarrow R_{HV,2} = 4.2 \,\Omega.$$

The load loss of each sub-coil of the HV winding at the minimum high voltage of 6270 V is:

$$LL_{HV,2,sub-coil} = 3 \cdot R_{HV,2} \cdot (I^p_{HV,2,sub-coil})^2 \cdot 1.06 = 3 \cdot (4.2 \,\Omega) \cdot (11.17 \text{ A})^2 \cdot 1.06 \Rightarrow$$

$$LL_{HV,2,sub-coil} = 1666 \text{ W}.$$

The load loss (of the three sub-coils) of the HV winding at the minimum high voltage of 6270 V is:

$$LL_{HV,2} = 3 \cdot LL_{HV,2,sub-coil} = 3 \cdot 1666 \Rightarrow LL_{HV,2} = 4998 \text{ W}.$$

6. The transformer load loss at the minimum high voltage of 6270 V is:

$$LL_2 = LL_{LV} + LL_{HV,2} + EdL_{LV} + EdL_{HV} = 4025 + 4998 + 399 + 266 \Rightarrow$$

$$LL_2 = 9688 \text{ W}.$$

In order for the IEC 60076-1 tolerance on the load loss at the minimum high voltage of 6270 V to be satisfied, the following constraint must be met:

$$LL_2 < 1.15 \cdot LL_g \Rightarrow 9688 < 1.15 \cdot 8900 \Rightarrow 9688 < 10235,$$

which is fulfilled.

2.12 Calculation of Impedance Voltage

7. In order for the IEC 60076-1 tolerance on the total loss at the minimum high voltage of 6270 V to be satisfied, the following constraint must be met:

$$(NLL + LL_2) < 1.1 \cdot (NLL_g + LL_g) \Rightarrow$$

$$(1195 + 9688) < 1.1 \cdot (1100 + 8900) \Rightarrow 10883 < 11000,$$

which is fulfilled.

8. The total weight of the LV winding is:

$$w_1 = CL_{LV} \cdot area_{LV} \cdot g_{LV} \cdot 1.05 \Rightarrow$$

$$w_1 = (42.74 \; m) \cdot (191.18 \cdot 10^{-6} \; m^2) \cdot (8856 \; kg/m^3) \cdot 1.05 \Rightarrow w_1 = 76 \; kg.$$

9. The total weight of the HV winding is:

$$w_2 = CL_{HV,1} \cdot \left[1 + Taps_{HV,\max}\right] \cdot area_{HV} \cdot g_{HV} \cdot 1.08 \Rightarrow$$

$$w_2 = (4875.6 \; m) \cdot (1 + 0.05) \cdot (2.54 \cdot 10^{-6} \; m^2) \cdot (8856 \; kg/m^3) \cdot 1.08 \Rightarrow$$

$$w_2 = 124.4 \; kg.$$

2.12 Calculation of Impedance Voltage

The ohmic part of the impedance voltage is:

$$IR = \frac{LL_1}{S_n}. \tag{2.62}$$

The impedance voltage is computed as follows:

$$U_k = \sqrt{IR^2 + IX^2}. \tag{2.63}$$

2.12.1 Example 2.7

It is desired to design a 630 kVA distribution transformer having the input data shown in Tables 2.6 to 2.12.

1. Compute the ohmic part of the impedance voltage.

2. Compute the impedance voltage.

3. Check if the IEC 60076-1 tolerance on the impedance voltage is satisfied.

Solution

1. The ohmic part of the impedance voltage is:

$$IR = \frac{LL_1}{S_n} = \frac{9388 \text{ W}}{630000 \text{ VA}} = 0.0149 \Rightarrow IR = 1.49\%.$$

2. The impedance voltage is:

$$U_k = \sqrt{IR^2 + IX^2} = \sqrt{0.0149^2 + 0.0565^2} = 0.0584 \Rightarrow U_k = 5.84\%.$$

3. In order for the impedance voltage tolerance given in IEC 60076-1 to be satisfied, the following constraint must be met:

$$0.9 \cdot U_{k,g} < U_k < 1.1 \cdot U_{k,g} \Rightarrow$$
$$0.9 \cdot (6\%) < 5.84\% < 1.1 \cdot (6\%) \Rightarrow 5.4\% < 5.84\% < 6.6\%,$$

which is fulfilled.

2.13 Calculation of Coil Length

Example 2.8 shows how the coil length is computed.

2.13 Calculation of Coil Length

2.13.1 Example 2.8

It is desired to design a 630 kVA distribution transformer having the input data shown in Tables 2.6 to 2.12. Compute the coil length.

Solution

The total thickness of the cooling ducts, *TCD*, is:

$$TCD = 2 \cdot (Ducts_{LV} + Ducts_{HV}) \cdot 3.33 = 2 \cdot (10 + 12) \cdot 3.33 \Rightarrow TCD = 146.52 \text{ mm}.$$

The overlap of the tube paper (thickness 1.5 mm) plus the LV layer insulation is:

$$OLI_{LV} = 1.5 + I_{LVL} = 1.5 + 0.28 \Rightarrow OLI_{LV} = 1.78 \text{ mm}.$$

The overlap of the layer insulation of the HV winding, OLI_{HV}, is:

$$OLI_{HV} = Layers_{HV} \cdot I_{HVL} = 16 \cdot 0.28 \Rightarrow OLI_{HV} = 4.48 \text{ mm}.$$

The total length of the coil, *LTC*, is calculated as follows:

$$TLC = ML + 2 \cdot F1 + TCD + TLT_{LV} + TLT_{HV} + OLI_{LV} + OLI_{HV} + TE \Rightarrow$$
$$TLC = 226 + 2 \cdot 69 + 146.52 + 12.48 + 14.2 + 1.78 + 4.48 + 38.1 \Rightarrow TLC = 582 \text{ mm}.$$

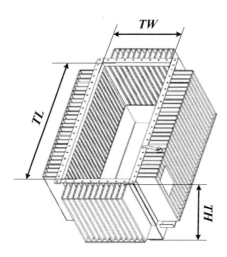

Fig. 2.11 Tank dimensions

2.14 Calculation of Tank Dimensions

The tank dimensions are shown in Fig. 2.11. Example 2.9 shows how the tank dimensions are computed.

2.14.1 Example 2.9

It is desired to design a 630 kVA distribution transformer having the input data shown in Tables 2.6 to 2.12.

1. Compute the tank length.
2. Compute the tank width.
3. Calculate the tank height and the oil height.

Solution

1. The tank length, TL, is:

$$TL = 2 \cdot (4 \cdot E_u + 3 \cdot F1 + K) + K + 108 = 2 \cdot (4 \cdot 96 + 3 \cdot 69 + 9) + 9 + 108 \Rightarrow$$

$$TL = 1317 \text{ mm} .$$

2. The tank width, TW, is:

$$TW = TLC + 38 = 582 + 38 \Rightarrow TW = 620 \text{ mm} .$$

3. The minimum tank height, TH_{min}, is:

$$TH_{min} = G + 2 \cdot E_u + 350 = 261 + 2 \cdot 96 + 350 \Rightarrow TH_{min} = 803 \text{ mm} .$$

Since a triple tap changer and a triple voltage regulator will be used, for constructional reasons, the tank height is selected to be:

$$TH = 1015 \text{ mm} .$$

This transformer has an oil conservator (a cylindrical tank that undergoes the oil volume fluctuation due to oil temperature variation), so the mineral oil height, OH, is:

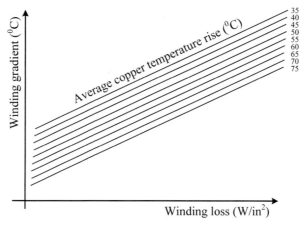

Fig. 2.12 Gradient curves based on average copper temperature rise

$$OH = TH \Rightarrow OH = 1015 \text{ mm}.$$

2.15 Calculation of Winding Gradient and Oil Gradient

The gradient (temperature rise) of the LV winding and the gradient of the HV winding are computed using gradient curves of the form shown in Fig. 2.12. The sequence of calculations is presented in Example 2.10.

2.15.1 Example 2.10

It is desired to design a 630 kVA distribution transformer having the input data shown in Tables 2.6 to 2.12.

1. Compute the average gradient between the oil and the LV winding.

2. Compute the average gradient between the oil and the HV winding.

3. Calculate the average oil temperature rise.

Solution

1. The area of the cooling ducts of the LV winding is:

$$DuctArea_{LV} = Ducts_{LV} \cdot 2 \cdot 2 \cdot MW \cdot TD_{LV} = 10 \cdot 2 \cdot 2 \cdot 201 \cdot 242 \Rightarrow$$

$$DuctArea_{LV} = 1945680 \text{ mm}^2 = \frac{1945680 \text{ mm}^2}{25.4^2 \frac{\text{mm}^2}{\text{in}^2}} \Rightarrow DuctArea_{LV} = 3016 \text{ in}^2.$$

The area of the cooling ducts of the LV gap is:

$$GapDuctArea_{LV} = D7 \cdot \pi \cdot TD_{LV} = 333.29 \cdot \pi \cdot 242 = 253389 \text{ mm}^2 \Rightarrow$$

$$GapDuctArea_{LV} = \frac{253389 \text{ mm}^2}{25.4^2 \frac{\text{mm}^2}{\text{in}^2}} \Rightarrow GapDuctArea_{LV} = 393 \text{ in}^2.$$

The total area of the cooling ducts of the LV winding is:

$$TotalDuctArea_{LV} = DuctArea_{LV} + GapDuctArea_{LV} = 3016 + 393 \Rightarrow$$

$$TotalDuctArea_{LV} = 3409 \text{ in}^2.$$

The loss in each one of the three LV windings is:

$$CoilLoss_{LV} = \frac{LL_{LV} + EdL_{LV}}{3} = \frac{4025 + 399}{3} \Rightarrow CoilLoss_{LV} = 1475 \text{ W}.$$

The LV winding loss per surface of the cooling ducts is:

$$LPS_{LV} = \frac{CoilLoss_{LV}}{TotalDuctArea_{LV}} = \frac{1475 \text{ W}}{3409 \text{ in}^2} \Rightarrow LPS_{LV} = 0.43 \frac{\text{W}}{\text{in}^2}.$$

The average copper temperature rise, ACR, is:

$$ACR = t_{w,\max} - t_{a,\max} = 105 - 45 \Rightarrow ACR = 60 \text{ °C}.$$

Since $ACR = 60 \text{ °C}$ and $LPS_{LV} = 0.43 \text{ W/in}^2$, from gradient curves it is found that the gradient of the LV winding, Gra_{LV}, is:

$$Gra_{LV} = 5.2 \text{ °C}.$$

The average gradient between the oil and the LV winding, $AvGra_{LV}$, is:

$$AvGra_{LV} = 2.09 \cdot Gra_{LV} = 2.09 \cdot 5.2 \Rightarrow AvGra_{LV} = 10.9 \text{ °C}.$$

2.15 Calculation of Winding Gradient and Oil Gradient

2. The area of the cooling ducts of the HV winding is:

$$DuctArea_{HV} = Ducts_{HV} \cdot 2 \cdot 2 \cdot LG_{HV} \cdot TD_{HV} = 12 \cdot 2 \cdot 2 \cdot 250.16 \cdot 177 \Rightarrow$$

$$DuctArea_{HV} = 2125359 \text{ mm}^2 = \frac{2125359 \text{ mm}^2}{25.4^2 \frac{\text{mm}^2}{\text{in}^2}} \Rightarrow DuctArea_{HV} = 3294 \text{ in}^2.$$

The area of the cooling ducts of the HV gap is:

$$GapDuctArea_{HV} = D7 \cdot \pi \cdot TD_{HV} = 333.29 \cdot \pi \cdot 177 = 185330 \text{ mm}^2 \Rightarrow$$

$$GapDuctArea_{HV} = \frac{185330 \text{ mm}^2}{25.4^2 \frac{\text{mm}^2}{\text{in}^2}} \Rightarrow GapDuctArea_{HV} = 287 \text{ in}^2.$$

The total area of the cooling ducts of the HV winding is:

$$TotalDuctArea_{HV} = DuctArea_{HV} + GapDuctArea_{HV} = 3294 + 287 \Rightarrow$$

$$TotalDuctArea_{HV} = 3581 \text{ in}^2.$$

The loss in each one of the three HV windings, at the minimum high voltage of 6270 V, is:

$$CoilLoss_{HV} = \frac{LL_{HV,2} + EdL_{HV}}{3} = \frac{4998 + 266}{3} \Rightarrow CoilLoss_{HV} = 1755 \text{ W}.$$

The HV winding loss per surface of the cooling ducts is:

$$LPS_{HV} = \frac{CoilLoss_{HV}}{TotalDuctArea_{HV}} = \frac{1755 \text{ W}}{3581 \text{ in}^2} \Rightarrow LPS_{HV} = 0.49 \frac{\text{W}}{\text{in}^2}.$$

Since $ACR = 60\ °C$ and $LPS_{HV} = 0.49 \text{ W/in}^2$, from gradient curves it is found that the gradient of the HV winding, Gra_{HV}, is:

$$Gra_{HV} = 5.8\ °C.$$

The average gradient between the oil and the HV winding, $AvGra_{HV}$, is:

$$AvGra_{HV} = 2.09 \cdot Gra_{HV} = 2.09 \cdot 5.8 \Rightarrow AvGra_{HV} = 12.1\ °C.$$

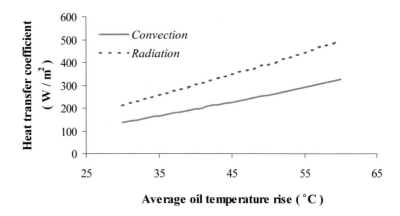

Fig. 2.13 Coefficients (W/m^2) of heat transfer by convection and radiation by transformer tank as a function of average oil temperature rise

3. The maximum gradient is:

$$MaxGra = \max(AvGra_{LV},\ AvGra_{HV}) = \max(10.9,\ 12.1) \Rightarrow MaxGra = 12.1\ °C.$$

The average oil temperature rise is:

$$AOR = ACR - MaxGra = 60 - 12.1 \Rightarrow AOR = 47.9\ °C.$$

2.16 Calculation of Heat Transfer

The transformer losses appear as heat in the core and coils. This heat must be dissipated without allowing the windings to reach a temperature that will cause excessive deterioration of the insulation (Bean et al. 1959). Rigorous mathematical treatment for expressing transformer heat transfer is quite difficult and hence designers mostly rely on empirical formulas (Flanagan 1993; Kulkarni and Khaparde 2004).

The heat transfer due to transformer tank convection is computed as follows:

$$TCL = TCA \cdot TCC, \tag{2.64}$$

where TCA is the tank convection area (m^2) and TCC is the tank convection coefficient (W/m^2), which is derived using Fig. 2.13.

The heat transfer due to transformer tank radiation is:

2.16 Calculation of Heat Transfer

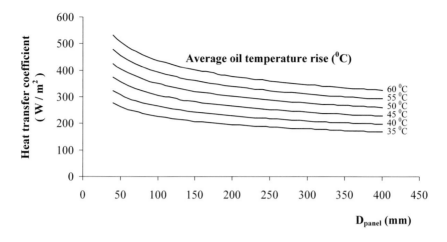

Fig. 2.14 Coefficients (W/m^2) of heat transfer by corrugated panels as a function of D_{panel} as well as a function of average oil temperature rise

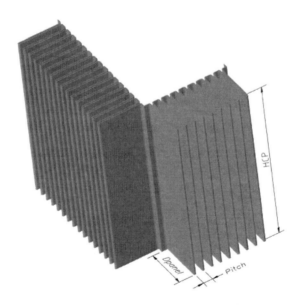

Fig. 2.15 Dimensions of corrugated panels

$$TRL = TRA \cdot TRC, \tag{2.65}$$

where *TRA* is the tank radiation area (m^2) and *TRC* is the tank radiation coefficient (W/m^2), which is derived using Fig. 2.13.

The heat transfer through the corrugated panels is computed as follows:

$$CPL = CPA \cdot CPC, \tag{2.66}$$

where *CPA* is the corrugated panels area (m^2) and *CPC* is the corrugated panels coefficient (W/m^2), which is derived using Fig. 2.14. The dimensions of the corrugated panels are shown in Fig. 2.15.

The total heat that the transformer can safely transfer is computed as follows:

$$TLRTT = TCL + TRL + CPL. \tag{2.67}$$

2.16.1 Example 2.11

It is desired to design a 630 kVA distribution transformer having the input data shown in Tables 2.6 to 2.12.

1. Compute the total heat that can be transferred by the transformer tank.

2. Check if the transformer tank is appropriate for transferring the heat arising from transformer total losses.

Solution

1. Since $AOR = 47.9\ ^0C$, Fig. 2.13 gives $TCC = 244.34\ W/m^2$ and $TRC = 371.32\ W/m^2$. Since $AOR = 47.9\ ^0C$ and $D_{Panel} = 220$ mm, Fig. 2.14 gives that $CPC = 280.88\ W/m^2$. Please note that the two digits accuracy of *TCC*, *TRC* and *CPC* is coming from computer simulation.

The tank convection area is:

$$TCA = 2 \cdot (TL + TW) \cdot TH + 2 \cdot TL \cdot TW = 2 \cdot (1317 + 620) \cdot 1015 + 2 \cdot 1317 \cdot 620 \Rightarrow$$

$$TCA = 5565190\ mm^2 \Rightarrow TCA = 5.57\ m^2.$$

The heat transfer due to transformer tank convection is computed using (2.64):

$$TCL = TCA \cdot TCC = 5.57 \cdot 244.34 \Rightarrow TCL = 1361\ W.$$

2.16 Calculation of Heat Transfer

The tank radiation area is computed as follows:

$$TRA = \left[2 \cdot (TL + TW) + 4 \cdot D_{Panel} \cdot \sqrt{2}\right] \cdot OH \Rightarrow$$

$$TRA = \left[2 \cdot (1317 + 620) + 4 \cdot 220 \cdot \sqrt{2}\right] \cdot 1015 = 5195286 \text{ mm}^2 \Rightarrow TRA = 5.2 \text{ m}^2.$$

The heat transfer due to transformer tank radiation is calculated using (2.65):

$$TRL = TRA \cdot TRC = 5.2 \cdot 371.32 \Rightarrow TRL = 1931 \text{ W}.$$

The transformer is constructed with corrugated panels around the four sides, i.e., across the tank length and tank width. The number of corrugated panels across the tank length, *NCPTL*, is:

$$NCPTL = \text{int}\left[\frac{TL - 60}{Pitch}\right] + 1 = \text{int}\left[\frac{1317 - 60}{44}\right] + 1 = \text{int}[28.57] + 1 \Rightarrow NCPTL = 29,$$

where int[*x*] is the integer part of *x*.

The number of corrugated panels across the tank width, *NCPTW*, is:

$$NCPTW = \text{int}\left[\frac{TW - 60}{Pitch}\right] + 1 = \text{int}\left[\frac{620 - 60}{44}\right] + 1 = \text{int}[12.73] + 1 \Rightarrow NCPTW = 13.$$

The total number of corrugated panels, *NCP*, is:

$$NCP = 2 \cdot (NCPTL + NCPTW) = 2 \cdot (29 + 13) \Rightarrow NCP = 84.$$

The corrugated panels area, *CPA*, is computed as follows:

$$CPA = 2 \cdot D_{Panel} \cdot HCP \cdot NCP = 2 \cdot 220 \cdot 800 \cdot 84 = 29568000 \Rightarrow CPA = 29.57 \text{ m}^2.$$

The heat transfer through the corrugated panels is calculated using (2.66):

$$CPL = CPA \cdot CPC = 29.57 \cdot 280.88 \Rightarrow CPL = 8306 \text{ W}.$$

The total heat that the transformer can safely transfer is computed using (2.67):

$$TLRTT = TCL + TRL + CPL = 1361 + 1931 + 8306 \Rightarrow TLRTT = 11598 \text{ W}.$$

2. The transformer total loss, TTL_2, at the minimum high voltage of 6270 V is:

$$TTL_2 = NLL + LL_2 = 1195 + 9688 \Rightarrow TTL_2 = 10883 \text{ W}.$$

The transformer tank is appropriate for safely transferring the transformer total losses because:

$$TLRTT > TTL_2 \Rightarrow 11598 \text{ W} > 10883 \text{ W}.$$

2.17 Calculation of the Weight of Insulating Materials

Example 2.12 shows how the weight of insulating materials is computed.

2.17.1 Example 2.12

It is desired to design a 630 kVA distribution transformer having the input data shown in Tables 2.6 to 2.12.

1. Compute the total area of each item of insulating material of the LV winding.

2. Compute the total area of each item of insulating material of the HV winding.

3. Calculate the total weight of insulating materials.

Solution

1. The LV winding has the following six insulating materials:
    - Tube paper with thickness 1.5 mm
    - Layer insulation with thickness 0.28 mm
    - End ducts insulation with thickness 0.15 mm
    - Perimetric duct insulation with thickness 0.15 mm
    - Extension paper insulation with thickness 0.41 mm
    - Gap insulation with thickness 0.28 mm

    The area of the tube paper (thickness 1.5 mm), $S_{LV,TP}$, is:

$$S_{LV,TP} = MW \cdot (G - 2 \cdot CCEE) + D3 \cdot \pi \cdot (G - 2 \cdot CCEE) \Rightarrow$$

$$S_{LV,TP} = 201 \cdot (261 - 2 \cdot 3) + 271.34 \cdot \pi \cdot (261 - 2 \cdot 3) = 268627 \text{ mm}^2 \Rightarrow$$

2.17 Calculation of the Weight of Insulating Materials

$$S_{LV,TP} = 0.27 \text{ m}^2.$$

The area of the layer insulation (thickness 0.28 mm), $S_{LV,LI}$, is:

$$S_{LV,LI} = MT_{LV} \cdot (G - 2 \cdot CCEE) \cdot Layers_{LV} = 949.75 \cdot (261 - 2 \cdot 3) \cdot 15 \Rightarrow$$

$$S_{LV,LI} = 3632793 \text{ mm}^2 \Rightarrow S_{LV,LI} = 3.63 \text{ m}^2.$$

The area of the end ducts insulation (thickness 0.15 mm), $S_{LV,EDI}$, is:

$$S_{LV,EDI} = MW \cdot (G - 2 \cdot CCEE) \cdot Ducts_{LV} \cdot 2 = 201 \cdot (261 - 2 \cdot 3) \cdot 10 \cdot 2 \Rightarrow$$

$$S_{LV,EDI} = 1025100 \text{ mm}^2 \Rightarrow S_{LV,EDI} = 1.03 \text{ m}^2.$$

The area of perimetric ducts insulation (thickness 0.15 mm), $S_{LV,PDI}$, is:

$$S_{LV,PDI} = D7 \cdot \pi \cdot (G - 2 \cdot CCEE) = 333.29 \cdot \pi \cdot (261 - 2 \cdot 3) \Rightarrow$$

$$S_{LV,PDI} = 267001 \text{ mm}^2 \Rightarrow S_{LV,PDI} = 0.27 \text{ m}^2.$$

Two extension papers are used. The area of extension paper insulation (thickness 0.41 mm), $S_{LV,EPI}$, is:

$$S_{LV,EPI} = 2 \cdot 2 \cdot \pi \cdot D13 \cdot \left[(G - 2 \cdot CCEE) + 1.8 \cdot BLD_{HV} \right] \Rightarrow$$

$$S_{LV,EPI} = 2 \cdot 2 \cdot \pi \cdot 449.35 \cdot \left[(261 - 2 \cdot 3) + 1.8 \cdot 36.47 \right] = 1810591 \text{ mm}^2 \Rightarrow$$

$$S_{LV,EPI} = 1.81 \text{ m}^2.$$

Two extension papers with 0.41 mm thickness are used, so the number of papers with thickness 0.28 mm that are used for gap insulation, *NPGI*, is:

$$NPGI = \frac{I_{HV-LV} - 3.3 - 2 \cdot 0.41}{I_{HVL}} = \frac{6.92 - 3.3 - 2 \cdot 0.41}{0.28} \Rightarrow NPGI = 10.$$

The area of gap insulation (thickness 0.28 mm), $S_{LV,GI}$, is:

$$S_{LV,GI} = D7 \cdot \pi \cdot (G - 2 \cdot CCEE) \cdot NPGI = 333.29 \cdot \pi \cdot (261 - 2 \cdot 3) \cdot 10 \Rightarrow$$

$$S_{LV,GI} = 2670007 \text{ mm}^2 \Rightarrow S_{LV,GI} = 2.67 \text{ m}^2.$$

2. The HV winding has the following six insulating materials:
   - Layer insulation with thickness 0.28 mm
   - End ducts insulation with thickness 0.15 mm
   - Perimetric duct insulation with thickness 0.15 mm
   - Insulation between HV sub-coils with thickness 0.28 mm
   - External gap insulation with thickness 0.28 mm
   - Insulating board paper with thickness 1.5 mm

The area of the layer insulation (thickness 0.28 mm), $S_{HV,LI}$, is:

$$S_{HV,LI} = MT_{HV} \cdot [(G - 2 \cdot CCEE) + 4 \cdot 19] \cdot Layers_{HV} \cdot 1.25 \Rightarrow$$

$$S_{HV,LI} = 1251.11 \cdot [(261 - 2 \cdot 3) + 4 \cdot 19] \cdot 16 \cdot 1.25 = 8282348 \text{ mm}^2 \Rightarrow$$

$$S_{HV,LI} = 8.28 \text{ m}^2.$$

The area of the end ducts insulation (thickness 0.15 mm), $S_{HV,EDI}$, is:

$$S_{HV,EDI} = LG_{HV} \cdot (G - 2 \cdot CCEE) \cdot Ducts_{HV} \cdot 2 = 250.16 \cdot (261 - 2 \cdot 3) \cdot 12 \cdot 2 \Rightarrow$$

$$S_{HV,EDI} = 1530979 \text{ mm}^2 \Rightarrow S_{HV,EDI} = 1.53 \text{ m}^2.$$

The area of perimetric ducts insulation (thickness 0.15 mm), $S_{HV,PDI}$, is:

$$S_{HV,PDI} = D13 \cdot \pi \cdot (G - 2 \cdot CCEE) = 449.35 \cdot \pi \cdot (261 - 2 \cdot 3) \Rightarrow$$

$$S_{HV,PDI} = 359977 \text{ mm}^2 \Rightarrow S_{HV,PDI} = 0.36 \text{ m}^2.$$

Five insulating papers with thickness 0.28 mm are used for insulation between HV sub-coils, so the area of these insulating papers, $S_{HV,Sub}$, is:

$$S_{HV,Sub} = MT_{HV} \cdot (G - 2 \cdot CCEE) \cdot 5 = 1251.11 \cdot (261 - 2 \cdot 3) \cdot 5 = 1595165 \text{ mm}^2 \Rightarrow$$

$$S_{HV,Sub} = 1.60 \text{ m}^2.$$

2.17 Calculation of the Weight of Insulating Materials

Two extension papers with 0.41 mm thickness are used, so the number of papers with thickness 0.28 mm that are used for HV gap insulation, *NPEGI*, is:

$$NPEGI = \frac{I_{HV-HV} - 3.3 - 2 \cdot 0.41}{I_{HVL}} = \frac{6.64 - 3.3 - 2 \cdot 0.41}{0.28} \Rightarrow NPEGI = 9.$$

The area of HV gap insulation (thickness 0.28 mm), $S_{HV,GI}$, is:

$$S_{HV,GI} = D13 \cdot \pi \cdot (G - 2 \cdot CCEE) \cdot NPEGI = 449.35 \cdot \pi \cdot (261 - 2 \cdot 3) \cdot 9 \Rightarrow$$

$$S_{HV,GI} = 3239793 \text{ mm}^2 \Rightarrow S_{HV,GI} = 3.24 \text{ m}^2.$$

The area of the insulating board paper, $S_{HV,IBP}$, with thickness 1.5 mm that covers a total space of 40 mm is:

$$S_{HV,IBP} = MT_{HV} \cdot 40 \cdot Layers_{HV} = 1251.11 \cdot 40 \cdot 16 = 800710 \text{ mm}^2 \Rightarrow$$

$$S_{HV,IBP} = 0.80 \text{ m}^2.$$

3. Insulating materials with four different thicknesses are used for the insulation of the LV and HV windings:
   - Insulating material with thickness 1.5 mm
   - Insulating material with thickness 0.41 mm
   - Insulating material with thickness 0.28 mm
   - Insulating material with thickness 0.15 mm

The weight of the insulating material with thickness 1.5 mm, $WIM_{1.5}$, is:

$$WIM_{1.5} = 3 \cdot 1.1 \cdot 1.95 \cdot (S_{LV,TP} + S_{HV,IBP}) = 3 \cdot 1.1 \cdot \left[1.95 \frac{\text{kg}}{\text{m}^2}\right] \cdot \left[(0.27 + 0.80) \text{ m}^2\right] \Rightarrow$$

$$WIM_{1.5} = 6.89 \text{ kg}.$$

The weight of the insulating material with thickness 0.41 mm, $WIM_{0.41}$, is:

$$WIM_{0.41} = 3 \cdot 1.1 \cdot 0.37 \cdot S_{LV,EPI} = 3 \cdot 1.1 \cdot \left[0.37 \frac{\text{kg}}{\text{m}^2}\right] \cdot (1.81 \text{ m}^2) \Rightarrow WIM_{0.41} = 2.21 \text{ kg}.$$

The weight of the insulating material with thickness 0.28 mm, $WIM_{0.28}$, is:

$$WIM_{0.28} = 3 \cdot 1.1 \cdot 0.26 \cdot (S_{LV,LI} + S_{LV,GI} + S_{HV,LI} + S_{HV,Sub} + S_{HV,GI}) \Rightarrow$$

$$WIM_{0.28} = 3 \cdot 1.1 \cdot \left[0.26 \frac{\text{kg}}{\text{m}^2} \right] \cdot \left[(3.63 + 2.67 + 8.28 + 1.60 + 3.24) \, \text{m}^2 \right] \Rightarrow$$

$$WIM_{0.28} = 16.66 \, \text{kg}.$$

The weight of the insulating material with thickness 0.15 mm, $WIM_{0.15}$, is:

$$WIM_{0.15} = 3 \cdot 1.1 \cdot 0.13 \cdot (S_{LV,EDI} + S_{LV,PDI} + S_{HV,EDI} + S_{HV,PDI}) \Rightarrow$$

$$WIM_{0.15} = 3 \cdot 1.1 \cdot \left[0.13 \frac{\text{kg}}{\text{m}^2} \right] \cdot \left[(1.03 + 0.27 + 1.53 + 0.36) \, \text{m}^2 \right] \Rightarrow WIM_{0.15} = 1.37 \, \text{kg}.$$

The total weight of the insulating materials, w_4, is:

$$w_4 = WIM_{1.5} + WIM_{0.41} + WIM_{0.28} + WIM_{0.15} = 6.89 + 2.21 + 16.66 + 1.37 \Rightarrow$$

$$w_4 = 27.13 \, \text{kg}.$$

2.18 Calculation of the Weight of Ducts

Example 2.13 shows how the weight of the ducts is computed.

2.18.1 Example 2.13

It is desired to design a 630 kVA distribution transformer having the input data shown in Tables 2.6 to 2.12. Compute the weight of duct strips.

Solution

The number of duct strips for the LV winding, N_1, is:

$$N_1 = \frac{MW}{DWPG_{LV}} \cdot Ducts_{LV} \cdot 2 = \frac{201}{25} \cdot 10 \cdot 2 \Rightarrow N_1 = 161.$$

The number of perimetric duct strips for the LV winding, N_2, is:

2.19 Calculation of the Weight of Oil

$$N_2 = \frac{\pi \cdot D7}{DWPG_{LV}} = \frac{\pi \cdot 333.29}{25} \Rightarrow N_2 = 42.$$

The number of duct strips for the HV winding, N_3, is:

$$N_3 = \frac{(MW + 2 \cdot BLD_{LV})}{DWPG_{HV}} \cdot Ducts_{HV} \cdot 2 = \frac{(201 + 2 \cdot 17.66)}{35} \cdot 12 \cdot 2 \Rightarrow N_3 = 162.$$

The number of perimetric duct strips for the HV winding, N_4, is:

$$N_4 = \frac{\pi \cdot D13}{DWPG_{HV}} = \frac{\pi \cdot 449.35}{35} \Rightarrow N_4 = 40.$$

The total weight of duct strips, w_5, is:

$$w_5 = 3 \cdot 1.1 \cdot g_{DS} \cdot \left[(N_1 + N_2 + N_3 + N_4) \cdot (G - 2 \cdot CCEE) \cdot D_W \cdot T_{DS} \right] \Rightarrow$$

$$w_5 = 3 \cdot 1.1 \cdot \left[1.25 \frac{kg}{m^3} \right] \cdot \left[(161 + 42 + 162 + 40) \cdot (261 - 2 \cdot 3) \cdot 15 \cdot 3 \right] \Rightarrow$$

$$w_5 = 3 \cdot 1.1 \cdot \left[1.25 \frac{kg}{m^3} \right] \cdot \left[4647375 \text{ mm}^3 \right] = 3 \cdot 1.1 \cdot \left[1.25 \frac{kg}{m^3} \right] \cdot \left[4.65 \text{ m}^3 \right] \Rightarrow$$

$$w_5 = 19.18 \text{ kg}.$$

2.19 Calculation of the Weight of Oil

Example 2.14 shows how the weight of the oil is computed.

2.19.1 Example 2.14

It is desired to design a 630 kVA distribution transformer having the input data shown in Tables 2.6 to 2.12. Compute the weight of the mineral oil.

Solution

The volume of corrugated panels, V_{CP}, is:

$$V_{CP} = 8 \cdot HCP \cdot D_{Panel} \cdot NCP = 8 \cdot 800 \cdot 220 \cdot 84 = 118272000 \text{ mm}^3 \Rightarrow V_{CP} = 118.27 \text{ L}.$$

The volume of the LV and HV windings, V_{Wd}, is:

$$V_{Wd} = \frac{w_1 + w_2}{g_{LV}} = \frac{(76 + 124.4) \text{ kg}}{8856 \frac{\text{kg}}{\text{m}^3}} = 0.02263 \text{ m}^3 \Rightarrow V_{Wd} = 22.63 \text{ L}.$$

The volume of magnetic material, V_{MM}, is:

$$V_{MM} = \frac{w_3}{g_{MM}} = \frac{639 \text{ kg}}{7650 \frac{\text{kg}}{\text{m}^3}} = 0.08353 \text{ m}^3 \Rightarrow V_{MM} = 83.53 \text{ L}.$$

The volume of the tank, V_T, is:

$$V_T = TL \cdot TW \cdot OH = 1317 \cdot 620 \cdot 1015 = 828788100 \text{ mm}^3 \Rightarrow V_T = 828.79 \text{ L}.$$

The volume of mineral oil, V_O, is:

$$V_O = V_T + V_{CT} + V_{CP} - V_{Wd} - V_{MM} = 828.79 + 25 + 118.27 - 22.63 - 83.53 \Rightarrow$$

$$V_O = 865.9 \text{ L} \Rightarrow V_O = 0.866 \text{ m}^3.$$

The total weight of mineral oil, w_6, is:

$$w_6 = 0.95 \cdot g_O \cdot V_O = 0.95 \cdot \left[870 \frac{\text{kg}}{\text{m}^3} \right] \cdot (0.866 \text{ m}^3) \Rightarrow w_6 = 715.75 \text{ kg}.$$

2.20 Calculation of the Weight of Sheet Steel

Sheet steel is used for the construction of:

1. The tank

2. The oil conservator

3. The frame that supports the active part of the transformer

An accurate estimate of the weight of the sheet steel can be obtained only after completion of the transformer constructional drawings. However, during the transformer design phase, approximate formulas are used to compute the total weight of the sheet steel.

2.20.1 Example 2.15

It is desired to design a 630 kVA distribution transformer having the input data shown in Tables 2.6 to 2.12. Compute the weight of the sheet steel.

Solution

Using approximate formulas, the total weight of the sheet steel, w_7, is:

$$w_7 = 217.2 \text{ kg}.$$

2.21 Calculation of the Weight of Corrugated Panels

Example 2.16 shows how the weight of the corrugated panels is computed.

2.21.1 Example 2.16

It is desired to design a 630 kVA distribution transformer having the input data shown in Tables 2.6 to 2.12. Compute the weight of the corrugated panels.

Solution

The total weight of corrugated panels, w_8, is:

$$w_8 = g_{CP} \cdot CPA = \left[9.87 \, \frac{\text{kg}}{\text{m}^2} \right] \cdot (29.57 \text{ m}^2) \Rightarrow w_8 = 291.86 \text{ kg}.$$

2.22 Calculation of the Cost of Transformer Main Materials

The cost of transformer main materials is computed as follows:

$$CMM = \sum_{i=1}^{8} uc_i \cdot w_i. \tag{2.68}$$

2.22.1 Example 2.17

It is desired to design a 630 kVA distribution transformer having the input data shown in Tables 2.6 to 2.12. Compute the cost of transformer main materials.

Solution

The cost of the LV winding, C_1, is:

$$C_1 = uc_1 \cdot w_1 = \left[12.01 \frac{\$}{kg}\right] \cdot (76 \text{ kg}) \Rightarrow C_1 = \$\,912.76.$$

The cost of the HV winding, C_2, is:

$$C_2 = uc_2 \cdot w_2 = \left[12.01 \frac{\$}{kg}\right] \cdot (124.4 \text{ kg}) \Rightarrow C_2 = \$\,1494.04.$$

The cost of magnetic material, C_3, is:

$$C_3 = uc_3 \cdot w_3 = \left[6.01 \frac{\$}{kg}\right] \cdot (639 \text{ kg}) \Rightarrow C_3 = \$\,3840.39.$$

The cost of insulating materials, C_4, is:

$$C_4 = uc_4 \cdot w_4 = \left[7.72 \frac{\$}{kg}\right] \cdot (27.13 \text{ kg}) \Rightarrow C_4 = \$\,209.44.$$

The cost of duct strips, C_5, is:

$$C_5 = uc_5 \cdot w_5 = \left[8.58 \frac{\$}{kg}\right] \cdot (19.18 \text{ kg}) \Rightarrow C_5 = \$\,164.56.$$

The cost of mineral oil, C_6, is:

$$C_6 = uc_6 \cdot w_6 = \left[1.72 \frac{\$}{kg}\right] \cdot (715.75 \text{ kg}) \Rightarrow C_6 = \$\,1231.09.$$

The cost of sheet steel, C_7, is:

$$C_7 = uc_7 \cdot w_7 = \left[1.03 \frac{\$}{kg}\right] \cdot (217.2 \text{ kg}) \Rightarrow C_7 = \$\,223.72.$$

2.23 Calculation of Transformer Manufacturing Cost

The cost of corrugated panels, C_8, is:

$$C_8 = uc_8 \cdot w_8 = \left[1.20 \frac{\$}{kg}\right] \cdot (291.86 \text{ kg}) \Rightarrow C_8 = \$350.23.$$

The cost of transformer main materials, CMM, is:

$$CMM = \sum_{i=1}^{8} C_i \Rightarrow$$

$$CMM = 912.76 + 1494.04 + 3840.39 + 209.44 + 164.56 + 1231.09 + 223.72 + 350.23 \Rightarrow$$

$$CMM = \$8426.23.$$

2.23 Calculation of Transformer Manufacturing Cost

The transformer materials cost is:

$$CM = CMM + CRM. \tag{2.69}$$

The transformer manufacturing cost is:

$$CTM = CM + C_{Lab}. \tag{2.70}$$

The transformer bid price (also called sales price or purchasing price) is:

$$BP = \frac{CTM}{1 - SM}. \tag{2.71}$$

The sales margin of the transformer is:

$$MS = BP - CTM. \tag{2.72}$$

The transformer total owning cost is:

$$TOC = BP + A \cdot NLL + B \cdot LL, \tag{2.73}$$

where LL is the transformer load loss.

2.23.1 Example 2.18

It is desired to design a 630 kVA distribution transformer having the input data shown in Tables 2.6 to 2.12.

1. Compute the cost of transformer materials.

2. Compute the transformer manufacturing cost.

3. Compute the transformer sales price.

4. Calculate the transformer total owning cost.

Solution

1. The cost of transformer materials, *CM*, is computed using (2.69):

$$CM = CMM + CRM = 8426.23 + 1236 \Rightarrow CM = \$\,9662.23.$$

2. The transformer manufacturing cost, *CTM*, is computed using (2.70):

$$CTM = CM + C_{Lab} = 9662.23 + 4541 \Rightarrow CTM = \$\,14203.23.$$

3. The transformer bid price, *BP*, is calculated using (2.71):

$$BP = \frac{CTM}{1-SM} = \frac{14203.23}{1-0.35} \Rightarrow BP = \$\,21851.12.$$

The sales margin of the transformer, *MS*, is:

$$MS = BP - CTM = 21851.12 - 14203.23 \Rightarrow MS = \$\,7647.89.$$

4. The transformer total owning cost, *TOC*, is:

$$TOC = BP + A \cdot NLL + B \cdot LL_1 = 21851.12 + 13.39 \cdot 1195 + 2.09 \cdot 9388 \Rightarrow$$

$$TOC = \$\,57473.09.$$

Figure 2.16 shows the transformer cost components and the transformer bid price.

2.23 Calculation of Transformer Manufacturing Cost

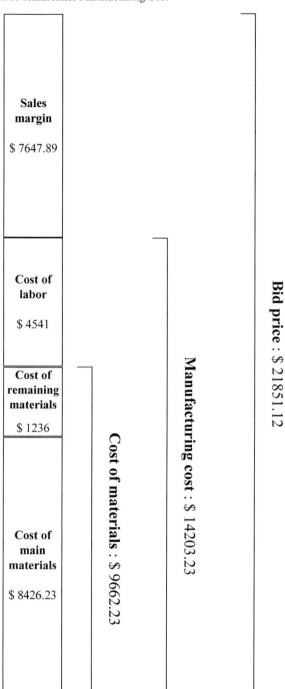

Fig. 2.16 Transformer cost components and bid price

References

Amoiralis EI, Tsili MA, Georgilakis PS, Kladas AG, Souflaris AT (2008) A parallel mixed integer programming–finite element method technique for global design optimization of power transformers. IEEE Transactions on Magnetics 44(6):1022–1025

Amoiralis EI, Georgilakis PS, Tsili MA, Kladas AG (2009) Global transformer optimization method using evolutionary design and numerical field computation. IEEE Transactions on Magnetics (accepted for publication)

Andersen OW (1991) Optimized design of electric power equipment. IEEE Computer Applications in Power 4(1):11–15

Bean RL, Chackan N, Moore HR, Wentz EC (1959) Transformers for the electric power industry. McGraw-Hill, New York

Del Vecchio RM, Poulin B, Feghali PT, Shah DM, Ahuja R (2002) Transformer design principles with applications to core-form power transformers. CRC Press, Boca Raton, Florida

Flanagan WM (1993) Handbook of transformer design & applications, second edition. McGraw-Hill, Boston

Georgilakis PS (2008) A recursive genetic algorithm–finite element method technique for the solution of transformer manufacturing cost minimization problem. Proc IEEE Conference on Electromagnetic Field Computation

Georgilakis PS (2009) A recursive genetic algorithm–finite element method technique for the solution of transformer manufacturing cost minimization problem. IET Electric Power Applications (accepted for publication)

Georgilakis PS, Tsili MA, Souflaris AT (2007) A heuristic solution to the transformer manufacturing cost optimization problem. Journal of Materials Processing Technology 181(1-3):260–266

Hatziargyriou N, Georgilakis P, Spiliopoulos D, Bakopoulos J (1998) Quality improvement of individual cores of distribution transformers using decision trees. International Journal of Engineering Intelligent Systems for Electrical Engineering and Communications 6(3):141–146

Jabr RA (2005) Application of geometric programming to transformer design. IEEE Transactions on Magnetics 41(11):4261–4269

Judd FF, Kressler DR (1977) Design optimization of small low-frequency power transformers. IEEE Transactions on Magnetics 13(4):1058–1069

Kulkarni SV and Khaparde SA (2004) Transformer engineering: design and practice. Marcel-Dekker, New York

MIT (1962) Magnetic circuits and transformers, 14th edn. John Wiley and Sons, New York

Mittle VN, Mittal A (1996) Design of electrical machines, 4th edn. Standard Publishers Distributors, Nai Sarak, Delhi

Odessey PH (1974) Transformer design by computer. IEEE Transactions on Manufacturing Technology 3(1):1–17

Rubaai A (1994) Computer aided instruction of power transformer design in the undergraduate power engineering class. IEEE Transactions on Power Systems 9(3):1174–1181

Part II
Evaluation and Optimization Methods

3 Numerical Analysis

Abstract Transformers involve magnetostatic problems. These problems can be solved by analytical and numerical techniques. The limitations of analytical techniques as well as the progress of computers facilitated the development of numerical techniques. Among the numerical techniques, the most popular method in the solution of magnetostatic problems is the finite element method. A very real advantage of the finite element method is its ability to deal with complex geometries. Another advantage is that it yields stable and accurate solutions. This chapter presents the finite element method for the solution of linear and nonlinear magnetostatic problems, the latter being very common in transformer design. Carefully selected arithmetic examples make clear the application of the finite element method to the solution of linear and nonlinear magnetostatic problems.

3.1 Introduction

3.1.1 Magnetostatic Problems

Transformers involve *magnetostatic problems*. In such problems, the magnetic field is practically constant in terms of time, so the following two *Maxwell equations* in differential form are applied in magnetostatics:

$$\nabla \times \mathbf{H} = \mathbf{J}, \tag{3.1}$$

$$\nabla \cdot \mathbf{B} = 0, \tag{3.2}$$

where **H** is the magnetic field intensity, **B** is the magnetic flux density, and **J** is the electric current density.

In the linear region of the magnetic material, the relationship between **B** and **H** is:

$$\mathbf{B} = \mu \cdot \mathbf{H}, \tag{3.3}$$

where μ is the permeability of the magnetic material that can be expressed as follows:

$$\mu = \mu_0 \cdot \mu_r = \frac{1}{v} = \frac{\mu_0}{v_r}, \tag{3.4}$$

where $\mu_0 = 4 \cdot \pi \cdot 10^{-7}$ H/m is the permeability of free space, μ_r is the relative permeability of the material, v_r is its relative reluctivity, and v is the reluctivity of the magnetic material.

In the case of ferromagnetic materials, e.g., those used in the construction of transformer magnetic circuits, the *B–H* characteristic curve is nonlinear and the permeability is a function of the magnetic field intensity:

$$\mathbf{B} = \mu(\mathbf{H}) \cdot \mathbf{H} . \tag{3.5}$$

Since $\nabla \cdot \mathbf{B} = 0$ according to (3.2), and using the vector identity:

$$\nabla \cdot (\nabla \times \mathbf{A}) = 0 , \tag{3.6}$$

it is concluded that there exists a *magnetic vector potential* **A** such that:

$$\mathbf{B} = \nabla \times \mathbf{A} . \tag{3.7}$$

Solving (3.3) for **H** and substituting into (3.1), we obtain:

$$\nabla \times \left(\frac{\mathbf{B}}{\mu} \right) = \mathbf{J} . \tag{3.8}$$

The substitution of (3.7) into (3.8) gives:

$$\frac{1}{\mu} \cdot \nabla \times (\nabla \times \mathbf{A}) = \mathbf{J} . \tag{3.9}$$

The following vector calculus relationship holds:

$$\nabla \times (\nabla \times \mathbf{A}) = \nabla (\nabla \cdot \mathbf{A}) - \nabla^2 \mathbf{A} . \tag{3.10}$$

By combining (3.9) and (3.10), we have:

$$\nabla (\nabla \cdot \mathbf{A}) - \nabla^2 \mathbf{A} = \mu \cdot \mathbf{J} . \tag{3.11}$$

Supposing that $\nabla \cdot \mathbf{A} = 0$ and assuming a constant permeability μ, (3.11) reduces to *Poisson's equation*:

$$\nabla^2 \mathbf{A} = -\mu \cdot \mathbf{J} . \tag{3.12}$$

3.1 Introduction

In the case of two-dimensional magnetostatic problems, the magnitude of the magnetic flux density, B, is computed as a function of its coordinates B_x and B_y at the x and y axis, respectively, as follows:

$$B = \sqrt{B_x^2 + B_y^2}, \tag{3.13}$$

where:

$$B_x = \frac{\partial \mathbf{A}}{\partial y}, \tag{3.14}$$

$$B_y = -\frac{\partial \mathbf{A}}{\partial x}. \tag{3.15}$$

Equations 3.14 and 3.15 follow from (3.7).

3.1.2 Methods for the Solution of Magnetostatic Problems

There are two main families of techniques for the analysis of electromagnetic fields and the solution of magnetostatic problems (Binns et al. 1992):

1. Analytical techniques

2. Numerical techniques

3.1.2.1 Analytical Techniques

The analytical techniques include the following approaches:

1. Separation of variables method (Zachmanoglou and Thoe 1976)

2. Laplace transformations method (Hameyer and Belmans 1999)

3. Rayleigh method (Rayleigh 1870)

4. Ritz method (Ritz 1909)

5. Galerkin method (Galerkin 1915)

Among the analytical techniques, the separation of variables and the Laplace transformations method can find the exact solution of the electromagnetic field problem, however, these methods can be applied only to problems with simple geometry. On the other hand, the simplified analytical techniques, i.e., Rayleigh, Ritz, and Galerkin methods are applied to very well known problems where simplifications can be done. Analytical techniques are limited to linear isotropic media, e.g., constant values of permeability and conductivity for magnetostatic and eddy current problems must be used (Chari and Salon 2000). The great advantage of analytical techniques is that, when these methods are applicable, the solution is obtained in a short time (Hameyer and Belmans 1999).

3.1.2.2 Numerical Techniques

The limitations of the analytical techniques as well as the progress of computers has facilitated the development of numerical techniques for the solution of electromagnetic field problems.

The most important numerical techniques are the following:

1. Finite element method (Courant 1943; Clough 1960; Hildebrand 1962; Martic and Carey 1973; Silvester and Ferrari 1996; Kwon and Bang 1997; Reece and Preston 2000; Bastos and Sadowski 2003)

2. Finite difference method (Gauss 1823; Erdelyi and Ahmed 1965)

3. Boundary element method (Peng and Salon 1982; Brebbia 1984; Salon et al.1989; Brebbia and Dominguez 1992)

4. Magnetic equivalent circuit (Hameyer and Belmans 1999)

5. Point mirroring method (Hameyer and Belmans 1999)

Among the numerical techniques, the most popular method for the solution of magnetostatic problems is the finite element method, which will be the subject of the rest of this chapter.

3.2 Finite Element Method

3.2.1 Introduction

The finite element method is a numerical technique for solving differential equations in many disciplines, e.g., electromagnetics, magnetostatics, thermal conduction, solid and structural mechanics, fluid dynamics, and acoustics. The core idea of the finite element method is to substitute the overall complicated problem of solving the differential equations into a series of simpler sub-problems that corresponds to an easily solved linear system of equations. Consequently, the problem has to be discretized in adequate sub-problems. The sub-problems are described geometrically by geometrically simple shaped elements such as triangles or rectangles for two-dimensional and mainly tetrahedrons for three-dimensional problems. These elements forming the numerical discretization, the *mesh*, are called the finite elements. On this discretization, the problem-describing differential equation is locally approximated by a simple *shape function*. Assembling all sub-problems into a system of equations and solving this obtains an approximated overall solution.

In general, the procedure for solving an electromagnetic field problem using the finite element method is divided into three steps:

1. *Pre-processing*: the electromagnetic field problem is defined and prepared to be solved.

2. *Processing*: the numerical solution of the physical problem is obtained.

3. *Post-processing*: the solution obtained is prepared to calculate the required electromagnetic field quantities or to evaluate forces and other macroscopic quantities.

A very strong advantage of the finite element method, and the main reason why it is a favorite method in many engineering branches, is its ability to deal with complex geometries. Other advantages of the finite element method are the following (Chari and Salon 2000):

1. It yields stable and accurate solutions

2. It can handle nonlinearity and eddy currents well

3. Natural boundary conditions are implicit in the functional formulation

3.2.2 Applications to Power Engineering

The finite element method has shown its capability in solving real-world problems, as reflected by the growing number of publications on its application to electrical power engineering problems, e.g., Demerdash and Nehl 1979; Ashtiani 1988; Labridis and Dokopoulos 1988; Pavlik et al. 1988; Moallem and Ong 1990; Wang and Demerdash 1991; Alhamadi and Demerdash 1994; Arjona and McDonald 1999; Bergeron and Trahar 1999; Bergeron et al. 1999; Hameyer and Belmans 1999; Minambres et al. 1999; Watson and Dorrell 1999; Yamazaki 1999; Papagiannis et al. 2000; Reece and Preston 2000; Kaehler and Henneberger 2004; Papazacharopoulos et al. 2004; Schlensok and Henneberger 2004; de Leon and Anders 2008, to quote only a few.

In transformer design, the finite element method has been applied for no-load and load loss evaluation (Holland et al. 1992; Pavlik et al. 1993; Pern and Yeh 1995; Enokizono and Soda 1997; Lin et al. 1998; Kefalas et al. 2008), impedance voltage evaluation (Andersen 1973; Kladas et al. 1994; Tsili et al. 2006), inrush current computation (Steurer and Fröhlich 2002), analysis of stresses and dynamic behavior under short-circuit (Renyuan et al. 1992; Ho et al. 2004; Abed and Mohammed 2007; Kumbhar and Kulkarni 2007), noise prediction (Rausch et al. 2002), insulation system analysis (Lesniewska 2002), thermal analysis and hot-spot temperature prediction (teNyenhuis et al. 2002; Hwang et al. 2005; Preis et al. 2006), and analysis of DC bias (Lu and Liu 1993; Viana et al.1999; Bíró et al. 2007).

3.2.3 Solution of Linear Magnetostatic Problems

The magnetostatic problem region is discretized by *triangular* node elements. The magnetic vector potential is approximated by the following linear shape function:

$$A = s_1 + s_2 \cdot x + s_3 \cdot y. \tag{3.16}$$

The unknowns s_1, s_2, s_3 in (3.16) are computed from the values of the magnetic vector potential A_1, A_2, and A_3 at the three nodes of an element shown in Fig. 3.1. This means that unknowns s_1, s_2, s_3 in (3.16) are found from solution of the following system of three linear equations:

$$A_1 = s_1 + s_2 \cdot x_1 + s_3 \cdot y_1, \tag{3.17}$$

$$A_2 = s_1 + s_2 \cdot x_2 + s_3 \cdot y_2, \tag{3.18}$$

$$A_3 = s_1 + s_2 \cdot x_3 + s_3 \cdot y_3. \tag{3.19}$$

The linear system (3.17), (3.18), and (3.19) is solved using the Cramer rule. For example, s_1 is computed using the following sequence of calculations that involve matrix determinants:

$$s_1 = \frac{\begin{vmatrix} A_1 & x_1 & y_1 \\ A_2 & x_2 & y_2 \\ A_3 & x_3 & y_3 \end{vmatrix}}{\begin{vmatrix} 1 & x_1 & y_1 \\ 1 & x_2 & y_2 \\ 1 & x_3 & y_3 \end{vmatrix}} \Rightarrow$$

3.2 Finite Element Method

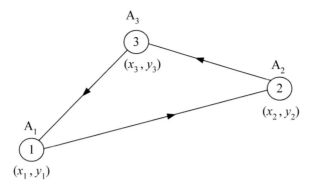

Fig. 3.1 A triangular finite element with its rectangular coordinates. The vertices of this finite element are 1, 2, and 3, given in counter clockwise order

$$s_1 = \frac{A_1 \cdot (x_2 \cdot y_3 - x_3 \cdot y_2) - A_2 \cdot (x_1 \cdot y_3 - x_3 \cdot y_1) + A_3 \cdot (x_1 \cdot y_2 - x_2 \cdot y_1)}{(x_2 \cdot y_3 - x_3 \cdot y_2) - (x_1 \cdot y_3 - x_3 \cdot y_1) + (x_1 \cdot y_2 - x_2 \cdot y_1)} \Rightarrow$$

$$s_1 = \frac{(x_2 \cdot y_3 - x_3 \cdot y_2) \cdot A_1 + (x_3 \cdot y_1 - x_1 \cdot y_3) \cdot A_2 + (x_1 \cdot y_2 - x_2 \cdot y_1) \cdot A_3}{2 \cdot \Delta}, \quad (3.20)$$

where Δ is the area of the triangular finite element of Fig. 3.1, which is computed as follows:

$$\Delta = \frac{|(x_2 - x_3) \cdot (y_3 - y_1) - (y_2 - y_3) \cdot (x_3 - x_1)|}{2}. \quad (3.21)$$

Similarly, s_2 and s_3 are computed from the following equations:

$$s_2 = \frac{(y_2 - y_3) \cdot A_1 + (y_3 - y_1) \cdot A_2 + (y_1 - y_2) \cdot A_3}{2 \cdot \Delta}, \quad (3.22)$$

$$s_3 = \frac{(x_3 - x_2) \cdot A_1 + (x_1 - x_3) \cdot A_2 + (x_2 - x_1) \cdot A_3}{2 \cdot \Delta}. \quad (3.23)$$

Substituting (3.20), (3.22), and (3.23) into (3.16), we obtain:

$$A = \frac{(x_2 \cdot y_3 - x_3 \cdot y_2) \cdot A_1 + (x_3 \cdot y_1 - x_1 \cdot y_3) \cdot A_2 + (x_1 \cdot y_2 - x_2 \cdot y_1) \cdot A_3}{2 \cdot \Delta}$$
$$+ \left[\frac{(y_2 - y_3) \cdot A_1 + (y_3 - y_1) \cdot A_2 + (y_1 - y_2) \cdot A_3}{2 \cdot \Delta} \right] \cdot x \quad (3.24)$$
$$+ \left[\frac{(x_3 - x_2) \cdot A_1 + (x_1 - x_3) \cdot A_2 + (x_2 - x_1) \cdot A_3}{2 \cdot \Delta} \right] \cdot y$$

Combining (3.14) and (3.24), the x-coordinate of the magnetic flux density is computed as follows:

$$B_x = \frac{(x_3 - x_2) \cdot A_1 + (x_1 - x_3) \cdot A_2 + (x_2 - x_1) \cdot A_3}{2 \cdot \Delta}. \qquad (3.25)$$

Combining (3.15) and (3.24), the y-coordinate of the magnetic flux density is calculated as follows:

$$B_y = \frac{(y_3 - y_2) \cdot A_1 + (y_1 - y_3) \cdot A_2 + (y_2 - y_1) \cdot A_3}{2 \cdot \Delta}. \qquad (3.26)$$

By combining (3.4) and (3.12), the two-dimensional Poisson equation for the magnetostatic problem is obtained as:

$$v \cdot \nabla^2 \mathbf{A} = -\mathbf{J}. \qquad (3.27)$$

It can be proved (Silvester and Ferrari 1996) that solving the Poisson equation (3.27), it is equivalent to minimizing the following energy-related *functional*:

$$F = \iint \left\{ \frac{v}{2} \cdot \left[\left(\frac{\partial A}{\partial x} \right)^2 + \left(\frac{\partial A}{\partial y} \right)^2 \right] - J \cdot A \right\} \cdot dx \cdot dy. \qquad (3.28)$$

The minimization of (3.28) with respect to each vertex value A_k, with $k = 1, 2, 3$ (Fig. 3.1), gives:

$$\frac{\partial F}{\partial A_k} = 0 \Rightarrow$$

$$\iint \left\{ v \cdot \left[\left(\frac{\partial A}{\partial x} \right) \cdot \frac{\partial}{\partial A_k} \left(\frac{\partial A}{\partial x} \right) + \left(\frac{\partial A}{\partial y} \right) \cdot \frac{\partial}{\partial A_k} \left(\frac{\partial A}{\partial y} \right) \right] - J \cdot \frac{\partial A}{\partial A_k} \right\} \cdot dx \cdot dy = 0 \Rightarrow$$

$$\iint v \cdot \left[\left(\frac{\partial A}{\partial x} \right) \cdot \frac{\partial}{\partial A_k} \left(\frac{\partial A}{\partial x} \right) + \left(\frac{\partial A}{\partial y} \right) \cdot \frac{\partial}{\partial A_k} \left(\frac{\partial A}{\partial y} \right) \right] \cdot dx \cdot dy =$$

$$= \iint J \cdot \frac{\partial A}{\partial A_k} \cdot dx \cdot dy, \quad \forall \, k = 1, 2, 3 \qquad (3.29)$$

Starting from (3.24), we find:

3.2 Finite Element Method

$$\frac{\partial A}{\partial x} = \frac{(y_2 - y_3) \cdot A_1 + (y_3 - y_1) \cdot A_2 + (y_1 - y_2) \cdot A_3}{2 \cdot \Delta}, \tag{3.30}$$

$$\frac{\partial A}{\partial y} = \frac{(x_3 - x_2) \cdot A_1 + (x_1 - x_3) \cdot A_2 + (x_2 - x_1) \cdot A_3}{2 \cdot \Delta}. \tag{3.31}$$

Differentiating (3.30) and (3.31) with respect to A_1, we obtain:

$$\frac{\partial}{\partial A_1}\left(\frac{\partial A}{\partial x}\right) = \frac{(y_2 - y_3)}{2 \cdot \Delta}, \tag{3.32}$$

$$\frac{\partial}{\partial A_1}\left(\frac{\partial A}{\partial y}\right) = \frac{(x_3 - x_2)}{2 \cdot \Delta}. \tag{3.33}$$

The first integral of (3.29), for the case $k=1$, is computed by substituting (3.30) to (3.33) into (3.29) as follows:

$$\iint v \cdot \left[\left(\frac{\partial A}{\partial x}\right) \cdot \frac{\partial}{\partial A_1}\left(\frac{\partial A}{\partial x}\right) + \left(\frac{\partial A}{\partial y}\right) \cdot \frac{\partial}{\partial A_1}\left(\frac{\partial A}{\partial y}\right)\right] \cdot dx \cdot dy =$$

$$= \iint v \cdot \left[\begin{array}{l}\left(\dfrac{(y_2 - y_3) \cdot A_1 + (y_3 - y_1) \cdot A_2 + (y_1 - y_2) \cdot A_3}{2 \cdot \Delta}\right) \cdot \left(\dfrac{y_2 - y_3}{2 \cdot \Delta}\right) + \\ \left(\dfrac{(x_3 - x_2) \cdot A_1 + (x_1 - x_3) \cdot A_2 + (x_2 - x_1) \cdot A_3}{2 \cdot \Delta}\right) \cdot \left(\dfrac{x_3 - x_2}{2 \cdot \Delta}\right)\end{array}\right] \cdot dx \cdot dy =$$

$$= \frac{v}{4 \cdot \Delta^2} \cdot \left\{\begin{array}{l}\left[(x_2 - x_3)^2 + (y_2 - y_3)^2\right] \cdot A_1 + \\ \left[(y_3 - y_1) \cdot (y_2 - y_3) + (x_3 - x_1) \cdot (x_2 - x_3)\right] \cdot A_2 + \\ \left[(y_1 - y_2) \cdot (y_2 - y_3) + (x_1 - x_2) \cdot (x_2 - x_3)\right] \cdot A_3\end{array}\right\} \cdot \iint dx \cdot dy \Rightarrow$$

$$\iint v \cdot \left[\left(\frac{\partial A}{\partial x}\right) \cdot \frac{\partial}{\partial A_1}\left(\frac{\partial A}{\partial x}\right) + \left(\frac{\partial A}{\partial y}\right) \cdot \frac{\partial}{\partial A_1}\left(\frac{\partial A}{\partial y}\right)\right] \cdot dx \cdot dy =$$

$$= \frac{1}{4 \cdot \mu_0 \cdot \mu_r \cdot \Delta} \cdot \left\{\begin{array}{l}\left[(x_2 - x_3)^2 + (y_2 - y_3)^2\right] \cdot A_1 + \\ \left[(y_3 - y_1) \cdot (y_2 - y_3) + (x_3 - x_1) \cdot (x_2 - x_3)\right] \cdot A_2 + \\ \left[(y_1 - y_2) \cdot (y_2 - y_3) + (x_1 - x_2) \cdot (x_2 - x_3)\right] \cdot A_3\end{array}\right\}. \tag{3.34}$$

where we used (a) the property that $\iint dx \cdot dy = \Delta$, where Δ is the area of the triangle of Fig. 3.1, and (b) the relation $v = 1/(\mu_0 \cdot \mu_r)$ from (3.4).

It can be proved (Chari and Salon 2000) that the second integral of (3.29) is computed as follows:

$$\iint J \cdot \frac{\partial A}{\partial A_k} \cdot dx \cdot dy = \frac{J \cdot \Delta}{3}, \quad \forall\, k = 1, 2, 3. \tag{3.35}$$

For the case $k=1$, (3.29) is simplified by substituting (3.34) and (3.35) into (3.29) as follows:

$$\iint v \cdot \left[\left(\frac{\partial A}{\partial x}\right) \cdot \frac{\partial}{\partial A_1}\left(\frac{\partial A}{\partial x}\right) + \left(\frac{\partial A}{\partial y}\right) \cdot \frac{\partial}{\partial A_1}\left(\frac{\partial A}{\partial y}\right) \right] \cdot dx \cdot dy = \iint J \cdot \frac{\partial A}{\partial A_1} \cdot dx \cdot dy \Rightarrow$$

$$\frac{1}{4 \cdot \mu_r \cdot \Delta} \cdot \begin{Bmatrix} \left[(x_2 - x_3)^2 + (y_2 - y_3)^2\right] \cdot A_1 + \\ \left[(y_3 - y_1) \cdot (y_2 - y_3) + (x_3 - x_1) \cdot (x_2 - x_3)\right] \cdot A_2 + \\ \left[(y_1 - y_2) \cdot (y_2 - y_3) + (x_1 - x_2) \cdot (x_2 - x_3)\right] \cdot A_3 \end{Bmatrix} = \frac{\mu_0 \cdot J \cdot \Delta}{3}. \tag{3.36}$$

Similarly, (3.29), for the case $k=2$, is equivalent to the following expression:

$$\frac{1}{4 \cdot \mu_r \cdot \Delta} \cdot \begin{Bmatrix} \left[(y_3 - y_1) \cdot (y_2 - y_3) + (x_3 - x_1) \cdot (x_2 - x_3)\right] \cdot A_1 + \\ \left[(x_1 - x_3)^2 + (y_1 - y_3)^2\right] \cdot A_2 + \\ \left[(y_3 - y_1) \cdot (y_1 - y_2) + (x_3 - x_1) \cdot (x_1 - x_2)\right] \cdot A_3 \end{Bmatrix} = \frac{\mu_0 \cdot J \cdot \Delta}{3}. \tag{3.37}$$

Also, (3.29), for the case $k=3$, is equivalent to the following formula:

$$\frac{1}{4 \cdot \mu_r \cdot \Delta} \cdot \begin{Bmatrix} \left[(y_1 - y_2) \cdot (y_2 - y_3) + (x_1 - x_2) \cdot (x_2 - x_3)\right] \cdot A_1 + \\ \left[(y_3 - y_1) \cdot (y_1 - y_2) + (x_3 - x_1) \cdot (x_1 - x_2)\right] \cdot A_2 + \\ \left[(x_1 - x_2)^2 + (y_1 - y_2)^2\right] \cdot A_3 \end{Bmatrix} = \frac{\mu_0 \cdot J \cdot \Delta}{3}. \tag{3.38}$$

Now, (3.36), (3.37), and (3.38) can be written in matrix form as follows:

$$\begin{bmatrix} K_{11} & K_{12} & K_{13} \\ K_{21} & K_{22} & K_{23} \\ K_{31} & K_{32} & K_{33} \end{bmatrix} \begin{bmatrix} A_1 \\ A_2 \\ A_3 \end{bmatrix} = \frac{\mu_0 \cdot J \cdot \Delta}{3} \cdot \begin{bmatrix} 1 \\ 1 \\ 1 \end{bmatrix}, \tag{3.39}$$

where:

$$\Delta = \frac{\left|(x_2 - x_3) \cdot (y_3 - y_1) - (y_2 - y_3) \cdot (x_3 - x_1)\right|}{2}, \tag{3.40}$$

3.2 Finite Element Method

$$K_{12} = K_{21} = \frac{1}{4 \cdot \mu_r \cdot \Delta} \cdot [(y_3 - y_1)\cdot(y_2 - y_3) + (x_3 - x_1)\cdot(x_2 - x_3)], \quad (3.41)$$

$$K_{23} = K_{32} = \frac{1}{4 \cdot \mu_r \cdot \Delta} \cdot [(y_3 - y_1)\cdot(y_1 - y_2) + (x_3 - x_1)\cdot(x_1 - x_2)], \quad (3.42)$$

$$K_{13} = K_{31} = \frac{1}{4 \cdot \mu_r \cdot \Delta} \cdot [(y_1 - y_2)\cdot(y_2 - y_3) + (x_1 - x_2)\cdot(x_2 - x_3)], \quad (3.43)$$

$$K_{11} = \frac{1}{4 \cdot \mu_r \cdot \Delta} \cdot [(x_2 - x_3)^2 + (y_2 - y_3)^2] = -K_{13} - K_{12}, \quad (3.44)$$

$$K_{22} = \frac{1}{4 \cdot \mu_r \cdot \Delta} \cdot [(x_1 - x_3)^2 + (y_1 - y_3)^2] = -K_{12} - K_{23}, \quad (3.45)$$

$$K_{33} = \frac{1}{4 \cdot \mu_r \cdot \Delta} \cdot [(x_1 - x_2)^2 + (y_1 - y_2)^2] = -K_{23} - K_{13}. \quad (3.46)$$

This means that instead of solving the differential equation (3.27), the much easier linear system (3.39) has to be solved, thanks to the finite element method.

3.2.3.1 Example 3.1

The mesh of the two-dimensional magnetostatic problem of Fig. 3.2 is composed of four triangles. There is a constant current density of 0.9 A/mm^2 in triangles 2 and 4, while the current density in triangles 1 and 3 is zero. The relative permeability in the triangles 2 and 4 is constant and equal to 1, while the relative permeability in triangles 1 and 3 is equal to 1000. The boundary conditions are such that the magnetic vector potential at nodes 1 and 2 is equal to zero. Compute the magnetic vector potential at nodes 3, 4, and 5. Calculate also the element magnetic flux densities.

Solution
The solution procedure is composed of the following four steps:

1. For each one of the four triangular finite elements, the components K_{ij} are computed using (3.41) to (3.46) and the matrix **KK** is calculated assuming that the four finite elements of Fig. 3.2 are disconnected.

2. The global matrix **KK** is formed taking into account the connections among the four finite elements of Fig. 3.2.

3. The boundary conditions are applied and the magnetic vector potentials are calculated.

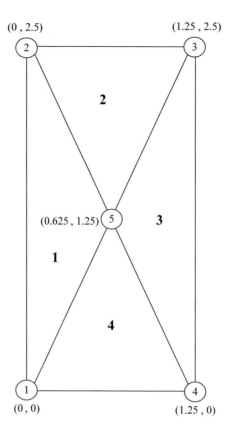

Fig. 3.2 The finite element mesh with four triangles and their rectangular coordinates in m

4. For each one of the four triangular finite elements, the magnetic flux density is computed.

Step 1 Computation of matrix **KK** for each one of the four finite elements

Computation of matrix **KK** *for the first finite element with nodes 2-1-5*

The first triangular finite element with code number 1 is composed of the vertices 2, 1, and 5, given in counter-clockwise order, as can be seen from Fig. 3.2. This means that:

- The first node is node 2 with rectangular coordinates $(x_1, y_1) = (0, 2.5)$.

- The second node is node 1 with rectangular coordinates $(x_2, y_2) = (0, 0)$.

- The third node is node 5 with rectangular coordinates $(x_3, y_3) = (0.625, 1.25)$.

3.2 Finite Element Method

The area of finite element 1 is computed using (3.40):

$$\Delta^{(1)} = \frac{|(x_2 - x_3) \cdot (y_3 - y_1) - (y_2 - y_3) \cdot (x_3 - x_1)|}{2} \Rightarrow$$

$$\Delta^{(1)} = \frac{|(0 - 0.625) \cdot (1.25 - 2.5) - (0 - 1.25) \cdot (0.625 - 0)|}{2} \Rightarrow \Delta^{(1)} = 0.78125 \text{ m}^2.$$

The components $K_{12}^{(1)}$, $K_{23}^{(1)}$, and $K_{13}^{(1)}$ of the first triangular finite element are computed using (3.41), (3.42), and (3.43), respectively:

$$K_{12}^{(1)} = K_{21}^{(1)} = \frac{1}{4 \cdot \mu_r^{(1)} \cdot \Delta^{(1)}} \cdot [(y_3 - y_1) \cdot (y_2 - y_3) + (x_3 - x_1) \cdot (x_2 - x_3)] \Rightarrow$$

$$K_{12}^{(1)} = \frac{1}{4 \cdot 1000 \cdot 0.78125} \cdot [(1.25 - 2.5) \cdot (0 - 1.25) + (0.625 - 0) \cdot (0 - 0.625)] \Rightarrow$$

$$K_{12}^{(1)} = K_{21}^{(1)} = 0.000375,$$

$$K_{23}^{(1)} = K_{32}^{(1)} = \frac{1}{4 \cdot \mu_r^{(1)} \cdot \Delta^{(1)}} \cdot [(y_3 - y_1) \cdot (y_1 - y_2) + (x_3 - x_1) \cdot (x_1 - x_2)] \Rightarrow$$

$$K_{23}^{(1)} = \frac{1}{4 \cdot 1000 \cdot 0.78125} \cdot [(1.25 - 2.5) \cdot (2.5 - 0) + (0.625 - 0) \cdot (0 - 0)] \Rightarrow$$

$$K_{23}^{(1)} = K_{32}^{(1)} = -0.001,$$

$$K_{13}^{(1)} = K_{31}^{(1)} = \frac{1}{4 \cdot \mu_r^{(1)} \cdot \Delta^{(1)}} \cdot [(y_1 - y_2) \cdot (y_2 - y_3) + (x_1 - x_2) \cdot (x_2 - x_3)] \Rightarrow$$

$$K_{13}^{(1)} = \frac{1}{4 \cdot 1000 \cdot 0.78125} \cdot [(2.5 - 0) \cdot (0 - 1.25) + (0 - 0) \cdot (0 - 0.625)] \Rightarrow$$

$$K_{13}^{(1)} = K_{31}^{(1)} = -0.001.$$

The components $K_{11}^{(1)}$, $K_{22}^{(1)}$, and $K_{33}^{(1)}$ of the first triangular finite element are computed using (3.44), (3.45), and (3.46), respectively:

$$K_{11}^{(1)} = -K_{13}^{(1)} - K_{12}^{(1)} = -(-0.001) - 0.000375 \Rightarrow K_{11}^{(1)} = 0.001375,$$

$$K_{22}^{(1)} = -K_{12}^{(1)} - K_{23}^{(1)} = -0.000375 - (-0.001) \Rightarrow K_{22}^{(1)} = 0.000625,$$

$$K_{33}^{(1)} = -K_{23}^{(1)} - K_{13}^{(1)} = -(-0.001) - (-0.001) \Rightarrow K_{33}^{(1)} = 0.002.$$

For the first triangular finite element (2-1-5):

- The first node is node 2, which implies the index substitution $1 \leftarrow 2$, i.e., the index 1 is substituted by index 2.

- The second node is node 1, which implies the index substitution $2 \leftarrow 1$.

- The third node is node 5, which implies the index substitution $3 \leftarrow 5$.

The above substitution of indices determines the contribution of the first triangular finite element to the global matrix **KK** as follows:

$$KK_{21}^{(1)} = KK_{12}^{(1)} = K_{12}^{(1)} = K_{21}^{(1)} = 0.000375,$$

$$KK_{15}^{(1)} = KK_{51}^{(1)} = K_{23}^{(1)} = K_{32}^{(1)} = -0.001,$$

$$KK_{25}^{(1)} = KK_{52}^{(1)} = K_{13}^{(1)} = K_{31}^{(1)} = -0.001.$$

$$KK_{22}^{(1)} = K_{11}^{(1)} = 0.000625,$$

$$KK_{11}^{(1)} = K_{22}^{(1)} = 0.000625,$$

$$KK_{55}^{(1)} = K_{33}^{(1)} = 0.002.$$

Consequently, the contribution of the first finite element 2-1-5 to the global matrix **KK** is:

$$\mathbf{KK}^{(1)} = \begin{bmatrix} KK_{11}^{(1)} & KK_{12}^{(1)} & KK_{13}^{(1)} & KK_{14}^{(1)} & KK_{15}^{(1)} \\ KK_{21}^{(1)} & KK_{22}^{(1)} & KK_{23}^{(1)} & KK_{24}^{(1)} & KK_{25}^{(1)} \\ KK_{31}^{(1)} & KK_{32}^{(1)} & KK_{33}^{(1)} & KK_{34}^{(1)} & KK_{34}^{(1)} \\ KK_{41}^{(1)} & KK_{42}^{(1)} & KK_{43}^{(1)} & KK_{44}^{(1)} & KK_{45}^{(1)} \\ KK_{51}^{(1)} & KK_{52}^{(1)} & KK_{53}^{(1)} & KK_{54}^{(1)} & KK_{55}^{(1)} \end{bmatrix} \Rightarrow$$

$$\mathbf{KK}^{(1)} = \begin{bmatrix} 0.000625 & 0.000375 & 0 & 0 & -0.001 \\ 0.000375 & 0.000625 & 0 & 0 & -0.001 \\ 0 & 0 & 0 & 0 & 0 \\ 0 & 0 & 0 & 0 & 0 \\ -0.001 & -0.001 & 0 & 0 & 0.002 \end{bmatrix}.$$

3.2 Finite Element Method

Computation of matrix **KK** *for the second finite element with nodes 3-2-5*

The second triangular finite element with code number 2 is composed of the vertices 3, 2, and 5, given in counter-clockwise order, as can be seen from Fig. 3.2. This means that:

- The first node is node 3 with rectangular coordinates $(x_1, y_1) = (1.25, 2.5)$.

- The second node is node 2 with rectangular coordinates $(x_2, y_2) = (0, 2.5)$.

- The third node is node 5 with rectangular coordinates $(x_3, y_3) = (0.625, 1.25)$.

The area of finite element 2 is:

$$\Delta^{(2)} = \frac{|(x_2 - x_3)\cdot(y_3 - y_1) - (y_2 - y_3)\cdot(x_3 - x_1)|}{2} \Rightarrow$$

$$\Delta^{(2)} = \frac{|(0 - 0.625)\cdot(1.25 - 2.5) - (2.5 - 1.25)\cdot(0.625 - 1.25)|}{2} \Rightarrow$$

$$\Delta^{(2)} = 0.78125 \text{ m}^2.$$

The components $K_{ij}^{(2)}$ of the second triangular finite element are computed as follows:

$$K_{12}^{(2)} = K_{21}^{(2)} = \frac{1}{4\cdot \mu_r^{(2)} \cdot \Delta^{(2)}}\cdot[(y_3 - y_1)\cdot(y_2 - y_3) + (x_3 - x_1)\cdot(x_2 - x_3)] \Rightarrow$$

$$K_{12}^{(2)} = \frac{1}{4\cdot 1\cdot 0.78125}\cdot[(1.25 - 2.5)\cdot(2.5 - 1.25) + (0.625 - 1.25)\cdot(0 - 0.625)] \Rightarrow$$

$$K_{12}^{(2)} = K_{21}^{(2)} = -0.375,$$

$$K_{23}^{(2)} = K_{32}^{(2)} = \frac{1}{4\cdot \mu_r^{(2)} \cdot \Delta^{(2)}}\cdot[(y_3 - y_1)\cdot(y_1 - y_2) + (x_3 - x_1)\cdot(x_1 - x_2)] \Rightarrow$$

$$K_{23}^{(2)} = \frac{1}{4\cdot 1\cdot 0.78125}\cdot[(1.25 - 2.5)\cdot(2.5 - 2.5) + (0.625 - 1.25)\cdot(1.25 - 0)] \Rightarrow$$

$$K_{23}^{(2)} = K_{32}^{(2)} = -0.25,$$

$$K_{13}^{(2)} = K_{31}^{(2)} = \frac{1}{4 \cdot \mu_r^{(2)} \cdot \Delta^{(2)}} \cdot [(y_1 - y_2) \cdot (y_2 - y_3) + (x_1 - x_2) \cdot (x_2 - x_3)] \Rightarrow$$

$$K_{13}^{(2)} = \frac{1}{4 \cdot 1 \cdot 0.78125} \cdot [(2.5 - 2.5) \cdot (2.5 - 1.25) + (1.25 - 0) \cdot (0 - 0.625)] \Rightarrow$$

$$K_{13}^{(2)} = K_{31}^{(2)} = -0.25,$$

$$K_{11}^{(2)} = -K_{13}^{(2)} - K_{12}^{(2)} = -(-0.25) - (-0.375) \Rightarrow K_{11}^{(2)} = 0.625,$$

$$K_{22}^{(2)} = -K_{12}^{(2)} - K_{23}^{(2)} = -(-0.375) - (-0.25) \Rightarrow K_{22}^{(1)} = 0.625,$$

$$K_{33}^{(2)} = -K_{23}^{(2)} - K_{13}^{(2)} = -(-0.25) - (-0.25) \Rightarrow K_{33}^{(2)} = 0.5.$$

For the second triangular finite element (3-2-5):

- The first node is node 3, which implies the index substitution $1 \leftarrow 3$, i.e., the index 1 is substituted by index 3.

- The second node is node 2, which implies the index substitution $2 \leftarrow 2$.

- The third node is node 5, which implies the index substitution $3 \leftarrow 5$.

The above substitution of indices determines the contribution of the second triangular finite element to the global matrix **KK** as follows:

$$KK_{32}^{(2)} = KK_{23}^{(2)} = K_{12}^{(2)} = K_{21}^{(2)} = -0.375,$$

$$KK_{25}^{(2)} = KK_{52}^{(2)} = K_{23}^{(2)} = K_{32}^{(2)} = -0.25,$$

$$KK_{35}^{(2)} = KK_{53}^{(2)} = K_{13}^{(2)} = K_{31}^{(2)} = -0.25.$$

$$KK_{33}^{(2)} = K_{11}^{(2)} = 0.625,$$

$$KK_{22}^{(2)} = K_{22}^{(2)} = 0.625,$$

$$KK_{55}^{(2)} = K_{33}^{(2)} = 0.5.$$

Consequently, the contribution of the second finite element 3-2-5 to the global matrix **KK** is:

3.2 Finite Element Method

$$\mathbf{KK}^{(2)} = \begin{bmatrix} KK_{11}^{(2)} & KK_{12}^{(2)} & KK_{13}^{(2)} & KK_{14}^{(2)} & KK_{15}^{(2)} \\ KK_{21}^{(2)} & KK_{22}^{(2)} & KK_{23}^{(2)} & KK_{24}^{(2)} & KK_{25}^{(2)} \\ KK_{31}^{(2)} & KK_{32}^{(2)} & KK_{33}^{(2)} & KK_{34}^{(2)} & KK_{34}^{(2)} \\ KK_{41}^{(2)} & KK_{42}^{(2)} & KK_{43}^{(2)} & KK_{44}^{(2)} & KK_{45}^{(2)} \\ KK_{51}^{(2)} & KK_{52}^{(2)} & KK_{53}^{(2)} & KK_{54}^{(2)} & KK_{55}^{(2)} \end{bmatrix} \Rightarrow$$

$$\mathbf{KK}^{(2)} = \begin{bmatrix} 0 & 0 & 0 & 0 & 0 \\ 0 & 0.625 & -0.375 & 0 & -0.25 \\ 0 & -0.375 & 0.625 & 0 & -0.25 \\ 0 & 0 & 0 & 0 & 0 \\ 0 & -0.25 & -0.25 & 0 & 0.5 \end{bmatrix}.$$

Computation of matrix **KK** *for the third finite element with nodes 4-3-5*

Similarly, the contribution of the third finite element 4-3-5 to the global matrix **KK** is:

$$\mathbf{KK}^{(3)} = \begin{bmatrix} 0 & 0 & 0 & 0 & 0 \\ 0 & 0 & 0 & 0 & 0 \\ 0 & 0 & 0.000625 & 0.000375 & -0.001 \\ 0 & 0 & 0.000375 & 0.000625 & -0.001 \\ 0 & 0 & -0.001 & -0.001 & 0.002 \end{bmatrix}.$$

Computation of matrix **KK** *for the fourth finite element with nodes 1-4-5*

Similarly, the contribution of the fourth finite element 1-4-5 to the global matrix **KK** is:

$$\mathbf{KK}^{(4)} = \begin{bmatrix} 0.625 & 0 & 0 & -0.375 & -0.25 \\ 0 & 0 & 0 & 0 & 0 \\ 0 & 0 & 0 & 0 & 0 \\ -0.375 & 0 & 0 & 0.625 & -0.25 \\ -0.25 & 0 & 0 & -0.25 & 0.5 \end{bmatrix}.$$

Step 2 Computation of global matrix **KK**

The global matrix **KK** is computed as follows:

$$\mathbf{KK} = \mathbf{KK}^{(1)} + \mathbf{KK}^{(2)} + \mathbf{KK}^{(3)} + \mathbf{KK}^{(4)} \Rightarrow$$

$$\mathbf{KK} = \begin{bmatrix} 0.625625 & 0.000375 & 0 & -0.375 & -0.251 \\ 0.000375 & 0.625625 & -0.375 & 0 & -0.251 \\ 0 & -0.375 & 0.625625 & 0.000375 & -0.251 \\ -0.375 & 0 & 0.000375 & 0.625625 & -0.251 \\ -0.251 & -0.251 & -0.251 & -0.251 & 1.004 \end{bmatrix}.$$

Step 3 Determination of magnetic vector potentials

Based on (3.39), the global set of equations is:

$$\mathbf{KK} \cdot \mathbf{A} = \frac{\mu_0 \cdot \mathbf{J} \cdot \Delta}{3} \Rightarrow$$

$$\begin{bmatrix} 0.625625 & 0.000375 & 0 & -0.375 & -0.251 \\ 0.000375 & 0.625625 & -0.375 & 0 & -0.251 \\ 0 & -0.375 & 0.625625 & 0.000375 & -0.251 \\ -0.375 & 0 & 0.000375 & 0.625625 & -0.251 \\ -0.251 & -0.251 & -0.251 & -0.251 & 1.004 \end{bmatrix} \cdot \begin{bmatrix} A_1 \\ A_2 \\ A_3 \\ A_4 \\ A_5 \end{bmatrix} = \frac{\mu_0 \cdot \mathbf{J} \cdot \Delta}{3} \Rightarrow$$

$$\begin{bmatrix} 0.625625 & 0.000375 & 0 & -0.375 & -0.251 \\ 0.000375 & 0.625625 & -0.375 & 0 & -0.251 \\ 0 & -0.375 & 0.625625 & 0.000375 & -0.251 \\ -0.375 & 0 & 0.000375 & 0.625625 & -0.251 \\ -0.251 & 0.251 & -0.251 & -0.251 & 1.004 \end{bmatrix} \cdot \begin{bmatrix} 0 \\ 0 \\ A_3 \\ A_4 \\ A_5 \end{bmatrix} = \frac{\mu_0 \cdot \Delta}{3} \cdot \begin{bmatrix} J_1 \\ J_2 \\ J_3 \\ J_4 \\ J_5 \end{bmatrix} \Rightarrow$$

$$\begin{bmatrix} 0.625625 & 0.000375 & -0.251 \\ 0.000375 & 0.625625 & -0.251 \\ -0.251 & -0.251 & 1.004 \end{bmatrix} \cdot \begin{bmatrix} A_3 \\ A_4 \\ A_5 \end{bmatrix} = \frac{\mu_0 \cdot \Delta}{3} \cdot \begin{bmatrix} J_3 \\ J_4 \\ J_5 \end{bmatrix}, \qquad (3.47)$$

where the input data $A_1 = 0$ and $A_2 = 0$ were used.

As can be seen from Fig. 3.2, node 3 belongs to triangles 2 and 3, so the current density J_3 at node 3 is computed as follows:

$$J_3 = J^{(2)} + J^{(3)} \Rightarrow \frac{\mu_0 \cdot J_3 \cdot \Delta}{3} = \frac{\mu_0}{3} \cdot \left[J^{(2)} \cdot \Delta^{(2)} + J^{(3)} \cdot \Delta^{(3)} \right] \Rightarrow$$

$$\frac{\mu_0 \cdot J_3 \cdot \Delta}{3} = \frac{4 \cdot \pi \cdot 10^{-7} \frac{H}{m}}{3} \cdot \left[\left(0.9 \cdot 10^6 \frac{A}{m^2} \right) \cdot (0.78125 \text{ m}^2) + \left(0 \frac{A}{m^2} \right) \cdot (0.78125 \text{ m}^2) \right] \Rightarrow$$

3.2 Finite Element Method

$$\frac{\mu_0 \cdot J_3 \cdot \Delta}{3} = 0.29452 \frac{H \cdot A}{m} \Rightarrow$$

$$\frac{\mu_0 \cdot J_3 \cdot \Delta}{3} = 0.29452 \frac{Wb}{m}, \quad (3.48)$$

where $J^{(2)} = 0.9 \, A/mm^2 = 0.9 \cdot 10^6 \, A/m^2$ is the current density in triangle 2, $J^{(3)}$ is the current density in triangle 3, $\Delta^{(2)}$ is the area of triangle 2, and $\Delta^{(3)}$ is the area of triangle 3.

Similarly, the current density J_4 at node 4 is computed as follows:

$$J_4 = J^{(3)} + J^{(4)} \Rightarrow \frac{\mu_0 \cdot J_4 \cdot \Delta}{3} = \frac{\mu_0}{3} \cdot \left[J^{(3)} \cdot \Delta^{(3)} + J^{(4)} \cdot \Delta^{(4)} \right] \Rightarrow$$

$$\frac{\mu_0 \cdot J_4 \cdot \Delta}{3} = 0.29452 \frac{Wb}{m}. \quad (3.49)$$

The current density J_5 at node 5 is:

$$J_5 = J^{(1)} + J^{(2)} + J^{(3)} + J^{(4)} \Rightarrow$$

$$\frac{\mu_0 \cdot J_5 \cdot \Delta}{3} = \frac{\mu_0}{3} \cdot \left[J^{(1)} \cdot \Delta^{(1)} + J^{(2)} \cdot \Delta^{(2)} + J^{(3)} \cdot \Delta^{(3)} + J^{(4)} \cdot \Delta^{(4)} \right] \Rightarrow$$

$$\frac{\mu_0 \cdot J_5 \cdot \Delta}{3} = 0.58905 \frac{Wb}{m}. \quad (3.50)$$

Substituting (3.48), (3.49), and (3.50) into (3.47) we obtain:

$$\begin{bmatrix} 0.625625 & 0.000375 & -0.251 \\ 0.000375 & 0.625625 & -0.251 \\ -0.251 & -0.251 & 1.004 \end{bmatrix} \cdot \begin{bmatrix} A_3 \\ A_4 \\ A_5 \end{bmatrix} = \begin{bmatrix} 0.29452 \\ 0.29452 \\ 0.58905 \end{bmatrix} \Rightarrow$$

$$\begin{bmatrix} A_3 \\ A_4 \\ A_5 \end{bmatrix} = \begin{bmatrix} 0.625625 & 0.000375 & -0.251 \\ 0.000375 & 0.625625 & -0.251 \\ -0.251 & -0.251 & 1.004 \end{bmatrix}^{-1} \cdot \begin{bmatrix} 0.29452 \\ 0.29452 \\ 0.58905 \end{bmatrix} \Rightarrow$$

$$\begin{bmatrix} A_3 \\ A_4 \\ A_5 \end{bmatrix} = \begin{bmatrix} 1.7987 & 0.1993 & 0.4995 \\ 0.1993 & 1.7987 & 0.4995 \\ 0.4995 & 0.4995 & 1.2458 \end{bmatrix} \cdot \begin{bmatrix} 0.29452 \\ 0.29452 \\ 0.58905 \end{bmatrix} \Rightarrow \begin{bmatrix} A_3 \\ A_4 \\ A_5 \end{bmatrix} = \begin{bmatrix} 0.8827 \\ 0.8827 \\ 1.0280 \end{bmatrix} \frac{\text{Wb}}{\text{m}}.$$

Step 4 Computation of magnetic flux density for each finite element

Computation of magnetic flux density for the first finite element

The coordinates of the first finite element are: $(x_1, y_1) = (0, 2.5)$, $(x_2, y_2) = (0, 0)$, and $(x_3, y_3) = (0.625, 1.25)$, and its area is $\Delta = 0.78125 \text{ m}^2$. The magnetic vector potential at its nodes 2, 1, and 5 is $A_2 = 0$, $A_1 = 0$, and $A_5 = 1.0280 \text{ Wb/m}$, respectively. The x and y coordinates of the magnetic flux density are computed using (3.25) and (3.26), respectively:

$$B_x = \frac{(x_3 - x_2) \cdot A_2 + (x_1 - x_3) \cdot A_1 + (x_2 - x_1) \cdot A_5}{2 \cdot \Delta} \Rightarrow$$

$$B_x = \frac{(0.625 - 0) \cdot 0 + (0 - 0.625) \cdot 0 + (0 - 0) \cdot 1.0280}{2 \cdot 0.78125} \Rightarrow B_x = 0$$

and

$$B_y = \frac{(y_3 - y_2) \cdot A_2 + (y_1 - y_3) \cdot A_1 + (y_2 - y_1) \cdot A_5}{2 \cdot \Delta} \Rightarrow$$

$$B_y = \frac{(1.25 - 0) \cdot 0 + (2.5 - 1.25) \cdot 0 + (0 - 2.5) \cdot 1.0280}{2 \cdot 0.78125} \Rightarrow B_y = -1.6449 \text{ T}.$$

The magnitude of the magnetic flux density at the first triangle is computed using (3.13):

$$B = \sqrt{B_x^2 + B_y^2} \Rightarrow B = \sqrt{0^2 + (-1.6449)^2} \Rightarrow B = 1.6449 \text{ T}.$$

Computation of magnetic flux density for the second finite element

The coordinates of the second finite element are: $(x_1, y_1) = (1.25, 2.5)$, $(x_2, y_2) = (0, 2.5)$, and $(x_3, y_3) = (0.625, 1.25)$, and its area is $\Delta = 0.78125 \text{ m}^2$. The magnetic vector potential at its nodes 3, 2, and 5 is $A_3 = 0.8827 \text{ Wb/m}$, $A_2 = 0$, and $A_5 = 1.0280 \text{ Wb/m}$, respectively. The x and y coordinates of the magnetic flux density are computed using (3.25) and (3.26), respectively:

3.2 Finite Element Method

$$B_x = \frac{(x_3 - x_2) \cdot A_3 + (x_1 - x_3) \cdot A_2 + (x_2 - x_1) \cdot A_5}{2 \cdot \Delta} \Rightarrow$$

$$B_x = \frac{(0.625 - 0) \cdot 0.8827 + (1.25 - 0.625) \cdot 0 + (0 - 1.25) \cdot 1.0280}{2 \cdot 0.78125} \Rightarrow$$

$$B_x = -0.4694 \text{ T}$$

and

$$B_y = \frac{(y_3 - y_2) \cdot A_3 + (y_1 - y_3) \cdot A_2 + (y_2 - y_1) \cdot A_5}{2 \cdot \Delta} \Rightarrow$$

$$B_y = \frac{(1.25 - 2.5) \cdot 0.8827 + (2.5 - 1.25) \cdot 0 + (2.5 - 2.5) \cdot 1.0280}{2 \cdot 0.78125} \Rightarrow$$

$$B_y = -0.7062 \text{ T}.$$

The magnitude of the magnetic flux density at the second triangle is computed using (3.13):

$$B = \sqrt{B_x^2 + B_y^2} \Rightarrow B = \sqrt{(-0.4694)^2 + (-0.7062)^2} \Rightarrow B = 0.8479 \text{ T}.$$

Computation of magnetic flux density for the third finite element

Similarly, the magnitude of the magnetic flux density of the third finite element is $B = 0.2326$ T.

Computation of magnetic flux density for the fourth finite element

The magnitude of the magnetic flux density of the fourth finite element is $B = 0.8479$ T.

Final results

The magnetic vector potentials at nodes 3, 4, and 5 are:

$$\begin{bmatrix} A_3 \\ A_4 \\ A_5 \end{bmatrix} = \begin{bmatrix} 0.8827 \\ 0.8827 \\ 1.0280 \end{bmatrix} \frac{\text{Wb}}{\text{m}}.$$

The magnetic flux densities of the four triangles are:

$$\begin{bmatrix} B_1 \\ B_2 \\ B_3 \\ B_4 \end{bmatrix} = \begin{bmatrix} 1.6449 \\ 0.8479 \\ 0.2326 \\ 0.8479 \end{bmatrix} \text{T}.$$

3.2.4 Solution of Nonlinear Magnetostatic Problems

Most transformer magnetostatic problems are inherently *nonlinear*. The analysis of transformer magnetic circuit requires knowledge of the physical properties of the magnetic materials used. Magnetic material manufacturers normally supply the magnetic material characteristic in the form of $B-H$ curves (e.g., Fig. 3.3).

The basic principles of linear finite element analysis, presented in Sect. 3.2.3 carry over to nonlinear magnetostatic problems almost without modification. A stationary functional is constructed and discretized over finite elements. The equations resulting from nonlinear magnetostatic problems are nonlinear too (Silvester and Ferrari 1996). They can be solved by several methods. Simple iterative methods are not always stable and can take a long time to converge. The most common approach to the solution of nonlinear equations derived for the nonlinear magnetostatic problem is to use *Newton iterative method* since its convergence is in principle much faster than that of simple iterative methods.

Newton's method requires a mathematical model describing the properties of the nonlinear magnetic material. The classical $B-H$ curve is not the best choice. It is a common practice to use the $v_r = v_r(B^2)$ curve to introduce ferromagnetic material properties in the finite element method (Silvester and Ferrari 1996). The $v_r - B^2$ curve (e.g., Fig. 3.4) can easily be constructed from the $B-H$ curve that is commonly supplied by magnetic material manufacturers. The relative reluctivity as a function of flux density squared, $v_r = v_r(B^2)$, is commonly modeled using cubic splines (Silvester and Ferrari 1996), however, there is a danger that unacceptable undulations may be produced in the fitted curve (Reece and Preston 2000). Some algebraic expressions have been found to be useful, e.g.:

$$v_r = v_0 + P \cdot e^{q \cdot B^2}, \tag{3.51}$$

and

$$v_r = v_0 + P \cdot B^{2q}, \tag{3.52}$$

where v_0, P, and q are chosen to fit the $v_r - B^2$ curve (Coulson 1981).

3.2 Finite Element Method

Newton's iterative method for the solution of nonlinear finite element equations is composed of the following steps:

1. We start with iteration 0. We assume a certain value for the magnetic vector potentials and compute the element magnetic flux densities. Next, using the $v_r - B^2$ curve, we find the relative reluctivity and afterwards the relative permeability for all the finite elements that correspond to the geometry covered by the nonlinear magnetic material. We then form the global matrix **KK** and compute the error vector in the solution of the set of equations $\mathbf{KK} \cdot \mathbf{A} = \dfrac{\mu_0 \cdot \mathbf{J} \cdot \Delta}{3}$. If the infinite norm of this error vector is smaller than a predefined tolerance (e.g., 10^{-4}), then the solution has been found, otherwise we proceed to step 2.

2. We enter the next iteration, i.e., the iteration number is increased by one. We compute the new values of the magnetic vector potential at all nodes by solving the set of linear equations $\mathbf{KK} \cdot \mathbf{A} = \dfrac{\mu_0 \cdot \mathbf{J} \cdot \Delta}{3}$ formed during the previous iteration. Next, we calculate the new values of the element magnetic flux densities. Next, using the $v_r - B^2$ curve, we find the relative reluctivity and then the relative permeability for all the finite elements that correspond to the geometry covered by the nonlinear magnetic material. We then form the new global matrix **KK** and compute the new error vector in the solution of the set of equations $\mathbf{KK} \cdot \mathbf{A} = \dfrac{\mu_0 \cdot \mathbf{J} \cdot \Delta}{3}$. If the infinite norm of this new error vector is smaller than a predefined tolerance (e.g., 10^{-4}), then the solution has been found, otherwise we repeat step 2.

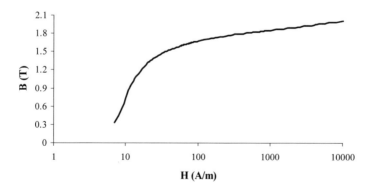

Fig. 3.3 $B - H$ curve for the magnetic material in triangles 1 and 3 of Fig. 3.2

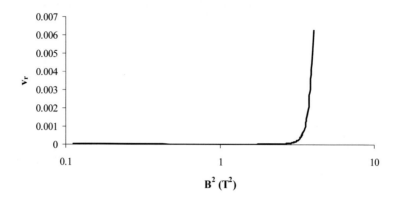

Fig. 3.4 $v_r - B^2$ curve for the magnetic material in triangles 1 and 3 of Fig. 3.2

3.2.4.1 Example 3.2

Solve Example 3.1, if the triangles 1 and 3 of Fig. 3.2 are composed of a nonlinear magnetic material with the $B-H$ characteristic shown in Fig. 3.3.

Solution

In order to solve the problem, it is convenient to construct the $v_r - B^2$ characteristic as follows:

1. From the $B-H$ curve of Fig. 3.3, we find N pairs of data (H_i, B_i), $\forall\, i = 1,...,N$.

2. For each pair (H_i, B_i), we compute the respective pair $(B_i^2, v_{r,i})$ using the formula $v_{r,i} = \mu_0 \cdot \dfrac{H_i}{B_i}$, $\forall\, i = 1,...,N$, so the $v_r - B^2$ characteristic is constructed. In particular, using $N = 93$ pairs of data, the $v_r - B^2$ curve that corresponds to the $B-H$ characteristic of Fig. 3.3 is shown in Fig. 3.4.

3. Using curve-fitting techniques, the $v_r - B^2$ curve of Fig. 3.4 can be expressed mathematically as follows:

3.2 Finite Element Method

$$v_r = \begin{cases} \begin{aligned} & 0.0002664 \cdot (B^2)^6 - 0.001015 \cdot (B^2)^5 + \\ & +0.001553 \cdot (B^2)^4 - 0.001222 \cdot (B^2)^3 + \\ & +0.0005382 \cdot (B^2)^2 - 0.000141 \cdot (B^2) + \\ & +3.681 \cdot 10^{-5} \end{aligned} & , \text{if } 0 \leq B^2 \leq 1.04 \\[1em] \begin{aligned} & 4.082 \cdot 10^{-5} \cdot (B^2)^6 - 0.0004697 \cdot (B^2)^5 + \\ & +0.002229 \cdot (B^2)^4 - 0.005564 \cdot (B^2)^3 + \\ & +0.007691 \cdot (B^2)^2 - 0.005576 \cdot (B^2) + \\ & +0.001671 \end{aligned} & , \text{if } 1.04 < B^2 \leq 3 \\[1em] 1 - e^{-0.001439 \cdot B^2 + 0.004295} & , \text{if } 3 < B^2 \leq 3.55 \\[1em] 1 - e^{-0.01083 \cdot B^2 + 0.03787} & , \text{if } B^2 > 3.55 \end{cases} \quad (3.53)$$

Having expressed mathematically the $v_r - B^2$ curve, the problem is solved as follows:

1. We start with *iteration* 0, where we assume that the magnetic vector potential at all nodes is zero, so the element magnetic flux densities are also zero. Next, we find the relative reluctivity in triangles 1 and 3. We form the global matrix **KK** as well as the set of equations $\mathbf{KK} \cdot \mathbf{A} = \dfrac{\mu_0 \cdot \mathbf{J} \cdot \Delta}{3}$. Next, we compute the error vector in the solution of the set of equations $\mathbf{KK} \cdot \mathbf{A} = \dfrac{\mu_0 \cdot \mathbf{J} \cdot \Delta}{3}$. If the infinite norm of this error vector is smaller than a predefined tolerance (e.g., 10^{-4}), then the solution has been found, otherwise we proceed to step 2.

2. We enter the next iteration (i.e., *iteration* 1), where we compute the new values of the magnetic vector potentials at nodes 3, 4, and 5 as well as the new values of the element magnetic flux densities. Next, we find the relative reluctivity in triangles 1 and 3. We form the global matrix **KK**, and compute the error vector in the solution of the set of equations $\mathbf{KK} \cdot \mathbf{A} = \dfrac{\mu_0 \cdot \mathbf{J} \cdot \Delta}{3}$. If the infinite norm of this error vector is smaller than a predefined tolerance (e.g., 10^{-4}), then the solution is found, otherwise we repeat step 2.

Iteration 0

We assume that the magnetic vector potential at all nodes is zero, so the element magnetic flux densities are also zero.

Since $B^2 = 0$ in triangle 1, using the first branch of (3.53), we find that $v_r = 3.681 \cdot 10^{-5}$, so $\mu_r^{(1)} = 1/v_r \Rightarrow \mu_r^{(1)} = 27166.53$. Similarly, in triangle 3, we find that $\mu_r^{(3)} = 27166.53$.

On the other hand, in triangles 2 and 4, the relative permeability is equal to 1, as given in Example 3.1.

Following the solution methodology of Example 3.1, we form the global matrix $\mathbf{KK}$ and the set of equations $\mathbf{KK} \cdot \mathbf{A} = \dfrac{\mu_0 \cdot \mathbf{J} \cdot \Delta}{3}$. The boundary conditions impose that $A_1 = A_2 = 0$, so the resulting set of equations is:

$$\mathbf{KK} \cdot \mathbf{A} = \frac{\mu_0 \cdot \mathbf{J} \cdot \Delta}{3} \Rightarrow$$

$$\begin{bmatrix} 0.625023 & 0.000014 & 0 & -0.375 & -0.25004 \\ 0.000014 & 0.625023 & -0.375 & 0 & -0.25004 \\ 0 & -0.375 & 0.625023 & 0.000014 & -0.25004 \\ -0.375 & 0 & 0.000014 & 0.625023 & -0.25004 \\ -0.25004 & -0.25004 & -0.25004 & -0.25004 & 1.000147 \end{bmatrix} \cdot \begin{bmatrix} A_1 \\ A_2 \\ A_3 \\ A_4 \\ A_5 \end{bmatrix} = \begin{bmatrix} 0.29452 \\ 0.29452 \\ 0.29452 \\ 0.29452 \\ 0.58905 \end{bmatrix} \Rightarrow$$

$$\begin{bmatrix} 0.625023 & 0.000014 & -0.25004 \\ 0.000014 & 0.625023 & -0.25004 \\ -0.25004 & -0.25004 & 1.000147 \end{bmatrix} \cdot \begin{bmatrix} A_3 \\ A_4 \\ A_5 \end{bmatrix} = \begin{bmatrix} 0.29452 \\ 0.29452 \\ 0.58905 \end{bmatrix}. \qquad (3.54)$$

Since at iteration 0, it is assumed that all magnetic vector potentials are zero, i.e., $A_3 = A_4 = A_5 = 0$, this assumption results in the following error vector in the solution of the set of equations (3.54):

$$\mathbf{err} = \begin{bmatrix} 0.625023 & 0.000014 & -0.25004 \\ 0.000014 & 0.625023 & -0.25004 \\ -0.25004 & -0.25004 & 1.000147 \end{bmatrix} \cdot \begin{bmatrix} A_3 \\ A_4 \\ A_5 \end{bmatrix} - \begin{bmatrix} 0.29452 \\ 0.29452 \\ 0.58905 \end{bmatrix} \Rightarrow$$

$$\mathbf{err} = \begin{bmatrix} 0.625023 & 0.000014 & -0.25004 \\ 0.000014 & 0.625023 & -0.25004 \\ -0.25004 & -0.25004 & 1.000147 \end{bmatrix} \cdot \begin{bmatrix} 0 \\ 0 \\ 0 \end{bmatrix} - \begin{bmatrix} 0.29452 \\ 0.29452 \\ 0.58905 \end{bmatrix} \Rightarrow$$

3.2 Finite Element Method

$$\mathbf{err} = \begin{bmatrix} -0.29452 \\ -0.29452 \\ -0.58905 \end{bmatrix}.$$

The infinite norm n_{err} of the error vector **err** is equal to the maximum value of the absolute value of its elements:

$$n_{err} = \|\mathbf{err}\|_\infty = \max(abs(\mathbf{err})) = \max \begin{bmatrix} 0.29452 \\ 0.29452 \\ 0.58905 \end{bmatrix} \Rightarrow n_{err} = 0.58905.$$

The tolerance is $tol = 10^{-4}$ and since $n_{err} > tol$, another iteration is required.

Iteration 1

Solving the system of linear equations (3.54), the new values of the magnetic vector potentials at nodes 3, 4, and 5 are:

$$\begin{bmatrix} A_3 \\ A_4 \\ A_5 \end{bmatrix} = \begin{bmatrix} 0.8835 \\ 0.8835 \\ 1.0307 \end{bmatrix} \frac{\text{Wb}}{\text{m}}. \tag{3.55}$$

Having calculated the new values of the magnetic vector potentials, the new values of the flux densities of the four triangles are computed using the sequence of calculations presented in Example 3.1 and the results are:

$$\begin{bmatrix} B_1 \\ B_2 \\ B_3 \\ B_4 \end{bmatrix} = \begin{bmatrix} 1.6492 \\ 0.8495 \\ 0.2355 \\ 0.8495 \end{bmatrix} \text{T}. \tag{3.56}$$

Since $B^2 = 1.6492^2 = 2.7198$ in triangle 1, using the second branch of (3.53), we find that $v_r = 0.0000449$, so $\mu_r^{(1)} = 1/v_r \Rightarrow \mu_r^{(1)} = 22271.01$.

Since $B^2 = 0.2355^2 = 0.0555$ in triangle 3, using the first branch of (3.53), we find that $v_r = 0.00003045$, so $\mu_r^{(3)} = 1/v_r \Rightarrow \mu_r^{(3)} = 32839.71$.

On the other hand, in triangles 2 and 4, the relative permeability is equal to 1, as given in Example 3.1.

Following the solution methodology of Example 3.1, we form the global matrix **KK** and the set of equations $\mathbf{KK} \cdot \mathbf{A} = \dfrac{\mu_0 \cdot J \cdot \Delta}{3}$. The boundary conditions impose that $A_1 = A_2 = 0$, so the resulting set of equations is:

$$\mathbf{KK} \cdot \mathbf{A} = \frac{\mu_0 \cdot J \cdot \Delta}{3} \Rightarrow$$

$$\begin{bmatrix} 0.625028 & 0.000017 & 0 & -0.375 & -0.250045 \\ 0.000017 & 0.625028 & -0.375 & 0 & -0.250045 \\ 0 & -0.375 & 0.625019 & 0.000011 & -0.250030 \\ -0.375 & 0 & 0.000011 & 0.625019 & -0.250030 \\ -0.250045 & -0.250045 & -0.250030 & -0.250030 & 1.000151 \end{bmatrix} \cdot \begin{bmatrix} A_1 \\ A_2 \\ A_3 \\ A_4 \\ A_5 \end{bmatrix} = \begin{bmatrix} 0.29452 \\ 0.29452 \\ 0.29452 \\ 0.29452 \\ 0.58905 \end{bmatrix} \Rightarrow$$

$$\begin{bmatrix} 0.625019 & 0.000011 & -0.250030 \\ 0.000011 & 0.625019 & -0.250030 \\ -0.250030 & -0.250030 & 1.000151 \end{bmatrix} \cdot \begin{bmatrix} A_3 \\ A_4 \\ A_5 \end{bmatrix} = \begin{bmatrix} 0.29452 \\ 0.29452 \\ 0.58905 \end{bmatrix}. \tag{3.57}$$

Using the solution of (3.55), the error vector in the solution of the set of equations (3.57) is:

$$\mathbf{err} = \begin{bmatrix} 0.625019 & 0.000011 & -0.250030 \\ 0.000011 & 0.625019 & -0.250030 \\ -0.250030 & -0.250030 & 1.000151 \end{bmatrix} \cdot \begin{bmatrix} A_3 \\ A_4 \\ A_5 \end{bmatrix} - \begin{bmatrix} 0.29452 \\ 0.29452 \\ 0.58905 \end{bmatrix} \Rightarrow$$

$$\mathbf{err} = \begin{bmatrix} 0.625019 & 0.000011 & -0.250030 \\ 0.000011 & 0.625019 & -0.250030 \\ -0.250030 & -0.250030 & 1.000151 \end{bmatrix} \cdot \begin{bmatrix} 0.8835 \\ 0.8835 \\ 1.0307 \end{bmatrix} - \begin{bmatrix} 0.29452 \\ 0.29452 \\ 0.58905 \end{bmatrix} \Rightarrow$$

$$\mathbf{err} = 10^{-4} \cdot \begin{bmatrix} -0.0003 \\ -0.0003 \\ 0.1668 \end{bmatrix}.$$

The infinite norm n_{err} of the error vector **err** is equal to the maximum value of the absolute value of its elements:

$$n_{err} = \|\mathbf{err}\|_\infty = \max\left\{ 10^{-4} \cdot \begin{bmatrix} 0.0003 \\ 0.0003 \\ 0.1668 \end{bmatrix} \right\} \Rightarrow n_{err} = 0.1668 \cdot 10^{-4}.$$

The tolerance is $tol = 10^{-4}$ and since $n_{err} < tol$, a solution is found, so no other iteration is required.

Final results

The magnetic vector potentials at nodes 3, 4, and 5 are:

$$\begin{bmatrix} A_3 \\ A_4 \\ A_5 \end{bmatrix} = \begin{bmatrix} 0.8835 \\ 0.8835 \\ 1.0307 \end{bmatrix} \frac{\text{Wb}}{\text{m}}.$$

The magnetic flux densities of the four triangles are:

$$\begin{bmatrix} B_1 \\ B_2 \\ B_3 \\ B_4 \end{bmatrix} = \begin{bmatrix} 1.6492 \\ 0.8495 \\ 0.2355 \\ 0.8495 \end{bmatrix} \text{T}.$$

References

Abed NY and Mohammed OA (2007) Modeling and characterization of transformers internal faults using finite element and discrete wavelet transforms. IEEE Transactions on Magnetics 43(4):1425–1428

Alhamadi MA and Demerdash NA (1994) Three dimensional magnetic field computation by a coupled vector-scalar potential method in brushless DC motors with skewed permanent magnet mounts – The formulation and FE grids. IEEE Transactions on Energy Conversion 9(1):1–10

Andersen OW (1973) Transformer leakage flux program based on the finite element method. IEEE Transactions on Power Apparatus and Systems 92(2):682–689

Arjona LMA and McDonald DC (1999) A new lumped steady-state synchronous machine model derived from finite element analysis. IEEE Transactions on Energy Conversion 14(1):1–7

Ashtiani CN (1988) Performance analysis of wound field synchronous alternators under load using finite elements. IEEE Transactions on Energy Conversion 3(2):330–334

Bastos JPA and Sadowski N (2003) Electromagnetic modeling by finite element methods. Marcel Dekker, New York

Bergeron DA and Trahar RE Jr (1999) A static finite element analysis of substation busbar structures. IEEE Transactions on Power Delivery 14(3):890–896

Bergeron DA, Trahar RE Jr, Dubinich MD, Opsetmoen A (1999) Verification of a dynamic finite element analysis of substation busbar structures. IEEE Transactions on Power Delivery 14(3):884–889

Binns KJ, Lawrenson PJ, Trowbridge CW (1992) The analytical and numerical solution of electric and magnetic fields. Wiley, New York

Bíró O, Auberhofer S, Burchgraber G, Preis K, Seitlinger W (2007) Prediction of magnetising current waveform in a single-phase power transformer under DC bias. IET Science, Measurement & Technology 1(1):2–5

Brebbia CA (1984) Boundary element techniques: theory and applications in engineering. Springer-Verlag, Berlin

Brebbia CA and Dominguez J (1992) Boundary elements: an introductory course. Computational Mechanics Publications, Southampton

Chari MVK and Salon SJ (2000) Numerical methods in electromagnetism. Academic Press, San Diego

Clough RW (1960) The finite element method in plane stress analysis. Proc American Society of Civil Engineers Conference on Electronic Computation

Coulson MA (1981) Magnetic non-linearity. PhD Thesis, University of Strathclyde, UK

Courant R (1943) Variational methods for the solution of problems of equilibrium and vibrations. Bulletin of the American Mathematical Society 49

de Leon F and Anders GJ (2008) Effects of backfilling on cable ampacity analyzed with the finite element method. IEEE Transactions on Power Delivery 23(2):537–543

Demerdash NA and Nehl TW (1979) Use of numerical analysis of nonlinear eddy current problems by finite element in the determination of parameters of electrical machines with solid iron cores. IEEE Transactions on Magnetics 15(6):1482–1484

Enokizono M and Soda N (1997) Finite element analysis of transformer model core with measured reluctivity tensor. IEEE Transactions on Magnetics 33(5):4110–4112

Erdelyi EA and Ahmed SV (1965) Flux distribution in saturated dc machines. IEEE Transactions on Power Apparatus and Systems 84(5):375–381

Galerkin BG (1915) Series solution of some problems of elastic equilibrium of rods and plates [In Russian]. Vestn. Inzh. Tekh. 19

Gauss CF (1823) Brief an gerling. Werke 9:278–281

Hameyer K and Belmans R (1999) Numerical modeling and design of electrical machines and devices. WIT Press, Southampton

Hildebrand FB (1962) Advanced calculus for applications. Prentice Hall, Englewood Cliffs

Ho SL, Li Y, Wong HC, Wang SH, Tang RY (2004) Numerical simulation of transient force and eddy current loss in a 720-MVA power transformer. IEEE Transactions on Magnetics 40(2):687–690

Holland SA, O'Connell GP, Haydock L (1992) Calculating stray losses in power transformers using surface impedance with finite elements. IEEE Transactions on Magnetics 28(2):1355–1358

Hwang CC, Tang PH, Jiang YH (2005) Thermal analysis of high-frequency transformers using finite elements coupled with temperature rise method. IEE Proc Electric Power Applications 152(4):832–836

Kaehler C and Henneberger G (2004) Transient 3-D FEM computation of eddy-current losses in the rotor of a claw-pole alternator. IEEE Transactions on Magnetics 40(2):1362–1365

Kefalas TD, Georgilakis PS, Kladas AG, Souflaris AT, Paparigas DG (2008) Multiple grade lamination wound core: a novel technique for transformer iron loss minimization using simulated annealing with restarts and an anisotropy model. IEEE Transactions on Magnetics 44(6):1082–1085

Kladas AG, Papadopoulos MP, Tegopoulos JA (1994) Leakage flux and force calculation on power transformer windings under short-circuit: 2D and 3D models based on the theory of images and the finite element method compared to measurements. IEEE Transactions on Magnetics 30(5):3487–3490

Kumbhar GB and Kulkarni SV (2007) Analysis of short-circuit performance of split-winding transformer using coupled field-circuit approach. IEEE Transactions on Power Delivery 22(2):936–943

Kwon YW and Bang H (1997) The finite element method using MATLAB. CRC Press, Boca Raton

References

Labridis D and Dokopoulos P (1988) Finite element computation of field, losses and forces in a three-phase gas cable with nonsymmetrical conductor arrangement. IEEE Transactions on Power Delivery 3(4):1326–1333

Lesniewska E (2002) The use of 3-D electric field analysis and the analytical approach for improvement of a combined instrument transformer insulation system. IEEE Transactions on Magnetics 38(2):1233–1236

Lin C, Xiang C, Yanlu Z, Zhingwang C, Guoqiang Z, Yinhan Z (1998) Losses calculation in transformer tie plate using the finite element method. IEEE Transactions on Magnetics 34(5):3644–3647

Lu S and Liu Y (1993) FEM analysis of DC saturation to assess transformer susceptibility to geomagnetically induced currents. IEEE Transactions on Power Delivery 8(3):1367–1376

Martic HC and Carey G (1973) Introduction to finite element analysis – Theory and applications. McGraw Hill, New York

Minambres JF, Barandiaran JJ, Alvarez-Isasi R, Zorrozua MA, Zamora I, Mazon AJ (1999) Radial temperature distribution in ACSR conductors applying finite elements. IEEE Transactions on Power Delivery 14(2):472–480

Moallem M and Ong CM (1990) Predicting the torque of a switched reluctance machine from its finite element field solution. IEEE Transactions on Energy Conversion 5(4):733–739

Papagiannis GK, Triantafyllidis DG, Labridis DP (2000) A one-step finite element formulation for the modeling of single and double-circuit transmission lines. IEEE Transactions on Power Systems 15(1):33–38

Papazacharopoulos ZK, Tatis KV, Kladas AG, Manias SN (2004) Dynamic model for harmonic induction motor analysis determined by finite elements. IEEE Transactions on Energy Conversion 19(1):102–108

Pavlik D, Garg VK, Repp JR, Weiss J (1988) A finite element technique for calculating the magnet sizes and inductances of permanent magnet machines. IEEE Transactions on Energy Conversion 3(1):116–122

Pavlik D, Johnson DC, Girgis RS (1993) Calculation and reduction of stray and eddy losses in core-form transformers using a highly accurate finite element modelling technique. IEEE Transactions on Power Delivery 8(1):239–244

Peng JP and Salon S (1982) A hybrid finite element boundary element formulation of Poisson's equation for axisymmetric vector potential problems. Journal of Applied Physics 53(11):8420–8422

Pern JF and Yeh SN (1995) Calculating the current distribution in power transformer windings using finite element analysis with circuit constraints. IEE Proc Science, Measurement and Technology 142(3):231–236

Preis K, Bíró O, Buchgraber G, Ticar I (2006) Thermal-electromagnetic coupling in the finite-element simulation of power transformers. IEEE Transactions on Magnetics 42(4):999–1002

Rausch M, Kaltenbacher M, Landes H, Lerch R, Anger J, Gerth J, Boss P (2002) Combination of finite and boundary element methods in investigation and prediction of load-controlled noise of power transformers. Journal of Sound and Vibration 250(2):323–338

Rayleigh L (1870) On the theory of resonance. Transactions of the Royal Society A161

Reece ABJ and Preston TW (2000) Finite element methods in electrical power engineering. Oxford University Press, Oxford

Renyuan T, Yan L, Dake L, Lijian T (1992) Numerical calculation of 3-D transient eddy current field and short circuit electromagnetic force in large transformers. IEEE Transactions on Magnetics 28(2):1418–1421

Ritz W (1909) Uber eine neue methode zur losung gewissen variations-probleme der mathematischen physik. J. Reine Angew. Math. 135

Salon S, Peaiyoung S, Mayergoyz I (1989) Some technical aspects of implementing boundary element equations. IEEE Transactions on Magnetics 25(4):2998–3000

Schlensok CS and Henneberger G (2004) Calculation of force excitations in induction machines with centric and excentric positioned rotor using 2-D transient FEM. IEEE Transactions on Magnetics 40(2):782–785

Silvester PP and Ferrari RL (1996) Finite elements for electrical engineers, 3rd edn. Cambridge University Press, Cambridge

Steurer M and Fröhlich K (2002) The impact of inrush currents on the mechanical stress of high voltage power transformer coils. IEEE Transactions on Power Delivery 17(1):155–160

teNyenhuis EG, Girgis RS, Mechler GF, Zhou G (2002) Calculation of core hot-spot temperature in power and distribution transformers. IEEE Transactions on Power Delivery 17(4):991–995

Tsili MA, Kladas AG, Georgilakis PS, Souflaris AT, Paparigas DG (2006) Advanced design methodology for single and dual voltage wound core power transformers based on a particular finite element model. Electric Power Systems Research 76:729–741

Viana WC, Micaleff RJ, Young S, Dawson FP (1999) Transformer design considerations for mitigating geomagnetic induced saturation. IEEE Transactions on Magnetics 35(5):3532–3534

Wang R and Demerdash NA (1991) A combined vector potential-scalar potential method for FE computation of 3D magnetic fields in electrical devices with iron cores. IEEE Transactions on Magnetics 27(5):3971–3977

Watson JF and Dorrell DG (1999) The use of finite element methods to improve techniques for the early detection of faults in 3-phase induction motors. IEEE Transactions on Energy Conversion 14(3):655–660

Yamazaki K (1999) Induction motor analysis considering both harmonics and end effects using combination of 2D and 3D finite element method. IEEE Transactions on Energy Conversion 14(3):698–703

Zachmanoglou EC and Thoe DW (1976) Introduction to partial differential equations with applications. The Williams and Wilkins Company, Baltimore

4 Classification and Forecasting

Abstract Classification aims at predicting the future class and forecasting aims at predicting the future value of a system that is intrinsically uncertain. This chapter briefly presents two artificial intelligence methods, namely decision trees and artificial neural networks. The decision tree methodology is a nonparametric inductive learning technique, able to produce classifiers for a given problem that can assess new, unseen situations and/or uncover the mechanisms driving this problem. The artificial neural network is a computer information processing system that is capable of sufficiently representing any nonlinear functions. The decision tree technique is appropriate for the solution of classification problems. The artificial neural network method is suitable for the solution of both classification and forecasting problems.

4.1 Introduction

Classification aims at predicting the future class of a system that is intrinsically uncertain. The number of classes is in principle arbitrary but generally rather small. In transformer design, it is necessary to classify the no-load losses into acceptable or non-acceptable quality prior to transformer manufacturing and quality control. Another example is the transformer winding material selection problem, where it is very important to classify the transformer winding material into one of the following two classes: copper or aluminum.

Forecasting aims at predicting the future value of a system that is intrinsically uncertain. In transformer design, it is necessary to forecast the values of transformer technical characteristics, e.g., no-load losses and impedance voltage, during the transformer design, i.e., prior to transformer manufacturing and quality control.

This chapter briefly presents decision trees and artificial neural networks. Both techniques are types of *automatic learning from examples*, i.e., they are able to extract high-level synthetic information from databases containing large amounts of low-level data. The decision trees technique is appropriate for the solution of classification problems. The artificial neural networks method is suitable for the solution of both classification and forecasting problems.

4.2 Automatic Learning

The term *automatic learning* is used to denote a highly multidisciplinary research field and set of methods to extract high-level synthetic information (knowledge) from databases containing large amounts of low-level data. Automatic learning encompasses statistical data analysis and modeling, artificial neural networks, and symbolic machine learning in artificial intelligence.

In the last two decades, automatic learning has progressed along many lines, in terms of theoretical understanding and actual applications. Probably the main reason for the important breakthrough was the tremendous increase in computing power. This makes possible the application of the often very computation-intensive automatic learning algorithms to practical large scale problems. Conversely, automatic learning algorithms allow one to make better use of existing computing power by exploiting more systematically the information contained in databases.

The generic problem of supervised learning from examples can be formulated as follows: given a set of examples (the learning set) of associated input/output pairs, derive a general rule representing the underlying input/output relationship that may be used to explain the observed pairs and/or predict output values for any new unseen input.

In automatic learning we use the term *attribute* to denote the parameters (or variables) used to describe the input information. The output information can be either symbolic (e.g., a classification) or numerical.

4.3 Data Mining

A database is a collection of examples. Each example is a vector of input/output pairs, i.e., inputs and their associated outputs. The inputs or *objects* or *patterns* are described by a certain number of attributes or input parameters.

Data mining is the application of automatic learning in order to discover interesting information from a database.

In general, data mining comprises the following five subtasks (Michie et al. 1994):

1. Representation

2. Attribute selection

3. Model selection

4. Interpretation and validation

5. Model use

4.3.1 Representation

Representation consists of:

1. Choosing appropriate input attributes to represent the practical problem instances

2. Defining the output information

3. Choosing a class of models (e.g., decision trees or artificial neural networks) suitable to represent input/output relations

Solving the representation problem is normally left to the engineer. Choosing an appropriate set of attributes is an iterative process during the first trials to apply an automatic learning technique to a new problem. Similarly, the selection of a suitable type of models is done by trial and error.

4.3.2 Attribute Selection

Attribute selection aims at reducing the dimensionality of the input space by dismissing attributes that do not carry useful information to forecast the output information considered.

4.3.3 Model Selection

Model selection identifies in the predefined class of models the one that best fits the learning states. This requires choice of model structure and parameters using an optimization technique adapted to the type of model considered.

The distinction between attribute selection and model selection is somewhat arbitrary. For example, some of the methods (e.g., decision trees) solve these two problems simultaneously.

4.3.4 Interpretation and Validation

Interpretation and validation aim at understanding the physical meaning of the synthesized model and at determining its range of validity. It consists of comparing the information that can be derived from the model with prior expertise, and testing it on a set of unseen test examples. It should be mentioned that some methods provide rather black-box information, difficult to interpret, while others provide more transparent explicit models, easier to compare with prior knowledge.

4.3.5 Model Use

Model use consists of applying the model to forecast outputs of new situations from the values assumed by the input parameters. Regarding the model use for fast decision making, it should be noted that speed variations of several orders of magnitude may exist between various techniques. This may reduce the usefulness of some methods in time-critical real-time applications.

4.4 Learning Set and Test Set

An example is a vector of input/output pairs belonging to the database. The learning set or training set is a subset composed of N examples drawn from the database. Similarly, the test set is a subset of M examples, used in order to evaluate the accuracy of the automatic learning technique. We assume always that the learning and test sets are disjoint and drawn randomly from the database.

4.4.1 Classification

Classification aims at predicting the future class of a system that is intrinsically uncertain. Without loss of generality, we will assume that each example is characterized by a certain number, say n, of ordered numerical attributes (the same number for each example), and that all the examples are classified into two classes only $\{1, 2\}$.

In the following, a learning set (LS) is a subset of N pre-classified examples (input/output pairs) that is defined by:

$$\mathbf{LS}(N) \triangleq \left\{ (\mathbf{v}^1, c^1), (\mathbf{v}^2, c^2), ..., (\mathbf{v}^N, c^N) \right\}, \tag{4.1}$$

where the input vector $\mathbf{v}^k$

$$\mathbf{v}^k = \left(v_1^k, v_2^k, ..., v_n^k \right)^T, \tag{4.2}$$

represents the attribute values of an object o_k, $\mathbf{x}^T$ represents the transpose of vector $\mathbf{x}$, and the output c^k

$$c^k \in \{1, 2\}, \tag{4.3}$$

is the class of the object o_k.

The test set (TS), is defined as a similar, but independent sample of size M:

$$\mathbf{TS}(M) \triangleq \{(\mathbf{v}^{N+1}, c^{N+1}), (\mathbf{v}^{N+2}, c^{N+2}), ..., (\mathbf{v}^{N+M}, c^{N+M})\}. \tag{4.4}$$

For example, in transformer winding material selection, objects may represent suitable technical characteristics of different transformers, while the output is the transformer winding material, which belongs to one of two classes: copper or aluminum.

4.4.2 Forecasting

Forecasting aims at predicting the future value of a system that is intrinsically uncertain. Figure 4.1 shows a forecasting problem with n inputs and m outputs. In general, supervised learning from examples can be used to represent the relationship between the inputs $\mathbf{x}$ and the outputs $\mathbf{y}$ and afterwards to solve the forecasting problem, i.e., for known inputs to forecast the outputs.

In the case of forecasting with supervised learning, a learning set (LS) is a subset of N learning examples (input/output pairs) that is defined by:

$$\mathbf{LS}(N) \triangleq \{(\mathbf{x}^1, \mathbf{y}^1), (\mathbf{x}^2, \mathbf{y}^2), ..., (\mathbf{x}^N, \mathbf{y}^N)\}, \tag{4.5}$$

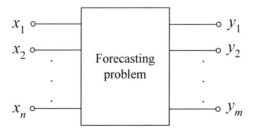

Fig. 4.1 Forecasting problem

where the input vector $\mathbf{x}^k$

$$\mathbf{x}^k = \left(x_1^k, x_2^k, ..., x_n^k\right)^T, \qquad (4.6)$$

represents the attribute values (inputs) of an object o_k and the vector $\mathbf{y}^k$

$$\mathbf{y}^k = \left(y_1^k, y_2^k, ..., y_m^k\right)^T, \qquad (4.7)$$

is the output of the object o_k.

The test set (TS), is defined as a similar, but independent sample of size M:

$$\mathbf{TS}(M) \triangleq \left\{(\mathbf{x}^{N+1}, \mathbf{y}^{N+1}), (\mathbf{x}^{N+2}, \mathbf{y}^{N+2}), ..., (\mathbf{x}^{N+M}, \mathbf{y}^{N+M})\right\}. \qquad (4.8)$$

For example, in transformer no-load loss forecasting, objects may represent technical characteristics of different transformers, while the output is the no-load loss value of these transformers.

4.5 Decision Trees

4.5.1 Introduction

The decision tree methodology belongs to the category of supervised learning from examples techniques. The decision tree is a nonparametric learning technique, independent of the statistical distribution of the analyzed population (the possible states of the examined system). The decision tree has the hierarchical form of a tree of rules built upside down. The decision tree methodology is used to draw inferences based on known and well recognized attributes.

There are three main families of decision tree induction:

1. *Crisp decision tree* induction. This is able to produce if-then decision rules that are used to classify into two classes the output of new, unobserved states of a system. This means that the output information (classification) is discrete, since it belongs exclusively to one of the two classes. The crisp decision tree (also called *binary decision tree*) is constructed off-line by an inductive inference similar to the ID3 (Quinlan 1983).

2. *Regression tree* induction. This produces if-then decision rules in order to conduct regression. The regression tree typically provides the mean value and the

standard deviation of the output variable (Wehenkel 1998). The regression tree is built using an appropriate algorithm (Breiman et al. 1984).

3. *Fuzzy decision tree* induction. This is a combination of fuzzy reasoning and automatic learning based on decision tree induction. It is a model concerned with the automatic design of fuzzy if-then rules in a tree structure. It is used in classification problems (Yuan and Shaw 1995; Janikow 1998; Boyen and Wehenkel 1999) in most cases but sometimes also in regression problems (Suarez and Lutsko 1999). The fuzzy decision tree is built using an appropriate algorithm (Boyen and Wehenkel 1999).

Because of the binary nature of classification problems in transformer design, this chapter is focused on crisp decision trees.

Decision trees offer significant advantages and valuable characteristics that are unavailable elsewhere. In particular:

1. The main strength of the decision trees is their interpretability. By merely looking at the test nodes of a tree one can easily sort out the most salient attributes (i.e., those that most strongly influence the output) and find out how they influence the output. Furthermore, at the tree growing stage the method provides a great deal of additional information, e.g., scores of candidate attributes, their correlations, and the overall information they provide to the decision tree.

2. Another very important asset is the ability of the decision tree method to automatically identify the most relevant attributes for each problem.

3. The third important feature of decision trees is their computational efficiency. Typically, tree growing computational complexity is linear in the number of learning examples, allowing one to tackle problems with a few hundred candidate attributes and a few thousand learning examples, with response times of only some minutes. The use of a decision tree to classify an unseen situation is ultrafast since only a few logical tests need to be computed.

4.5.2 Applications to Power Systems

Decision trees have shown their capabilities in solving real-world problems, as reflected by the growing number of publications on their applications to power system problems, e.g., Wehenkel and Pavella 1991; Van Cutsem et al. 1993; Wehenkel and Pavella 1993; Wehenkel et al. 1994; Hatziargyriou et al. 1994; Rovnyak et al. 1994; Yang and Hsu 1994; Hatziargyriou et al. 1995; Karapidakis and Hatziargyriou 2002; Ugedo et al. 2005; Leonidaki et al. 2006; Senroy et al. 2006, to name only a few.

Particularly in transformer design, decision trees have been applied for no-load loss classification (Hatziargyriou et al. 1998; Georgilakis et al. 1999; Georgilakis 2000) as well as for solution of the transformer winding material selection problem (Amoiralis et al. 2007; Georgilakis and Amoiralis 2007; Georgilakis et al. 2007).

4.5.3 General Characteristics

The first node of each decision tree is called root node or top node. A non-trivial binary decision tree is a tree with at least three nodes. On the other hand, a trivial binary decision tree has only one node, i.e., the root node.

Except for the root node, every node of a non-trivial binary decision tree is a successor of exactly one other node, called its parent node or successor node. There is exactly one path from the root towards any other node of the decision tree. Except for the root node, every node of a non-trivial binary decision tree can be either a non-terminal or terminal node.

Non-terminal nodes are also called interior nodes or *test nodes*. Each test node contains a suitable test, called the node splitting test, which produces exactly two successor nodes. In particular, if the test of the test node is satisfied, then the decision tree is directed to the first successor node (left node), otherwise it is directed to the second successor node (right node).

Nodes that have no successor nodes are called *terminal nodes*. These nodes classify the case analyzed into one of the two classes.

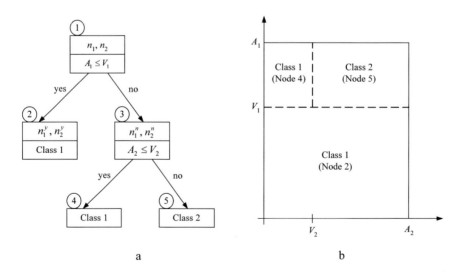

Fig. 4.2 a Hypothetical decision tree; **b** geometric interpretation of that decision tree

Table 4.1 Decision rules of the decision tree of Fig. 4.2

Node	Decision rule
2	If $A_1 \leq V_1 \Rightarrow$ class 1
4	If $A_1 > V_1$ and $A_2 \leq V_2 \Rightarrow$ class 1
5	If $A_1 > V_1$ and $A_2 > V_2 \Rightarrow$ class 2

Figure 4.2 illustrates a hypothetical decision tree and its geometric interpretation in its attribute space. At each test node, the number of examples that belongs to each one of the two classes as well as the suitable test are given. Each terminal node is labeled class 1 or class 2, depending upon the majority class of its learning examples. In Fig. 4.2, n_i denotes the number of learning examples of node 1 that belong to class i with $i \in \{1, 2\}$, and n_i^y (respectively n_i^n) denotes the number of learning examples of node 2 (respectively node 3) that belong to class i and verify (respectively do not verify) the test of node 1. In order to classify an example, one starts at the top node and applies sequentially the dichotomous tests encountered to select the appropriate successor. When a terminal node is reached, the output information (class) stored there is retrieved. Thus, at the terminal node 2 of the decision tree of Fig. 4.2, the majority of the learning examples belongs to class 1, i.e., $n_1^y > n_2^y$, while the ideal situation would be all the learning examples to belong to class 1, i.e., $n_2^y = 0$.

The decision tree of Fig. 4.2 comprises only five nodes: two test nodes and three terminal nodes. Nodes 1 and 3 are test nodes. Nodes 2, 4, and 5 are terminal nodes. The number of decision rules that can be extracted from a decision tree is equal to the number of its terminal nodes. Thus, the decision tree of Fig. 4.2 produces three decision rules, and using these rules, each example of the database is classified either in class 1 or in class 2. These decision rules are shown in Table 4.1.

4.5.4 Top Down Induction

Inductive inference is a subfield of automatic learning concerned with the automatic design of rules similar to those used by human experts, e.g., if-then rules. We will describe only top down induction of decision trees (TDIDT), which is one of the most successful classes of such methods (Breiman et al. 1984; Quinlan 1993).

The goal of TDIDT is to produce an as simple as possible tree, providing a maximum amount of information about the classification of the learning examples. For instance, the objective of the initial version of ID3 (interactive dichotomizer) method was to build the simplest tree of minimum classification error rate in the learning set (Quinlan 1993).

The process of decision tree building aims at producing a near optimal decision tree that reaches a good compromise between complexity (i.e., number of nodes) and classification accuracy.

The decision tree building generally decomposes into two subtasks (Wehenkel 1998):

1. *Tree growing.* This aims at deriving the tree structure and tests. During tree growing, the test nodes of the tree are progressively developed, by choosing appropriate tests that separate the learning examples of each node into two subsets as class-pure as possible, i.e., with the majority of the learning examples belonging to one of the two classes. The recursive tree growing algorithm starts with the complete learning set at the top node of the tree. At each step a test is selected that splits the current set of examples into subsets, corresponding to the current node's successors. This process stops when no further nodes need to be developed. This is a locally rather than globally optimal hill-climbing search that leads to an efficient algorithm, the computational complexity of which is at most of order $N \cdot \log N$, where N is the number of learning examples, and of order n, where n is the number of candidate attributes (Wehenkel 1998). During tree growing, two rules are needed:

   – *Optimal splitting rule.* This defines the criterion and search procedure in order to choose the best candidate test to split the current node.

   – *Stop splitting rule.* This allows one to decide whether one should further develop a node, depending on the information provided in the current learning subset

2. *Tree pruning.* It aims at determining the appropriate complexity of a decision tree. Too large trees will over-fit the data, whereas too small trees will under-exploit the information contained in the learning set. Therefore, some smoothing strategy is required to control the complexity of the tree and ensure that the learning samples at its terminal nodes remain sufficiently representative. For that purpose, various tree pruning methods have been proposed (Henrichon and Fu 1969; Friedman 1977; Rounds 1980; Breiman et al. 1984; Kononenko et al. 1984; Quinlan 1986, 1987; Mingers 1989; Wehenkel et al. 1989a' Wehenkel 1993).

4.5.5 Optimal Splitting Rule

The optimal splitting rule consists of a search for a locally optimal test maximizing a given score function. This implies finding for each candidate attribute its own optimal split and identifying the attribute that is, overall, optimal.

The optimal splitting rule is an important part of the tree building algorithm, which is composed of the following steps (Wehenkel et al. 1989b; Wehenkel and Pavella 1991):

1. Starting at the root node of the tree, with the list of candidate attributes and with the whole learning set, the learning examples are analyzed in order to select a test that allows a maximum increase in purity or, equivalently, which provides a maximum amount of information about their classification. The selection proceeds in two steps:

    a. For each attribute, say a_i, it finds the optimal test on its values, i.e., the test with the maximum score by computing and comparing the score of this candidate attribute on its values for the different learning examples. In our case of ordered numerical attributes, this step provides an *optimal threshold value* u_i^* and defines the test:

    $$a_i \leq u_i^* \ ? \tag{4.9}$$

    b. Among the different candidate attributes, it chooses the *optimal attribute*, a^*, i.e., the one with the maximum score, along with its optimal value, u^*, to split the node.

    In short, step 1.b defines the optimal attribute and step 1.a its optimal threshold value.

2. The selected test is applied to the learning set of the node and splits it into two subsets, corresponding to the two successors of the node. Starting with the root node of the tree and the entire learning set, the two subsets:

    $$LS_Y \triangleq \left\{ \mathbf{v}^k \in LS \,\middle|\, a^* \leq u^* \right\} \tag{4.10}$$

    $$LS_N \triangleq \left\{ \mathbf{v}^k \in LS \,\middle|\, a^* > u^* \right\} \tag{4.11}$$

    correspond to the two successors of the root node.

3. The successors are labeled terminal or not on the basis of the stop splitting rule described in Sect. 4.5.6.

4. For the non-terminal nodes, the overall procedure is called recursively in order to build the corresponding subtrees.

5. For the terminal nodes, the class label of the majority class is attached.

In addition to the above tree building algorithm, we need to specify the evaluation function or score used to select the best split. This score is calculated from the normalized information gain ((4.31) presented in the following) provided by a candidate partition at a node (Wehenkel 1998).

Each node n of the decision tree possesses a subset of learning examples with the following characteristics:

E_n : the learning examples subset of node n of the decision tree

N : size (number of learning examples) of E_n

n_1 : number of learning examples in E_n that belong to class 1

n_2 : number of learning examples in E_n that belong to class 2

The relative frequencies f_1 and f_2 of the learning examples of node n that belong to class 1 and class 2, respectively, will be:

$$f_1 = \frac{n_1}{n_1 + n_2} = \frac{n_1}{N}, \tag{4.12}$$

$$f_2 = \frac{n_2}{n_1 + n_2} = \frac{n_2}{N}. \tag{4.13}$$

The *prior classification entropy* estimated in the subset E_n at node n with respect to the class partition of the learning examples of E_n is calculated by:

$$H_C(E_n) = -(f_1 \cdot \log_2 f_1 + f_2 \cdot \log_2 f_2). \tag{4.14}$$

The prior classification entropy, $H_C(E_n)$, is a measure of class purity of node subset E_n and, consequently, of the uncertainty of the classification of a learning example by this node. The following relations hold for $H_C(E_n)$:

$$0 \le H_c(E_n) \le 1, \tag{4.15}$$

$$H_c(E_n) = 0 \Leftrightarrow (f_1 = 1 \text{ and } f_2 = 0) \text{ or } (f_1 = 0 \text{ and } f_2 = 1), \tag{4.16}$$

$$H_c(E_n) = 1 \Leftrightarrow f_1 = f_2 = 0.5. \tag{4.17}$$

4.5 Decision Trees

Equation 4.16 shows that $H_c(E_n) = 0$ corresponds to a perfectly pure subset (i.e., all the learning examples of the subset E_n belong either to class 1 or to class 2), whereas (4.17) indicates that $H_c(E_n) = 1$ corresponds to maximum uncertainty (i.e., 50% of the learning examples of the subset E_n belong to class 1 and the remaining 50% belong to class 2).

A test T is defined at node n as:

$$T : A_i \leq t, \qquad (4.18)$$

where A_i is the value of attribute i of a particular learning example and t is a threshold value.

By applying the test T to all learning examples of node n, E_n is split into two subsets E_{nY} and E_{nN}:

$$E_{nY} = \{\text{learning examples} \in E_n : A_i \leq t\}, \qquad (4.19)$$
$$E_{nN} = \{\text{learning examples} \in E_n : A_i > t\}. \qquad (4.20)$$

Let us denote as n_Y and n_N the number of learning examples in subsets E_{nY} and E_{nN}, respectively. We also denote as n_{Y1} and n_{Y2} the number of learning examples in subset E_{nY} that belong to class 1 and class 2, respectively. In addition, we denote as n_{N1} and n_{N2} the number of learning examples in subset E_{nN} that belong to class 1 and class 2, respectively. The corresponding frequencies are:

$$f_Y = \frac{n_Y}{n_Y + n_N} = \frac{n_Y}{N} \quad \text{and} \quad f_N = \frac{n_N}{n_Y + n_N} = \frac{n_N}{N}, \qquad (4.21)$$

$$f_{Y1} = \frac{n_{Y1}}{n_{Y1} + n_{Y2}} = \frac{n_{Y1}}{n_Y} \quad \text{and} \quad f_{Y2} = \frac{n_{Y2}}{n_{Y1} + n_{Y2}} = \frac{n_{Y2}}{n_Y}, \qquad (4.22)$$

$$f_{N1} = \frac{n_{N1}}{n_{N1} + n_{N2}} = \frac{n_{N1}}{n_N} \quad \text{and} \quad f_{N2} = \frac{n_{N2}}{n_{N1} + n_{N2}} = \frac{n_{N2}}{n_N}. \qquad (4.23)$$

The entropy of E_n with respect to the partition induced by T is:

$$H_T(E_n) = -(f_Y \cdot \log_2 f_Y + f_N \cdot \log_2 f_N). \qquad (4.24)$$

$H_T(E_n)$ is a measure of the uncertainty of the outcome of test T and has similar properties to $H_C(E_n)$:

$$0 \leq H_T(E_n) \leq 1, \tag{4.25}$$

$$H_T(E_n) = 0 \Leftrightarrow (f_Y = 1 \text{ and } f_N = 0) \text{ or } (f_Y = 0 \text{ and } f_N = 1), \tag{4.26}$$

$$H_T(E_n) = 1 \Leftrightarrow f_Y = f_N = 0.5. \tag{4.27}$$

The *mean conditional entropy* or posterior classification entropy of E_n, given the outcome of test T, corresponds to residual entropy after the application of T and is defined as:

$$H_C(E_n|T) = f_Y \cdot H_C(E_{nY}) + f_N \cdot H_C(E_{nN}). \tag{4.28}$$

The *information* gained from the application of test T is expressed by the reduction achieved in the learning subset entropy:

$$I(E_n; T) = H_C(E_n) - H_C(E_n|T). \tag{4.29}$$

The following relation holds:

$$0 \leq H_C(E_n|T) \leq H_C(E_n) \Leftrightarrow 0 \leq I(E_n; T) \leq H_C(E_n). \tag{4.30}$$

A more objective (less biased) estimator of the merit of test T is provided by the normalized information gain or *score* measure, defined as (Wehenkel 1998):

$$SCORE(E_n; T) \triangleq \frac{2 \cdot I(E_n; T)}{H_C(E_n) + H_T(E_n)} \in [0, 1]. \tag{4.31}$$

Equation 4.31 defines the score measure to select the best split during the tree building algorithm.

4.5.6 Stop Splitting Rule

The stop splitting rule aims at detecting two types of nodes:

1. *Leaf node.* A node is LEAF if it corresponds to a pure enough learning subset, i.e., if the classification entropy of the node is lower than a minimum preset value H_{min}:

$$\text{If } H_C(E_n) < H_{min} \Rightarrow \text{LEAF}. \tag{4.32}$$

4.5 Decision Trees

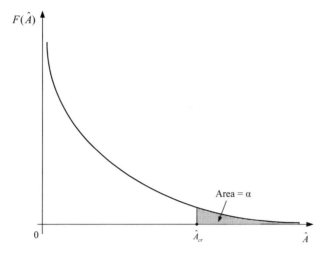

Fig. 4.3 χ^2 probability density function of $\hat{A}$ with one degree of freedom

2. *Deadend node.* A node is DEADEND if the optimal split found at that node leads to a score that is not significantly larger than zero (in the statistical sense).

Deadend detection amounts to applying a hypothesis test to the information quantity. More precisely, under the hypothesis of zero score, the quantity:

$$\hat{A} = 2 \cdot N \cdot \ln 2 \cdot I(E_n ; T), \qquad (4.33)$$

is distributed according to a χ-square (χ^2) distribution with one degree of freedom, where N is the number of learning examples of E_n. Figure 4.3 sketches such a χ-square probability density function.

Thus, the deadend detection rule amounts to fixing a priori a value of the non-detection risk a of the hypothesis test, and to comparing the value of $\hat{A}$ obtained for the optimal test, with the threshold value $\hat{A}_{cr}$ obtained from the χ^2 distribution table. For example, using a value of $a = 0.0001$ (a good choice in practice) yields a threshold value $\hat{A}_{cr} = 15.2$. If $\hat{A}$ is smaller than or equal to the tabulated value $\hat{A}_{cr}$ the node will become a deadend:

$$\text{If } \hat{A} \leq \hat{A}_{cr} \Rightarrow \text{DEADEND}. \qquad (4.34)$$

On the other hand, if $\hat{A} > \hat{A}_{cr}$, then the node is split into two successor nodes.

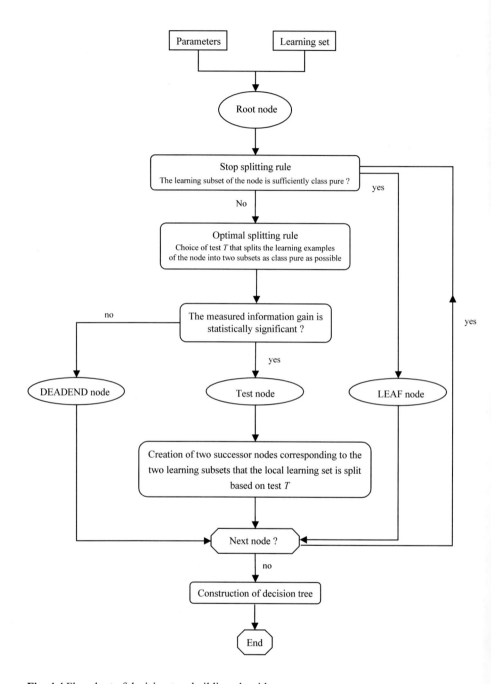

Fig. 4.4 Flowchart of decision tree building algorithm

The stop splitting rule is the most efficient way to avoid over-fitting in decision trees. It is easy to understand, and allows, in general, a reduction in tree size by a factor of 2 to 3, and in computational burden by a factor of 2. All in all, it increases the interpretability and the accuracy of decision trees.

4.5.7 Overview of Decision Tree Building Algorithm

The parameters involved in the tree building algorithm are:

- H_{min} : the *minimum node entropy*, below which the node is declared a leaf.

- a : the assumed *risk level*.

The values of the above parameters are defined before starting the tree building procedure. The risk level affects the structure of the decision much more than the minimum node entropy.

Figure 4.4 shows the flowchart of the decision tree building algorithm. The steps followed by the tree building algorithm are summarized in the following:

1. The procedure starts from the top node with the whole learning set.

2. It is examined if the node should be further split by applying the stop splitting rule:

    – If the node subset is sufficiently class-pure $\Rightarrow$ LEAF node.
    – If the node subset is not sufficiently class-pure $\Rightarrow$ proceed to Step 3.

3. Selection of the optimal splitting test (optimal splitting rule):

    – Selection of the test T that divides the learning examples of the node into two subsets, mostly purified.

4. Statistical significance testing of the optimal test:

    – If the measured information gain is statistically significant $\Rightarrow$ test node. Proceed to Step 5.
    – If there is no statistically significant way to expand the node $\Rightarrow$ DEADEND node.

5. Splitting of the node:

    – Two successor nodes are created, corresponding to the two learning subsets for the optimal splitting test.

6. Steps 2 to 5 are recursively applied to the successor nodes.

Table 4.2 Learning set for the Example 4.1

LEC	I_1	I_2	O_1	LEC	I_1	I_2	O_1
1	0.32	0.41	NA	14	0.38	0.34	NA
2	0.62	0.42	NA	15	0.62	0.34	A
3	0.86	0.41	A	16	0.38	0.32	NA
4	0.62	0.39	A	17	0.62	0.32	A
5	0.86	0.39	A	18	0.38	0.31	NA
6	0.32	0.37	NA	19	0.62	0.31	A
7	0.32	0.48	NA	20	0.62	0.31	A
8	0.86	0.37	A	21	0.62	0.37	A
9	0.32	0.36	NA	22	0.62	0.29	A
10	0.38	0.36	NA	23	0.62	0.35	A
11	0.32	0.34	NA	24	0.86	0.29	A
12	0.38	0.34	NA	25	0.62	0.39	A
13	0.62	0.34	A				

4.5.8 Example 4.1

The quality (output O_1) of an industrial design process is classified as acceptable (A) or non-acceptable (NA) and is described by two attributes (input parameters) I_1 and I_2. Table 4.2 presents the 25 learning examples of the learning set for the above industrial design process. In Table 4.2, LEC denotes the learning example code.

1. Describe the process to find the optimal splitting test for node 1.

2. Using appropriate decision tree software with the parameters $a = 0.0001$ and $H_{min} = 0.1$, we find that the optimal splitting test for node 1 is $I_1 \leq 0.3848$. We also see that the decision tree has in total three nodes. Design the decision tree for this learning set.

3. Calculate the classification success rate of the decision tree on the learning set.

4. Compute the information and the score of the optimal splitting test $I_1 \leq 0.3848$ for node 1.

4.5 Decision Trees

Solution

1. Initially, the candidate threshold values for each one of the two attributes are computed by dividing the interval between the minimum and the maximum value of each attribute into a constant number of, say, 50 equal sub-intervals.

   Node 1 contains the whole learning set.
   The attribute I_1 varies from 0.32 (minimum value) to 0.86 (maximum value), as can be seen from Table 4.2. If we divide the interval [0.32, 0.86] into 50 equal sub-intervals, the step is:

   $$Step\_I_1 = \frac{0.86 - 0.32}{50} = 0.0108.$$

   As a result, the 51 candidate threshold values t_1 for attribute I_1 are:

   $$t_1 = \{0.3200, 0.3200 + 0.0108, 0.3200 + 2 \cdot 0.0108, ..., 0.8600\} \Rightarrow$$

   $$t_1 = \{0.3200, 0.3308, 0.3416, ..., 0.8600\}.$$

   Consequently, the 51 candidate tests for attribute I_1 are:

   $$I_1 \leq 0.3200,$$

   $$I_1 \leq 0.3308,$$

   $$I_1 \leq 0.3416,$$

   $$...$$

   $$I_1 \leq 0.8600.$$

   The attribute I_2 varies from 0.29 (minimum value) to 0.48 (maximum value), as can be seen from Table 4.2. If we divide the interval [0.29, 0.48] into 50 equal sub-intervals, the step is:

   $$Step\_I_2 = \frac{0.48 - 0.29}{50} = 0.0038.$$

   As a result, the 51 candidate threshold values t_2 for attribute I_2 are:

   $$t_2 = \{0.2900, 0.2900 + 0.0038, 0.2900 + 2 \cdot 0.0038, ..., 0.4800\} \Rightarrow$$

$$t_2 = \{0.2900, 0.2938, 0.2976, ..., 0.4800\}.$$

Consequently, the 51 candidate tests for attribute I_2 are:

$$I_2 \leq 0.2900,$$

$$I_2 \leq 0.2938,$$

$$I_2 \leq 0.2976,$$

...

$$I_2 \leq 0.4800.$$

Next, for each one of the above 51 candidate tests for attribute I_1 and the 51 candidate tests for attribute I_2 we calculate the score using (4.31). Finally, the optimal splitting test is the one with the highest score.

2. Node 1 has the whole learning set, which has 14 acceptable and 11 non-acceptable learning samples (as can be seen from Table 4.2), so the *acceptability index* of node 1 is 14/25, or 0.56.

   Node 1 has the whole learning set of Table 4.2. It is given that the optimal splitting test for node 1 is $I_1 \leq 0.3848$. Using the optimal test $I_1 \leq 0.3848$ of node 1, its two successor nodes 2 and 3 are created as follows:
   - Node 2 is composed of the 10 learning examples with LEC { 1, 6, 7, 9, 10, 11, 12, 14, 16, 18 } of the learning set of Table 4.2 that verify the test $I_1 \leq 0.3848$. All these 10 learning examples are of non-acceptable quality, so the classification entropy of node 2 is $H_C(\text{node 2}) = 0.0$ ((4.16) was used). Since $H_C(\text{node 2}) < H_{min}$, node 2 is LEAF according to (4.32). The acceptability index, i.e., the ratio of acceptable learning examples to total learning examples of node 2 is 0.0, since all learning examples are of non-acceptable quality.
   - Node 3 is composed of the remaining 15 learning examples with LEC { 2, 3, 4, 5, 8, 13, 15, 17, 19, 20, 21, 22, 23, 24, 25 } of the learning set of Table 4.2 that do not verify the test $I_1 \leq 0.3848$. Among these 15 learning examples, only the learning example with code 2 is of non-acceptable quality, while the remaining 14 learning examples are of acceptable quality, so the acceptability index of node 3 is 14/15 or 0.9333. The classification entropy of node 3 is:

4.5 Decision Trees

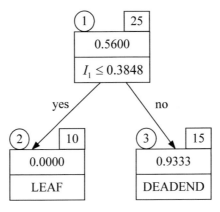

Fig. 4.5 Decision tree for the Example 4.1

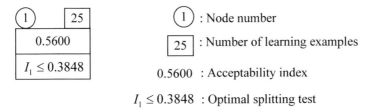

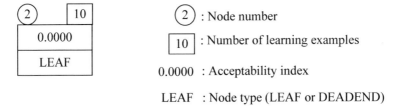

Fig. 4.6 Notation of the decision tree nodes

$$H_C(\text{node 3}) = -\left[\frac{14}{15} \cdot \log_2 \frac{14}{15} + \frac{1}{15} \cdot \log_2 \frac{1}{15}\right] = 0.3534.$$

At this point we see that $H_C(\text{node 3}) > H_{\min}$, which means that node 3 is either DEADEND or node 3 has to be split into two new nodes. However, since it is given that the decision tree has in total three nodes, the conclusion is that node 3 is of DEADEND type.

Figure 4.5 shows the decision tree. The notation used for the decision tree nodes is explained in Fig. 4.6.

3. Only the learning example with code 2 is incorrectly classified by the decision tree of Fig. 4.5. More specifically, since the learning example with code 2 has $I_1 = 0.62$, i.e., $I_1 > 0.3848$, we are led to node 3 that corresponds to acceptable quality, since the acceptability index of node 3 is 93.33%, i.e., greater than 50%. However, the learning example with code 2 is of non-acceptable quality as can be seen from Table 4.2.

All the other learning examples of Table 4.2 are correctly classified, which means that the decision tree of Fig. 4.5 classifies correctly 24 out of the 25 learning examples, i.e., the decision tree classification success rate is 24/25 or 96%.

4. Since the 14 learning examples of node 1 are of acceptable quality and the remaining 11 are of non-acceptable quality, the classification entropy of node 1 is:

$$H_C(\text{node 1}) = -\left[\frac{14}{25} \cdot \log_2 \frac{14}{25} + \frac{11}{25} \cdot \log_2 \frac{11}{15}\right] = 0.9896.$$

Because all the learning examples of node 2 are of acceptable quality, the entropy of node 2 with respect to the partition induced by test T ($I_1 \leq 0.3848$) is:

$$H_T(\text{node 2}) = 0.0.$$

Since 14 of the 15 learning examples of node 3 are of acceptable quality, the entropy of node 3 with respect to the partition induced by test T is:

$$H_T(\text{node 3}) = -\left[\frac{14}{15} \cdot \log_2 \frac{14}{15} + \frac{1}{15} \cdot \log_2 \frac{1}{15}\right] = 0.3534.$$

Since node 2 has 10 learning examples and node 3 is composed of 15 learning examples, the mean conditional entropy of node 1, given the outcome of test T, is:

$$H_C(\text{node 1} \mid T) = \frac{10}{25} \cdot H_T(\text{node 2}) + \frac{15}{25} \cdot H_T(\text{node 3}) \Rightarrow$$

$$H_C(\text{node 1} \mid T) = \frac{10}{25} \cdot 0.0 + \frac{15}{25} \cdot 0.3534 \Rightarrow H_C(\text{node 1} \mid T) = 0.2120.$$

4.5 Decision Trees

The information gained from the application of test T at node 1 is:

$$I(\text{node } 1; T) = H_C(\text{node } 1) - H_C(\text{node } 1 | T) = 0.9896 - 0.2120 \Rightarrow$$

$$I(\text{node } 1; T) = 0.7776.$$

The entropy of node 1 with respect to the partition induced by T is:

$$H_T(\text{node } 1) = -\left[\frac{10}{25} \cdot \log_2 \frac{10}{25} + \frac{15}{25} \cdot \log_2 \frac{15}{15}\right] = 0.9710.$$

The score of test T at node 1 is:

$$SCORE(\text{node } 1; T) = \frac{2 \cdot I(\text{node } 1; T)}{H_C(\text{node } 1) + H_T(\text{node } 1)} = \frac{2 \cdot 0.7776}{0.9896 + 0.9710} = 0.7932.$$

At this point we can confirm that node 1 is a test node by applying the stop splitting rule. We have:

$$\hat{A} = 2 \cdot N \cdot \ln 2 \cdot I(\text{node } 1; T) = 2 \cdot 25 \cdot \ln 2 \cdot 0.7776 = 26.95.$$

Since $a = 0.0001$, from the χ^2 distribution table it is determined that $\hat{A}_{cr} = 15.2$, and because $\hat{A} > \hat{A}_{cr}$, node 1 is split into two successor nodes.

4.5.9 Example 4.2

The quality (output O_1) of an industrial design process is classified as acceptable (A) or non-acceptable (NA) and is described by two attributes (input parameters) I_1 and I_2. Table 4.3 presents the 40 learning examples of the learning set for the above industrial design process. In Table 4.3, LEC denotes the learning example code.

Using appropriate decision tree software with parameters $a = 0.0001$ and $H_{min} = 0.1$, we find that the decision tree has in total five nodes, of which two are test nodes. The software also gives that the optimal splitting test for node 1 is $I_2 \leq 155.680$, while the optimal splitting test for the second test node is $I_1 \leq 36.952$.

Table 4.3 Learning set for the Example 4.2

LEC	I_1	I_2	O_1	LEC	I_1	I_2	O_1
1	34.1	154.4	NA	21	36.7	156.7	A
2	34.2	154.4	NA	22	36.7	156.7	A
3	34.3	154.4	NA	23	36.7	156.3	A
4	34.1	154.4	NA	24	36.7	156.3	A
5	34.1	154.4	NA	25	36.7	156.3	A
6	34.1	152.4	NA	26	36.7	156.7	A
7	34.1	152.4	NA	27	36.9	156.7	A
8	34.1	153.4	NA	28	36.9	156.7	A
9	34.1	153.4	NA	29	36.9	156.7	A
10	34.1	153.4	NA	30	36.9	156.7	A
11	34.1	153.4	NA	31	36.9	156.4	A
12	34.1	152.4	NA	32	36.9	156.4	A
13	36	155.6	NA	33	36.9	156.4	A
14	36.2	155.2	NA	34	37.2	156.7	NA
15	36.3	155.2	NA	35	37.2	156.7	NA
16	36.4	155.2	NA	36	37.2	156.7	NA
17	36.9	155.2	NA	37	37.2	156.7	NA
18	36.7	156.7	A	38	37.2	156.7	NA
19	36.7	156.7	A	39	37.2	156.7	NA
20	36.7	156.7	A	40	37.2	156.7	NA

1. Design the decision tree for this learning set.

2. Compute the classification success rate of the decision tree on the learning set.

3. Compute the information of the optimal splitting test $I_2 \leq 155.680$ for node 1.

4. Calculate the information of the optimal splitting test $I_1 \leq 36.952$ for the second test node.

5. Compute the total information of the decision tree.

4.5 Decision Trees

Solution

1. Node 1 has the whole learning set, which has 16 acceptable and 24 non-acceptable learning samples, so the acceptability index of node 1 is 16/40, or 0.4.

   Node 1 has the whole learning set of Table 4.3. The optimal splitting test for node 1 is $I_2 \leq 155.680$. Using the optimal test $I_2 \leq 155.680$ of node 1, its two successor nodes 2 and 3 are created as follows:
   - Node 2 is composed of the 17 learning examples with LEC {1, 2, 3, 4, 5, 6, 7, 8, 9, 10, 11, 12, 13, 14, 15, 16, 17} of the learning set of Table 4.3 that verify the test $I_2 \leq 155.680$. All these 17 learning examples are of non-acceptable quality, so the classification entropy of node 2 is $H_C(\text{node 2}) = 0.0$. Since $H_C(\text{node 2}) < H_{\min}$, node 2 is LEAF according to (4.32). The acceptability index, i.e., the ratio of the acceptable learning examples over the whole learning examples of node 2 is 0.0, since all learning examples are of non-acceptable quality.
   - Node 3 is composed of the remaining 23 learning examples with LEC {18, 19, 20, 21, 22, 23, 24, 25, 26, 27, 28, 29, 30, 31, 32, 33, 34, 35, 36, 37, 38, 39, 40} of the learning set of Table 4.3 that do not verify the test $I_2 \leq 155.680$. Among these 23 learning examples, the seven learning examples with LEC {34, 35, 36, 37, 38, 39, 40} are of non-acceptable quality, while the remaining 16 learning examples are of acceptable quality, so the acceptability index of node 3 is 16/23 or 0.6957. The classification entropy of node 3 is:

$$H_C(\text{node 3}) = -\left[\frac{16}{23} \cdot \log_2 \frac{16}{23} + \frac{7}{23} \cdot \log_2 \frac{7}{23}\right] = 0.8865.$$

   At this point we see that $H_C(\text{node 3}) > H_{\min}$, which means that node 3 is either DEADEND or node 3 has to be split into two new nodes. However, since it is given that the decision tree has in total five nodes, the conclusion is that node 3 is a test node.

   As we have seen above, node 3 is a test node that has the learning examples of Table 4.3 with LEC {18, 19, 20, 21, 22, 23, 24, 25, 26, 27, 28, 29, 30, 31, 32, 33, 34, 35, 36, 37, 38, 39, 40}. The optimal splitting test for node 3 is $I_1 \leq 36.952$. Using the optimal test $I_1 \leq 36.952$ of node 3, its two successor nodes 4 and 5 are created as follows:
   - Node 4 is composed of the 16 learning examples with LEC {18, 19, 20, 21, 22, 23, 24, 25, 26, 27, 28, 29, 30, 31, 32, 33} of the learning set of Table 4.3 that verify the test $I_1 \leq 36.952$. All these 16 learning examples are of acceptable quality, so the classification entropy of node 4 is

$H_C(\text{node } 4) = 0.0$. Since $H_C(\text{node } 4) < H_{min}$, node 4 is LEAF according to (4.32). The acceptability index of node 4 is 1.0, since all its learning examples are of acceptable quality.

- Node 5 is composed of the seven learning examples with LEC {34, 35, 36, 37, 38, 39, 40} of the learning set of Table 4.3 that do not verify the test $I_1 \leq 36.952$. All these seven learning examples are of non-acceptable quality, so the classification entropy of node 5 is $H_C(\text{node } 5) = 0.0$. Since $H_C(\text{node } 5) < H_{min}$, node 5 is LEAF according to (4.32). The acceptability index of node 5 is 0.0, since all its learning examples are of non-acceptable quality.

Figure 4.7 shows the decision tree. The notation used for the decision tree nodes is explained in Fig. 4.6.

2. The learning examples with LEC {1, 2, 3, 4, 5, 6, 7, 8, 9, 10, 11, 12, 13, 14, 15, 16, 17} are led to node 2 and are classified as NA (non-acceptable) by the decision tree of Fig. 4.7, since the acceptability index of node 2 is 0.0. As can be seen from Table 4.3, all these 17 learning examples are of NA quality, so all 17 examples are corrected classified, which means that the classification success rate of node 2 is 17/17 or 100%.

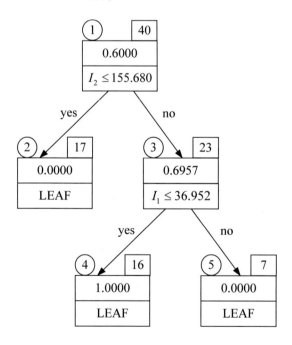

Fig. 4.7 Decision tree for the Example 4.2

4.5 Decision Trees

Similarly, the learning examples with LEC {18, 19, 20, 21, 22, 23, 24, 25, 26, 27, 28, 29, 30, 31, 32, 33} are led to node 4 and are classified as A (acceptable), since the acceptability index of node 4 is 1.0. As can be seen from Table 4.3, all these 16 learning examples are of A quality, so all 16 examples are corrected classified, which means that the classification success rate of node 4 is 16/16 or 100%.

The learning examples with LEC {34, 35, 36, 37, 38, 39, 40} are led to node 4 and are classified as NA (non-acceptable), since the acceptability index of node 5 is 0.0. As can be seen from Table 4.3, all these seven learning examples are of A quality, so all seven examples are corrected classified, which means that the classification success rate of node 5 is 7/7 or 100%.

From the above it is concluded that the decision tree of Fig. 4.7 correctly classifies all 40 learning examples of the learning set of Table 4.3, so the total classification success rate of the decision tree is 40/40 or 100%.

3. Since the 16 learning examples of node 1 are of acceptable quality and the remaining 24 are of non-acceptable quality, the classification entropy of node 1 is:

$$H_C(\text{node 1}) = -\left[\frac{16}{40} \cdot \log_2 \frac{16}{40} + \frac{24}{40} \cdot \log_2 \frac{24}{40}\right] = 0.9710.$$

Because all the learning examples of node 2 are of non-acceptable quality, the entropy of node 2 with respect to the partition induced by test T ($I_2 \leq 155.680$) is:

$$H_T(\text{node 2}) = 0.0.$$

Since the 16 of 23 learning examples of node 3 are of acceptable quality, the entropy of node 3 with respect to the partition induced by test T is:

$$H_T(\text{node 3}) = -\left[\frac{16}{23} \cdot \log_2 \frac{16}{23} + \frac{7}{23} \cdot \log_2 \frac{7}{23}\right] = 0.8865.$$

Since node 2 is composed of 17 learning examples and node 3 has 23 learning examples, the mean conditional entropy of node 1, given the outcome of test T, is:

$$H_C(\text{node 1}|T) = \frac{17}{40} \cdot H_T(\text{node 2}) + \frac{23}{40} \cdot H_T(\text{node 3}) \Rightarrow$$

$$H_C(\text{node 1}|T) = \frac{17}{40} \cdot 0.0 + \frac{23}{40} \cdot 0.8865 \Rightarrow H_C(\text{node 1}|T) = 0.5097.$$

The information gained from the application of test T at node 1 is:

$$I(\text{node }1;T) = H_C(\text{node }1) - H_C(\text{node }1|T) = 0.9710 - 0.5097 \Rightarrow$$

$$I(\text{node }1;T) = 0.4613.$$

4. Since the 16 learning examples of node 3 are of acceptable quality and the remaining seven are of non-acceptable quality, the classification entropy of node 3 is:

$$H_C(\text{node }3) = -\left[\frac{16}{23} \cdot \log_2 \frac{16}{23} + \frac{7}{23} \cdot \log_2 \frac{7}{23}\right] = 0.8865.$$

Because all 16 learning examples of node 4 are of acceptable quality, the entropy of node 4 with respect to the partition induced by test T ($I_1 \leq 36.952$) is:

$$H_T(\text{node }4) = 0.0.$$

Because all seven learning examples of node 5 are of non-acceptable quality, the entropy of node 5 with respect to the partition induced by test T ($I_1 \leq 36.952$) is:

$$H_T(\text{node }5) = 0.0.$$

The mean conditional entropy of node 3, given the outcome of test T, is:

$$H_C(\text{node }3 | T) = \frac{16}{23} \cdot H_T(\text{node }4) + \frac{7}{23} \cdot H_T(\text{node }5) \Rightarrow$$

$$H_C(\text{node }3 | T) = \frac{16}{23} \cdot 0.0 + \frac{7}{23} \cdot 0.0 \Rightarrow H_C(\text{node }3 | T) = 0.0.$$

The information gained from the application of test T at node 3 is:

$$I(\text{node }3;T) = H_C(\text{node }3) - H_C(\text{node }3|T) = 0.8865 - 0.0 \Rightarrow$$

$$I(\text{node }3;T) = 0.8865.$$

5. Node 1 has 40 learning examples, so the total information of node 1 is:

$$I_{\text{node 1}} = 40 \cdot I(\text{node 1}; T) = 40 \cdot 0.4613 \Rightarrow I_{\text{node 1}} = 18.5.$$

Node 3 has 23 learning examples, so the total information of node 3 is:

$$I_{\text{node 3}} = 23 \cdot I(\text{node 3}; T) = 23 \cdot 0.8865 \Rightarrow I_{\text{node 3}} = 20.4.$$

The total information of the decision tree of Fig. 4.7 is:

$$I_{tot} = I_{\text{node 1}} + I_{\text{node 3}} = 18.5 + 20.4 \Rightarrow I_{tot} = 38.9.$$

Node 1 possesses 18.5/38.9 or 48% of the information of the decision tree and node 3 possesses the rest 52% of decision tree information.

4.6 Artificial Neural Networks

4.6.1 Introduction

The artificial neural network (ANN) or simply neural network is a computer information processing system that is capable of sufficiently representing any nonlinear functions (Haykin 1994). The techniques based on ANNs are especially effective in the solution of high-complexity problems for which a traditional mathematical model is difficult to build, where the nature of the input/output relationship is neither well defined nor easily computable.

The ANN is composed of simple elements, called *neurons*, operating in parallel. ANNs are inspired by biological nervous systems. As in nature, the ANN function is determined largely by the connections between neurons. An ANN can be trained to perform a particular function by adjusting the values of the connections (weights) between neurons.

Traditional engineering practices suggest that any input/output relationship, regardless of how complex, can be developed from physical principles (e.g., electric circuit theory) starting at the individual component level. Unfortunately, there are two major problems with this assertion when real problems are considered:

1. The problem may be so large that an exact mathematical model is not realistic, so we use approximate analysis, based on sound engineering assumptions, on a regular basis to avoid this problem.

2. The complexity of the problem may introduce uncertainty for which we must generate suitable approximations.

In either case, the relationship between input and output variables is usually only approximately known, and much work is required to generate approximate relationships. ANNs have the ability to automatically learn approximate relationships between inputs and outputs without being overcome by the size and complexity of the problem. These approximate relationships are often superior to those generated by engineers because they are usually based on actual inputs and outputs (e.g., measured data) and are not biased based on preconceived ideas of engineers. Furthermore, the number of approximation can be gradually reduced as more information becomes available. In theory, the input/output relationship learned by an ANN can become exact.

ANNs are used for classification, prediction, and function estimation.

ANNs offer significant advantages and valuable characteristics that are unavailable elsewhere (Hammerstrom 1993). In particular:

1. ANNs can infer subtle, unknown relationships from data. This characteristic is useful because gathering data does not require explaining it.

2. ANNs can generalize, meaning that they can respond correctly to new input data that are only broadly similar to the original training data. Generalization is useful because real-world data is noisy, distorted, and often incomplete.

3. ANNs are nonlinear, that is, they can solve complex problems more accurately than linear techniques do. It should be noted that nonlinear behavior is common, but can be difficult to handle mathematically.

4. ANNs contain many identical, independent operations that can be executed simultaneously. This is useful because parallel hardware can run ANNs quickly, often making them faster than alternative methods.

4.6.2 Applications to Power Systems

Artificial neural networks have shown their capabilities in solving real-world problems, as reflected by the growing number of publications on their applications to power system problems (Sobajic and Pao 1989; Santoso and Tan 1990; Park et al. 1991; Peng et al. 1992; Papalexopoulos et al. 1994; Piras et al. 1996; Halpin and Burch 1997; Gubina and Halilcevic 1998; Halilcevic and Gubina 1999; Amjady and Ehsan 1999; Szkuta et al. 1999; Chen and Maun 2000; Naresh and Sharma 2000; Jensen et al. 2001; Hippert et al. 2001; Jain et al. 2003; Stankovic et al. 2003; Halilcevic et al. 2003; Guo and Luh 2004; Gerbec et al. 2005; Thalassinakis et al. 2006; Huang and Wang 2007; Georgilakis 2007), and electrical machines (Mohammed et al. 1992; Hoole 1993; Hoole and Haldar 1995; Wishart and Harley 1995; Filippeti et al. 1995; Tsekouras et al. 2001; Kiartzis and Kladas 2001), to name only a few.

In transformer engineering, in particular, ANNs have been applied for the evaluation of transformer technical characteristics (Georgilakis et al. 1998, Georgilakis et al. 1999, Georgilakis 2000, Nussbaum et al. 2000; He et al. 2000; Georgilakis et al. 2001; Yu et al. 2001; Mao and Aggarwal 2001; Adly 2001; Doulamis et al. 2002; Khorashadi-Zadeh and Sanaye-Pasand 2006; Georgilakis and Amoiralis 2007; Amoiralis et al. 2007; Rebizant and Bejmert 2007) as well as for transformer fault diagnosis (Perez et al. 1994; Zhang et al. 1996; Pihler et al. 1997; Zaman and Rahman 1998; Wang et al. 1998; Yang and Huang 1998; Orille-Fernandez et al. 2001; Farag et al. 2001; Guardado et al. 2001; Yang et al. 2001; Huang and Huang 2002; De and Chatterjee 2002; Huang 2003; Wang 2005; Miranda and Castro 2005; Castro and Miranda 2005; Segatto and Coury 2006; Pylvanainen et al. 2007; Hao and Cai-Xin 2007).

4.6.3 ANN Types

Many different ANN types are described in the literature (Haykin 1994). The main types of ANN that that have been used in power systems are classified as follows, with an approximate percentage of all reported applications given in parentheses following each major category (El-Sharkawi and Niebur 1996):

1. Multilayer perceptrons (81.2%):
   – Three-layer neural networks
   – Four-layer neural networks
   – Recurrent neural networks
   – Functional-link networks
   – Radial basis function neural networks
   – Fuzzy neural networks

2. Hopfield networks (5.4%):
   – Boltzman machine
   – Gaussian machine
   – Chaotic neural networks

3. Kohonen networks (8.3%):
   – Rectangular (two-dimensional) grid networks
   – Hexagonal (two-dimensional) grid networks

4. Other networks (5.1%).

In this chapter, we concentrate on the design, construction, and application of three-layer and four-layer neural networks, due to their very widespread application.

4.6.4 Neuron Mathematical Model

The basic element in an ANN is called a neuron and its mathematical model is depicted in Fig. 4.8. It can be seen from Fig. 4.8 that one neuron can accept at its input n inputs x_1, x_2, ..., x_n and one input for the bias b and produces one output y. The bias b of the neuron can be either non-zero or zero.

As can be seen from Fig. 4.8, each one of the n inputs x_1, x_2, ..., x_n, has a corresponding weight w_1, w_2, ..., w_n, respectively. The total input z of the neuron is calculated as follows:

$$z = w_1 \cdot x_1 + w_2 \cdot x_2 + ... + w_n \cdot x_n + b. \tag{4.35}$$

The output y of the neuron is calculated as follows:

$$y = f(z) = f(w_1 \cdot x_1 + w_2 \cdot x_2 + ... + w_n \cdot x_n + b). \tag{4.36}$$

The function f is called the *activation function* or the *transfer function*. Two typical activation functions that are usually used in ANNs are the sigmoid transfer function:

$$f(z) \equiv \text{logsig}(z) = \frac{1}{1+e^{-z}}, \tag{4.37}$$

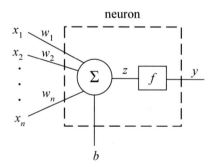

Fig. 4.8 Mathematical model for neuron

4.6 Artificial Neural Networks

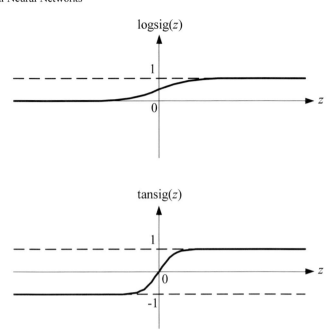

Fig. 4.9 Sigmoid and tan-sigmoid transfer functions

and the tan-sigmoid transfer function:

$$f(z) \equiv \operatorname{tansig}(z) = \frac{2}{1+e^{-2 \cdot z}} - 1. \tag{4.38}$$

These two transfer functions are shown in Fig. 4.9. The above two transfer functions introduce a nonlinearity to the neurons that further enhances the neural network ability to model complex functions.

4.6.5 ANN Architectures

A multilayer neural network is constructed by combining neurons in series and parallel. A small three-layer ANN is shown in Fig. 4.10. As is evident in Fig. 4.10, an ANN consists of multiple series connections (from left to right in Fig. 4.10) of multiple parallel connections (from top to bottom in Fig. 4.10). A *layer* is defined to be a set of parallel (top to bottom in Fig. 4.10) connected neurons or nodes. Data enters the neural network through the input layer. The nodes in the input layer are passive, not computational, i.e., each node simply broadcasts a single data value over weighted connections to the hidden nodes. The hidden layer and the output layer are identical in both form and functionality, and it is these layers that give the ANN its ability to learn complex nonlinear relationships between inputs and outputs. The mathematical model of each hidden and output neuron is the one depicted in Fig. 4.8.

The neural network of Fig. 4.10 has three neurons in the input layer, three neurons in the hidden layer, and two neurons in the output layer, and is said to be a neural network with architecture $3-3-2$, from the number of input, hidden, and output neurons, respectively. The neural network of Fig. 4.10 has $3 \cdot 3 = 9$ connections between the three input and the three hidden neurons and $3 \cdot 2 = 6$ connections between the three hidden and the two output neurons. This means that the neural network of Fig. 4.10 has in total 15 connections. A weight value $w_{pj,qk}$ corresponds to each connection that links the neuron p of layer j with the neuron q of layer k. It is supposed that the hidden and output neurons of the ANN of Fig. 4.10 have zero bias value. Moreover, it is assumed that the sigmoid transfer function is used in the hidden and output neurons of the ANN of Fig. 4.10.

Figure 4.11 shows the input and output signals of the second neuron of the second layer (hidden layer) that correspond to the ANN of Fig. 4.10. The total input, INP_{22}, of the second neuron of the second layer is:

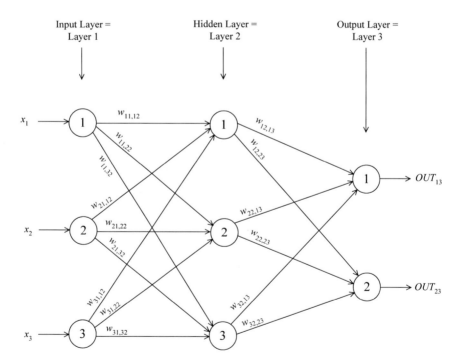

Fig. 4.10 A three-layer neural network with architecture 3-3-2

4.6 Artificial Neural Networks

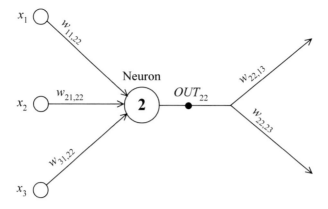

Fig. 4.11 Input/output signals of the second neuron of the second layer for the ANN of Fig. 4.10

$$INP_{22} = w_{11,22} \cdot x_1 + w_{21,22} \cdot x_2 + w_{31,22} \cdot x_3. \tag{4.39}$$

Since in the hidden and output layers the sigmoid transfer function is used, the output, OUT_{22}, of the second neuron of the second layer is:

$$OUT_{22} = \frac{1}{1+e^{-INP_{22}}}. \tag{4.40}$$

The input/output signals of the remaining hidden and output neurons of the ANN of Fig. 4.10 are similarly produced.

4.6.6 ANN Training

The major justification for the use of ANNs is their ability to learn relationships in complex data sets that may not be easily perceived by engineers. An ANN performs this function as a result of *training* that is a process of repetitively presenting a set of training data (typically a representative subset of the complete set of data available) to the network and adjusting the weights so that each input data set produces the desired output.

There are a number of methods that can be used to adjust the weights in an ANN. Generally speaking, there are two categories of methods: (1) unsupervised and (2) supervised learning. Unsupervised learning is a learning process that requires only input pairs to be included in the training set (learning set). On the other hand, supervised learning is a learning process that requires both input/output pairs to be included in the training set. Unsupervised learning can be characterized as a fast, but potentially inaccurate, method of adjusting the weights. On the other hand, supervised learning typically requires longer learning times and can be more accurate. There is no way to tell beforehand which learning method will work best for a given application. For this reason, we concentrate on the very popular supervised learning approach based on the *backpropagation training algorithm*, which has been shown to produce good results for a large number of different problems (El-Sharkawi and Niebur 1996).

The backpropagation training algorithm is a method of iteratively adjusting the neural network weights until the desired accuracy level is achieved. It is based on a gradient-search optimization method applied to an error function. Typical error functions include the *mean square error* shown in (4.41), where N is the total number of input/output pairs (which can be vector quantities) used for training:

$$mse = \frac{1}{N} \cdot \sum_{i=1}^{N} \left[OUT_{forecast,i} - OUT_{actual,i} \right]^2, \qquad (4.41)$$

where $OUT_{forecast,i}$ and $OUT_{actual,i}$ are the output forecast by the neural network and the actual (desired) output, respectively, of the ith training example. The set of N training examples (input/output pairs) defines the training set or learning set. For best results, the training set should adequately represent all expected variations in the complete set of data.

A recursive algorithm for adjusting the weights can be developed, such that the error defined by (4.41) is minimized. The equations (4.42) and (4.43) are recursive training equations based on the generalized delta rule (Wasserman 1989; Pao 1989; Kosko 1993) and the corresponding algorithm is called *gradient descent backpropagation*:

$$\Delta w_{pj,qk}(n+1) = lr \cdot \delta_{qk} \cdot OUT_{pj} + m \cdot \Delta w_{pj,qk}(n), \qquad (4.42)$$

$$w_{pj,qk}(n+1) = w_{pj,qk}(n) + \Delta w_{pj,qk}(n+1), \qquad (4.43)$$

where:

n	:	the number of the current iteration of the training algorithm
$w_{pj,qk}(n)$	:	the value of the weight that connects the neuron p of layer j with the neuron q of layer k during the iteration n
$\Delta w_{pj,qk}(n)$	:	the variation in the value of the weight $w_{pj,qk}(n)$ during the iteration n
δ_{qk}	:	the value of δ (delta coefficient) for the neuron q of layer k
OUT_{pj}	:	the output for the neuron p of layer j
lr	:	the learning rate
m	:	the momentum

4.6 Artificial Neural Networks

The value of δ is calculated differently depending on the specific location of the weight under consideration: (4.44) is the formula for calculating δ for any weight connected from a hidden layer neuron to an output layer neuron:

$$\delta_{qk} = \frac{2}{N} \cdot OUT_{qk} \cdot (1 - OUT_{qk}) \cdot (OUT_{\text{actual }qk} - OUT_{qk}), \tag{4.44}$$

where layer k is the output layer, $OUT_{\text{actual }qk}$ is the actual (desired) output of any neuron q of the output layer k, and N is the number of training examples of the training set. The values $OUT_{\text{actual }qk}$ in (4.44) are known from the training set. The calculated output of the network is compared to the actual value to generate an error signal. The error signal is propagated back through the neural network to adjust the weights, as shown in (4.42) and (4.43).

For neurons in any other than an output layer, however, an error value is not directly obtainable because no desired output value is given for these internal neurons as a part of the training set. The error values for any neurons other than the output neurons are calculated as weighted sums of the output layer errors:

$$\delta_{pj} = OUT_{pj} \cdot (1 - OUT_{pj}) \cdot \sum_{q=1}^{Q} \delta_{qk} \cdot w_{pj,qk}, \tag{4.45}$$

where Q is the number of neurons of the output layer.

The coefficient *lr* in (4.42) is called *learning rate* and directly controls how much the calculated error values are allowed to change the weights. The learning rate is typically selected between 0.01 and 1.0. The coefficient m in (4.42) is called *momentum* and allows the weight updates at one iteration to utilize information from previous error values. The momentum term helps avoid settling into a local minimum and is selected between 0.01 and 1.0.

The recursive training algorithm (set $n = n+1$) is executed until the network satisfactorily predicts the output values. Common stopping criteria for the training algorithm involve monitoring either the mean square error or the maximum error or both and stopping when the value is less than a specified tolerance. The selected tolerance is very problem dependent and may or may not be actually achievable. There is no mathematical proof that the backpropagation training algorithm will ever converge within a given tolerance. The only guarantee is that any changes of the weights will not increase the total error. Note that the inclusion of the momentum term may allow the error as defined in (4.41) to temporarily increase if the optimization process is moving away from the local minimum.

Table 4.4 Training set for the Example 4.3

Training example code	Input I_1	I_2	Output O_1
1	0	0	0
2	0	1	1
3	1	0	1
4	1	1	0

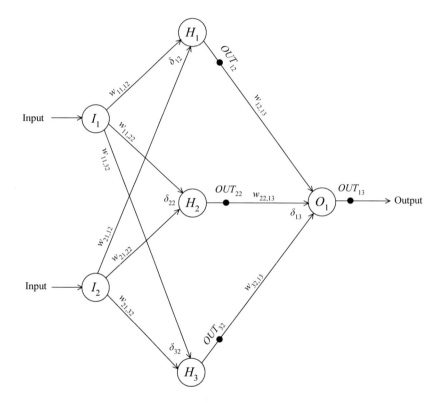

Fig. 4.12 Artificial neural network for the solution of Example 4.3

4.6.6.1 Example 4.3

An ANN will be used to solve the two-bit exclusive-or (XOR) problem. The training set for the XOR problem is given in Table 4.4. Show how the training process

4.6 Artificial Neural Networks

works by making one iteration of the gradient descent backpropagation algorithm. It is supposed that the learning rate is equal to one ($lr = 1$) and the momentum is equal to zero ($m = 0$). It is also assumed that the hidden and output neurons have zero bias values and at these neurons the sigmoid transfer function is used.

Solution

As can be seen from Table 4.4, the output (O_1) can take only two values, 0 or 1, so the XOR problem is a classification problem. Since the training set of Table 4.4 has two inputs (I_1, I_2) and one output (O_1), the neural network will have two input neurons and one output neuron. We will use the very simple three-layer neural network of Fig. 4.12 with architecture 2-3-1 (i.e., two input neurons, three hidden neurons, and one output neuron) for the solution of the XOR classification problem. It should be noted that conventional classification techniques, e.g., nearest-neighbor methods, cannot correctly classify these training examples.

As requested, one iteration of the backpropagation algorithm will be presented. More specifically, the steps to be followed for application of the backpropagation algorithm are:

1. *Iteration 0.* Initial weights are assigned to the neural network. Next, the output and the performance of the neural network are calculated.

2. *Iteration 1.* The value of delta for the output neuron is calculated using (4.44). The values of delta for the three hidden neurons is computed using (4.45). The new weights are calculated using (4.42) and (4.43). Finally, the performance of the neural network is computed.

Table 4.5 Initial random weights

Neuron From	Neuron To	Weight Symbol	Value
I_1	H_1	$w_{11,12}$	−0.07514
I_1	H_2	$w_{11,22}$	−0.05542
I_1	H_3	$w_{11,32}$	−0.53124
I_2	H_1	$w_{21,12}$	0.85124
I_2	H_2	$w_{21,22}$	−0.18542
I_2	H_3	$w_{21,32}$	0.47563
H_1	O_1	$w_{12,13}$	−0.01235
H_2	O_1	$w_{22,13}$	0.53121
H_3	O_1	$w_{32,13}$	0.18155

Iteration 0

The gradient descent backpropagation algorithm begins with the assignment of small ($-1 < x < 1$) random numbers to the neural network weights. Such a random weight assignment for the neural network of Fig. 4.12 is given in Table 4.5.

The computation of the output of the neural network is done in two steps:

1. *Simulation of the hidden layer.* The net inputs (INP_{12}, INP_{22}, INP_{32}) to the three hidden neurons are computed for the four training examples of Table 4.4. Next, the outputs (OUT_{12}, OUT_{22}, OUT_{32}) of the three hidden neurons are calculated for the four training examples.

2. *Simulation of the output layer.* The net input (INP_{13}) to the output neuron is computed for the four training examples. Next, the output (OUT_{13}) of the output neuron is calculated for the four training examples. This output (OUT_{13}) is in fact the output of the neural network, as can be seen from Fig. 4.12.

Initially, we will show the simulation of the hidden layer by presenting the calculation of the output of the first hidden neuron ($OUT_{12}^{(4)}$) for the fourth training example. It can be seen from Table 4.4 that the fourth training example is $(I_1, I_2, O_1) = (I_1^{(4)}, I_2^{(4)}, O_1^{(4)}) = (1, 1, 0)$, so $I_1^4 = 1$ and $I_2^{(4)} = 1$. The net input for the neuron H_1, i.e., the first neuron of the second layer is:

$$INP_{12}^{(4)} = w_{11,12} \cdot I_1^{(4)} + w_{21,12} \cdot I_2^{(4)} = (-0.07514) \cdot 1 + 0.85124 \cdot 1 \Rightarrow$$

$$INP_{12}^{(4)} = 0.7761,$$

where the initial weight values of Table 4.5 were used.

The output for the neuron H_1 is:

$$OUT_{12}^{(4)} = logsig\left[INP_{12}^{(4)}\right] = \frac{1}{1+e^{-INP_{12}^{(4)}}} = \frac{1}{1+e^{-0.7761}} \Rightarrow OUT_{12}^{(4)} = 0.6848.$$

Using matrix algebra, the net inputs for the three hidden neurons and for the four training examples are computed as follows:

$$[\mathbf{INP}_{hidden}] \equiv \begin{bmatrix} INP_{12}^{(1)} & INP_{12}^{(2)} & INP_{12}^{(3)} & INP_{12}^{(4)} \\ INP_{22}^{(1)} & INP_{22}^{(2)} & INP_{22}^{(3)} & INP_{22}^{(4)} \\ INP_{32}^{(1)} & INP_{32}^{(2)} & INP_{32}^{(3)} & INP_{32}^{(4)} \end{bmatrix} \Rightarrow$$

4.6 Artificial Neural Networks

$$[\mathbf{INP}_{hidden}] = [\mathbf{w}_{hidden}(0)] \cdot [\mathbf{OUT}_{input}] \Rightarrow$$

$$[\mathbf{INP}_{hidden}] = \begin{bmatrix} w_{11,12} & w_{21,12} \\ w_{11,22} & w_{21,22} \\ w_{11,32} & w_{21,32} \end{bmatrix} \cdot \begin{bmatrix} I_1^{(1)} & I_1^{(2)} & I_1^{(3)} & I_1^{(4)} \\ I_2^{(1)} & I_2^{(2)} & I_2^{(3)} & I_2^{(4)} \end{bmatrix} \Rightarrow$$

$$[\mathbf{INP}_{hidden}] = \begin{bmatrix} -0.07514 & 0.85124 \\ -0.05542 & -0.18542 \\ -0.53124 & 0.47563 \end{bmatrix} \cdot \begin{bmatrix} 0 & 0 & 1 & 1 \\ 0 & 1 & 0 & 1 \end{bmatrix} \Rightarrow$$

$$[\mathbf{INP}_{hidden}] = \begin{bmatrix} 0 & 0.8512 & -0.0751 & 0.7761 \\ 0 & -0.1854 & -0.0554 & -0.2408 \\ 0 & 0.4756 & -0.5312 & -0.0556 \end{bmatrix},$$

where $I_1^{(1)}$ denotes the input I_1 for the first training example and $INP_{12}^{(1)}$ denotes the net input of the first neuron of the second layer (hidden layer) for the first training example.

The outputs for the three hidden neurons and for the four training examples are computed as follows:

$$[\mathbf{OUT}_{hidden}] \equiv \begin{bmatrix} OUT_{12}^{(1)} & OUT_{12}^{(2)} & OUT_{12}^{(3)} & OUT_{12}^{(4)} \\ OUT_{22}^{(1)} & OUT_{22}^{(2)} & OUT_{22}^{(3)} & OUT_{22}^{(4)} \\ OUT_{32}^{(1)} & OUT_{32}^{(2)} & OUT_{32}^{(3)} & OUT_{32}^{(4)} \end{bmatrix} = logsig\{[\mathbf{INP}_{hidden}]\} \Rightarrow$$

$$[\mathbf{OUT}_{hidden}] = \begin{bmatrix} logsig(0) & logsig(0.8512) & logsig(-0.0751) & logsig(0.7761) \\ logsig(0) & logsig(-0.1854) & logsig(-0.0554) & logsig(-0.2408) \\ logsig(0) & logsig(0.4756) & logsig(-0.5312) & logsig(-0.0556) \end{bmatrix} \Rightarrow$$

$$[\mathbf{OUT}_{hidden}] = \begin{bmatrix} 0.5 & 0.7008 & 0.4812 & 0.6848 \\ 0.5 & 0.4538 & 0.4861 & 0.4401 \\ 0.5 & 0.6167 & 0.3702 & 0.4861 \end{bmatrix},$$

where $OUT_{12}^{(1)}$ is the output of the first neuron of the second layer (hidden layer) for the first training example.

The net input for the output neuron and for the four training examples is computed as follows:

$$[\mathbf{INP}_{output}] \equiv \begin{bmatrix} INP_{13}^{(1)} & INP_{13}^{(2)} & INP_{13}^{(3)} & INP_{13}^{(4)} \end{bmatrix} = \begin{bmatrix} \mathbf{w}_{output}(0) \end{bmatrix}^T \cdot [\mathbf{OUT}_{hidden}] \Rightarrow$$

$$[\mathbf{INP}_{output}] = \begin{bmatrix} w_{12,13} & w_{22,13} & w_{32,13} \end{bmatrix} \cdot [\mathbf{OUT}_{hidden}] \Rightarrow$$

$$[\mathbf{INP}_{output}] = \begin{bmatrix} -0.01235 & 0.53121 & 0.18155 \end{bmatrix} \cdot \begin{bmatrix} 0.5 & 0.7008 & 0.4812 & 0.6848 \\ 0.5 & 0.4538 & 0.4861 & 0.4401 \\ 0.5 & 0.6167 & 0.3702 & 0.4861 \end{bmatrix} \Rightarrow$$

$$[\mathbf{INP}_{output}] = \begin{bmatrix} 0.3502 & 0.3444 & 0.3195 & 0.3136 \end{bmatrix}.$$

The output for the neuron of the output layer and for the four training examples is computed as follows:

$$[\mathbf{OUT}_{output}] \equiv \begin{bmatrix} OUT_{13}^{(1)} & OUT_{13}^{(2)} & OUT_{13}^{(3)} & OUT_{13}^{(4)} \end{bmatrix} = logsig\{[\mathbf{INP}_{output}]\} \Rightarrow$$

$$[\mathbf{OUT}_{output}] = \begin{bmatrix} logsig(0.3502) & logsig(0.3444) & logsig(0.3195) & logsig(0.3136) \end{bmatrix} \Rightarrow$$

$$[\mathbf{OUT}_{output}] = \begin{bmatrix} 0.5867 & 0.5852 & 0.5792 & 0.5778 \end{bmatrix}.$$

As can be seen from Table 4.4, the desired output for the four training examples is $[\mathbf{OUT}_{desired}] = \begin{bmatrix} 0 & 1 & 1 & 1 \end{bmatrix}$, so the error in the prediction of the output by the neural network is:

$$[error] = [\mathbf{OUT}_{desired}] - [\mathbf{OUT}_{output}] \Rightarrow$$

$$[error] = \begin{bmatrix} 0 & 1 & 1 & 0 \end{bmatrix} - \begin{bmatrix} 0.5867 & 0.5852 & 0.5792 & 0.5778 \end{bmatrix} \Rightarrow$$

$$[error] = \begin{bmatrix} -0.5867 & 0.4148 & 0.4208 & -0.5778 \end{bmatrix}.$$

The performance of the neural network is given by the mean square error function, which is calculated using (4.41):

$$mse = \frac{1}{N} \cdot \sum_{i=1}^{N} [error_i] = \frac{1}{4} \cdot \left[(-0.5867)^2 + 0.4148^2 + 0.4208^2 + (-0.5778)^2 \right] \Rightarrow$$

$$mse = 0.2568.$$

Iteration 1

The value of delta for the output neuron and for all the training examples is calculated. We will show this calculation for the fourth training example, for which the delta for the output neuron, $\delta_{13}^{(4)}$, is calculated using (4.44):

$$\delta_{13}^{(4)} = \frac{2}{N} \cdot OUT_{13}^{(4)} \cdot \left[1 - OUT_{13}^{(4)}\right] \cdot \left[OUT_{actual\,13}^{(4)} - OUT_{13}^{(4)}\right] \Rightarrow$$

$$\delta_{13}^{(4)} = \frac{2}{4} \cdot 0.5778 \cdot [1 - 0.5778] \cdot [0 - 0.5778] \Rightarrow \delta_{13}^{(4)} = -0.0705 \,,$$

where $OUT_{actual\,13}^{(4)}$ is the desired output of the output neuron for the fourth training example. The value of $OUT_{actual\,13}^{(4)}$ is retrieved from Table 4.4.

The calculation of the delta value for the output neuron for the remaining three training examples is performed similarly and the results obtained are:

$$\left[\boldsymbol{\delta}_{output}\right] = \left[\delta_{13}^{(1)} \quad \delta_{13}^{(2)} \quad \delta_{13}^{(3)} \quad \delta_{13}^{(4)}\right] = \left[-0.0711 \quad 0.0503 \quad 0.0513 \quad -0.0705\right].$$

Next, the values of delta for the three hidden neurons and for the four training examples will be computed. We will show this calculation for the first hidden neuron (H_1) for the fourth training example, for which the delta for the output neuron, $\delta_{12}^{(4)}$, is calculated using (4.45):

$$\delta_{12}^{(4)} = OUT_{12}^{(4)} \cdot (1 - OUT_{12}^{(4)}) \cdot \sum_{q=1}^{1} \delta_{13}^{(4)} \cdot w_{12,13} \Rightarrow$$

$$\delta_{12}^{(4)} = 0.6848 \cdot (1 - 0.6848) \cdot (-0.0705) \cdot (-0.01235) \Rightarrow \delta_{12}^{(4)} = 0.0002 \,.$$

The remaining 11 calculations of delta value for all the hidden neurons and for all the training examples are performed similarly and the results obtained are:

$$\left[\boldsymbol{\delta}_{hidden}\right] = \begin{bmatrix} \delta_{12}^{(1)} & \delta_{12}^{(2)} & \delta_{12}^{(3)} & \delta_{12}^{(4)} \\ \delta_{22}^{(1)} & \delta_{22}^{(2)} & \delta_{22}^{(3)} & \delta_{22}^{(4)} \\ \delta_{32}^{(1)} & \delta_{32}^{(2)} & \delta_{32}^{(3)} & \delta_{32}^{(4)} \end{bmatrix} \Rightarrow$$

$$\left[\boldsymbol{\delta}_{hidden}\right] = \begin{bmatrix} 0.0002 & -0.0001 & -0.0002 & 0.0002 \\ -0.0094 & 0.0066 & 0.0068 & -0.0092 \\ -0.0032 & 0.0022 & 0.0022 & -0.0032 \end{bmatrix}.$$

Next, the new weights are calculated using (4.42) and (4.43), and (4.42) is used to compute the variation in the value of the weights linking the input neurons with the hidden neurons:

$$\left[\Delta \mathbf{w}_{hidden}(1)\right] = lr \cdot \left[\boldsymbol{\delta}_{hidden}\right] \cdot \left[\mathbf{OUT}_{input}\right]^T + m \cdot \left[\Delta \mathbf{w}_{hidden}(0)\right] \Rightarrow$$

$$\left[\Delta \mathbf{w}_{hidden}(1)\right] = 1 \cdot \begin{bmatrix} 0.0002 & -0.0001 & -0.0002 & 0.0002 \\ -0.0094 & 0.0066 & 0.0068 & -0.0092 \\ -0.0032 & 0.0022 & 0.0022 & -0.0032 \end{bmatrix} \cdot \begin{bmatrix} 0 & 0 \\ 0 & 1 \\ 1 & 0 \\ 1 & 1 \end{bmatrix} + 0 \Rightarrow$$

$$\left[\Delta \mathbf{w}_{hidden}(1)\right] = \begin{bmatrix} \Delta w_{11,12} & \Delta w_{21,12} \\ \Delta w_{11,22} & \Delta w_{21,22} \\ \Delta w_{11,32} & \Delta w_{21,32} \end{bmatrix} = \begin{bmatrix} 0.0000 & 0.0001 \\ -0.0024 & -0.0026 \\ -0.0010 & -0.0010 \end{bmatrix},$$

where $\left[\mathbf{OUT}_{input}\right]^T$ denotes the transpose of matrix $\left[\mathbf{OUT}_{input}\right]$. The values of matrix $\left[\mathbf{OUT}_{input}\right]$ are retrieved from Table 4.4.

The updated values (iteration 1) for the weights linking the input neurons with the hidden neurons are computed using (4.43):

$$\left[\mathbf{w}_{hidden}(1)\right] \equiv \begin{bmatrix} w_{11,12} & w_{21,12} \\ w_{11,22} & w_{21,22} \\ w_{11,32} & w_{21,32} \end{bmatrix} = \left[\mathbf{w}_{hidden}(0)\right] + \left[\Delta \mathbf{w}_{hidden}(1)\right] \Rightarrow$$

$$\left[\mathbf{w}_{hidden}(1)\right] = \begin{bmatrix} -0.07514 & 0.85124 \\ -0.05542 & -0.18542 \\ -0.53124 & 0.47563 \end{bmatrix} + \begin{bmatrix} 0.0000 & 0.0001 \\ -0.0024 & -0.0026 \\ -0.0010 & -0.0010 \end{bmatrix} \Rightarrow$$

$$\left[\mathbf{w}_{hidden}(1)\right] = \begin{bmatrix} -0.0751 & 0.8513 \\ -0.0578 & -0.1880 \\ -0.5323 & 0.4746 \end{bmatrix},$$

where the values of the matrix $\left[\mathbf{w}_{hidden}(0)\right]$ are retrieved from Table 4.5 (values of the weights linking the input neurons with the hidden neurons during iteration 0).

The variation in the value of the weights linking the hidden neurons with the output neuron is:

4.6 Artificial Neural Networks

Table 4.6 Neural network weights computed during the first iteration

Neuron		Weight	
From	To	Symbol	Value
I_1	H_1	$w_{11,12}$	-0.0751
I_1	H_2	$w_{11,22}$	-0.0578
I_1	H_3	$w_{11,32}$	-0.5323
I_2	H_1	$w_{21,12}$	0.8513
I_2	H_2	$w_{21,22}$	-0.1880
I_2	H_3	$w_{21,32}$	0.4746
H_1	O_1	$w_{12,13}$	-0.0362
H_2	O_1	$w_{22,13}$	0.5124
H_3	O_1	$w_{32,13}$	0.1618

$$[\Delta \mathbf{w}_{output}(1)] \equiv \begin{bmatrix} \Delta w_{12,13} \\ \Delta w_{22,13} \\ \Delta w_{32,13} \end{bmatrix} = lr \cdot [\boldsymbol{\delta}_{output}] \cdot [\mathbf{OUT}_{hidden}]^T + m \cdot [\Delta \mathbf{w}_{output}(0)] =$$

$$= 1 \cdot [-0.0711 \quad 0.0503 \quad 0.0513 \quad -0.0705] \cdot \begin{bmatrix} 0.5 & 0.7008 & 0.4812 & 0.6848 \\ 0.5 & 0.4538 & 0.4861 & 0.4401 \\ 0.5 & 0.6167 & 0.3702 & 0.4861 \end{bmatrix}^T + 0 \Rightarrow$$

$$[\Delta \mathbf{w}_{output}(1)] = \begin{bmatrix} -0.0239 \\ -0.0188 \\ -0.0198 \end{bmatrix}.$$

The updated values (iteration 1) for the weights linking the hidden neurons with the output neuron are computed using (4.43):

$$\left[\mathbf{w}_{output}(1)\right] \equiv \begin{bmatrix} w_{12,13} \\ w_{22,13} \\ w_{32,13} \end{bmatrix} = \left[\mathbf{w}_{output}(0)\right] + \left[\Delta\mathbf{w}_{output}(1)\right] = \begin{bmatrix} -0.01235 \\ 0.53121 \\ 0.18155 \end{bmatrix} + \begin{bmatrix} -0.0239 \\ -0.0188 \\ -0.0198 \end{bmatrix} \Rightarrow$$

$$\left[\mathbf{w}_{output}(1)\right] = \begin{bmatrix} -0.0362 \\ 0.5124 \\ 0.1618 \end{bmatrix},$$

where the values of the matrix $\left[\mathbf{w}_{output}(0)\right]$ are retrieved from Table 4.5 (values of the weights linking the hidden neurons with the output neuron during iteration 0).

The neural network weights that have been computed during iteration 1 are shown in Table 4.6.

Next the simulation of the hidden layer and the output layer take place. The calculations are done in the same way as during iteration 0 with the only difference that the new weights of Table 4.6 are used. The simulation results of the output layer are:

$$[\mathbf{OUT}_{output}] = \begin{bmatrix} 0.5791 & 0.5760 & 0.5723 & 0.5692 \end{bmatrix}.$$

The error in the prediction of the output by the neural network is:

$$[\mathbf{error}] = [\mathbf{OUT}_{desired}] - [\mathbf{OUT}_{output}] \Rightarrow$$

$$[\mathbf{error}] = \begin{bmatrix} 0 & 1 & 1 & 0 \end{bmatrix} - \begin{bmatrix} 0.5791 & 0.5760 & 0.5723 & 0.5692 \end{bmatrix} \Rightarrow$$

$$[\mathbf{error}] = \begin{bmatrix} -0.5791 & 0.4240 & 0.4277 & -0.5692 \end{bmatrix}.$$

The mean square error performance of the neural network is:

$$mse = \frac{1}{N} \cdot \sum_{i=1}^{N} [error_i] = \frac{1}{4} \cdot \left[(-0.5791)^2 + 0.4240^2 + 0.4277^2 + (-0.5692)^2\right] \Rightarrow$$

$$mse = 0.2555.$$

The improvement in the mean square error performance (reduction in error) of the neural network during iteration 1 in comparison with iteration 0 is:

$$Impr = \frac{(0.2555 - 0.2568)}{0.2568} \cdot 100\% \Rightarrow Impr = -0.5\%.$$

4.6 Artificial Neural Networks

Table 4.7 Neural network weights

Neuron		Weight	
From	To	Symbol	Value
I_1	H_1	$w_{11,12}$	57.7
I_1	H_2	$w_{11,22}$	-136.4
I_1	H_3	$w_{11,32}$	209.6
I_2	H_1	$w_{21,12}$	-89.6
I_2	H_2	$w_{21,22}$	229.5
I_2	H_3	$w_{21,32}$	-346.6
H_1	O_1	$w_{12,13}$	23.2
H_2	O_1	$w_{22,13}$	-64.0
H_3	O_1	$w_{32,13}$	93.1

Table 4.8 Neural network biases

	Bias	
Neuron	Symbol	Value
H_1	b_{12}	-2.1
H_2	b_{22}	-9.6
H_3	b_{32}	12.3
O_1	b_{13}	-15.0

This small improvement in performance means that there is a need for sufficient iterations of the gradient descent backpropagation algorithm to converge to a solution with small mean square error. Using a computer program, we found that after 246 iterations the mean square error was 0.01, which is a satisfactory error value for the specific problem.

4.6.6.2 Example 4.4

A very simple neural network with two input neurons (I_1 and I_2), three hidden neurons (H_1, H_2, and H_3) and one output neuron (O_1) has been trained with the learning set of Table 4.2 so as to solve the classification problem of Example 4.1, and the weights and biases of the trained neural network are shown in Tables 4.7

and 4.8, respectively. The output neuron O_1 takes the value 1 in the case of acceptable quality (class A) and the value 0 in the case of non-acceptable quality (class NA). The activation function for the hidden and the output layer is $f(x) = 1/(1+e^{-0.1 \cdot x})$. For the unknown input example $(I_1, I_2) = (0.5, 0.4)$, which does not belong to the learning set of Table 4.2, calculate the output provided (1) by the neural network, and (2) by the decision tree of Fig. 4.5.

Solution

1. The net input for neuron H_1, i.e., the first neuron of the second layer is:

$$INP_{12} = w_{11,12} \cdot I_1 + w_{21,12} \cdot I_2 + b_{12} = 57.7 \cdot 0.5 - 89.6 \cdot 0.4 - 2.1 \Rightarrow INP_{12} = -9.09.$$

The output for neuron H_1 is:

$$OUT_{12} = f(INP_{12}) = \frac{1}{1+e^{-0.1 \cdot INP_{12}}} = \frac{1}{1+e^{-0.1 \cdot (-9.09)}} \Rightarrow OUT_{12} = 0.2872.$$

Similarly, we calculate the net inputs and the outputs of neurons H_2 and H_3, and obtain the following results:

$$INP_{22} = 14.0, \ INP_{23} = -21.54, \ OUT_{22} = 0.8022, \ OUT_{32} = 0.1040.$$

The net input for neuron O_1 is:

$$INP_{13} = w_{12,13} \cdot OUT_{12} + w_{22,13} \cdot OUT_{22} + w_{32,13} \cdot OUT_{32} + b_{13} \Rightarrow$$

$$INP_{13} = 23.2 \cdot 0.2872 - 64 \cdot 0.8022 + 93.1 \cdot 0.1040 - 15 \Rightarrow INP_{13} = -49.9981.$$

The output of neuron O_1, which is also the output of the neural network, is:

$$OUT_{13} = f(INP_{13}) = \frac{1}{1+e^{-0.1 \cdot INP_{13}}} = \frac{1}{1+e^{-0.1 \cdot 49.9981}} \Rightarrow OUT_{13} = 0.0067.$$

Since the output of the neural network, i.e., $OUT_{13} = 0.0067$, is closer to zero than to one, the neural network classifies the input example $(I_1, I_2) = (0.5, 0.4)$ as non-acceptable.

2. Since the input example $(I_1, I_2) = (0.5, 0.4)$ has $I_1 = 0.5$, the test $I_1 \leq 0.3848$ of the root node of the decision tree of Fig. 4.5 is true, so the input example is led to node 2, which has acceptability index equal to zero, so the input example $(I_1, I_2) = (0.5, 0.4)$ is classified as non-acceptable.

4.6 Artificial Neural Networks

Table 4.9 Neural network types

Name	Description
network	Custom neural network (Demuth and Beale 2001)
newcf	Cascade-forward backpropagation network
newelm	Elman backpropagation network (Elman 1990)
newff	Feedforward backpropagation network
newfftd	Feedforward input-delay backpropagation network
newgrnn	Generalized regression neural network (Wasserman 1993)
newhop	Hopfield recurrent network (Li et al. 1989)
newlvq	Learning vector quantization network (Kohonen 1987)
newp	Perceptron (Rosenblatt 1961)
newpnn	Probabilistic neural network (Wasserman 1993)
newrb	Radial basis network (Chen et al. 1991)
newsom	Self-organizing map (Kohonen 1987)

4.6.7 ANN Configuration

Designing a neural network can be as simple as selecting commercially available neural network software and configuring it to agree with the data. It can also be as complex as coding a fully custom made neural network software.

Neural network design choices include defining the behavior of the neurons (e.g., transfer function), the training algorithm (e.g., backpropagation), the topology of the network (e.g., number of hidden layers, numbers of neurons per layer), and the values of the training parameters (e.g., learning rate, momentum).

Assuming standard commercial software, the following steps are needed so as to configure the ANN:

1. The first choice is which neural network to use. This decision is closely tied to the data and application, making it difficult to offer general advice. One idea is to investigate the best-known neural networks. Another idea is to study previous neural network applications similar to the task at hand.

2. After choosing a neural network, the next step to configure it by setting the number of input and output neurons to agree with the number of inputs and outputs in the data. The selection of the training algorithm, the topology of the network, and the values of the training parameters are also important.

Table 4.10 Training functions

Name	Description
trainb	Batch training with weight and bias learning rules
trainbfg	BFGS quasi-Newton backpropagation (Gill et al. 1981)
trainbr	Bayesian regularization (Foresee and Hagan 1997; MacKay 1992)
trainc	Cyclical order incremental training
traincgb	Powell–Beale conjugate gradient backpropagation (Powell 1977)
traincgf	Fletcher–Powell conjugate gradient backpropagation (Scales 1985)
traincgp	Polak–Ribiere conjugate gradient backpropagation (Scales 1985)
traingd	Gradient descent backpropagation (Hagan et al. 1996)
traingdm	Gradient descent with momentum backpropagation (Hagan et al. 1996)
traingda	Gradient descent with adaptive learning rate (lr) backpropagation (Hagan et al. 1996)
traingdx	Gradient descent with momentum and adaptive lr backpropagation (Hagan et al. 1996)
trainlm	Levenberg–Marquardt backpropagation (Hagan and Menhaj 1994)
trainoss	One step secant backpropagation (Battiti 1992)
trainr	Random order incremental training with learning functions
trainrp	Resilient backpropagation (Reidmiller and Braun 1993)
trains	Sequential order incremental training with learning functions
trainscg	Scaled conjugate gradient backpropagation (Moller 1993)

3. The final step is training and testing the network. Training is an interactive process: the trainer tries a configuration, evaluates a result, makes a change, tries it again, and so on until satisfied. During testing, the neural network passes the test set forward through itself and calculates a performance index such as the mean square error without changing the weights. The real goal during training is the accuracy on the test set, not on the training set.

Among the various commercial neural network software packages, MATLAB neural network toolbox is a choice that offers flexibility in prototyping neural network systems (Demuth and Beale 2001). Table 4.9 presents the different neural network types, Table 4.10 shows the different training functions and Table 4.11 presents the different transfer functions that are included in MATLAB neural network toolbox for the benefit of the user.

4.6 Artificial Neural Networks

Table 4.11 Transfer functions (Demuth and Beale 2001)

Name	Description
compet	Competitive transfer function
hardlim	Hard limit transfer function
hardlims	Symmetric hard limit transfer function
logsig	Log sigmoid transfer function
poslin	Positive linear transfer function
purelin	Linear transfer function
radbas	Radial basis transfer function
satlin	Saturating linear transfer function
satlins	Symmetric saturating linear transfer function
softmax	Soft max transfer function
tansig	Hyperbolic tangent sigmoid transfer function
tribas	Triangular basis transfer function

4.6.8 Example 4.5

Compare the speed to train the neural network of Example 4.3 using different training functions of MATLAB neural network toolbox if the targeted error performance is (1) $mse = 0.01$, and (2) $mse = 0.001$. The neural network has two input neurons, three hidden neurons and one output neuron. The sigmoid transfer function is used for the hidden neurons and the output neuron.

Solution

The tests in this example were performed on a laptop computer with Intel Pentium M 1.5 GHz processor, 512 MB RAM memory. MATLAB version 6.1 and MATLAB Neural Network Toolbox version 4.0 were used.

1. For targeted error performance $mse = 0.01$, we obtained the results shown in Table 4.12 by examining the performance of 10 training functions. Each entry in the table represents 30 different trials, where different random initial weights and biases were used in each trial. The maximum epochs (iterations) of the training algorithm was set to 1000 in each trial. The column "Successes" on Table 4.12 shows the number of times the algorithm successfully converged out of the 30 trials. This table also shows the mean value and the standard deviation (St_dev) of the time required for convergence, the epochs (iterations of the training algorithm), and the actual mean square error (MSE).

Table 4.12 Training speed comparison of different training functions for $mse = 0.01$

Training function	Successes	Time (s) Mean	St_dev	Epochs Mean	St_dev	MSE Mean	St_dev
trainbfg	23	0.36	0.12	19	17	0.0050	0.0029
traincgb	22	0.26	0.03	7	5	0.0051	0.0030
traincgf	24	0.33	0.11	16	13	0.0051	0.0027
traincgp	22	0.31	0.11	8	5	0.0052	0.0031
traingd	16	1.29	0.46	435	184	0.0100	0.0000
traingdx	26	0.65	0.32	160	125	0.0091	0.0010
trainlm	29	0.25	0.09	7	6	0.0052	0.0025
trainoss	26	0.58	0.58	42	73	0.0060	0.0029
trainrp	26	0.32	0.09	30	20	0.0082	0.0014
trainscg	27	0.32	0.08	19	15	0.0079	0.0012

Table 4.13 Training speed comparison of different training functions for $mse = 0.001$

Training function	Successes	Time (s) Mean	St_dev	Epochs Mean	St_dev	MSE Mean	St_dev
trainbfg	21	0.42	0.11	28	16	0.0005	0.0004
traincgb	21	0.31	0.08	11	6	0.0005	0.0003
traincgf	25	0.35	0.11	20	16	0.0005	0.0003
traincgp	23	0.31	0.07	13	10	0.0006	0.0003
traingd	0						
traingdx	27	0.69	0.37	170	141	0.0010	0.0000
trainlm	30	0.26	0.08	8	6	0.0005	0.0003
trainoss	26	0.75	0.36	62	43	0.0006	0.0004
trainrp	23	0.37	0.12	51	34	0.0007	0.0002
trainscg	26	0.35	0.10	26	17	0.0008	0.0002

It is concluded from Table 4.12 that the fastest algorithm for this problem is the Levenberg–Marquardt backpropagation algorithm (transfer function trainlm) since it requires on average 0.25 s to converge. This algorithm converged 29 times out of the 30 trials.

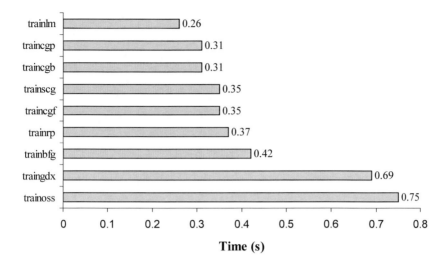

Fig. 4.13 Speed comparison of different training functions for $mse = 0.001$

2. Table 4.13 presents results obtained for targeted error performance $mse = 0.001$. Figure 4.13 compares the speed of the different training algorithms.

   Analyzing the results of Table 4.13, the following conclusions are drawn:

- Among the 10 training algorithms, the best performing one is the Levenberg–Marquardt backpropagation algorithm (training function trainlm) because:
  - It converges faster than all other algorithms, on average in 0.26 s
  - It converged 30 times out of 30 trials, while the second best performing algorithm (traingdx) converged 27 times out of 30 trials

- The gradient descent backpropagation algorithm (training function traingd) did not manage to converge in any of the 30 trials performed.

- Although the studied problem is not complex and the neural network used is simple, there are differences in terms of speed performance and convergence among the 10 different training algorithms considered. These differences are more obvious when larger problems are studied (Demuth and Beale 2001).

4.7 Hybrid Decision Tree–Neural Network Classifier

A binary decision tree induces a hierarchical partitioning over the decision space. Starting with the root node, each internal (test) node partitions its associated decision region into two half spaces. It is obvious that all the conditions along any particular path from the root to the terminal node of the decision tree must be satisfied in order to reach the particular terminal node. Thus, each path of a decision tree implements an AND operation on a set of half spaces. If two or more terminal nodes result in the same class, then the corresponding paths are in an OR relationship.

Multilayer perceptrons (MLPs) are feedforward neural networks consisting of one input layer, one or more hidden layers and one output layer. A multilayer perceptron with two hidden layers used for classification performs the following functions. The first hidden layer is the partitioning layer that divides the entire feature space into several regions. The second hidden layer is the ANDing layer that performs ANDing of partitioned regions to yield convex decision regions for each class. The output layer is the ORing layer that combines the results of the previous layer to produce disjoint regions of arbitrary shape.

It is concluded from the above that a decision tree and a four-layer perceptron are equivalent in terms of input/output mapping. In addition, a decision tree can be reformulated as a neural network, called an *entropy network* (EN), as follows (Sethi 1990):

1. The input layer (IL) consists of one neuron per attribute selected and tested by the decision tree.

2. The first hidden layer called the partitioning or test layer (TL) consists of one neuron per decision tree test node.

3. The second hidden layer called the ANDing Layer (AL) consists of one neuron per decision tree terminal node.

4. The output layer or ORing Layer (OL) consists of one neuron per decision tree class.

The connections between the neurons of the above four layers implement the hierarchy of the decision tree. In particular, each neuron of the TL is connected to the neuron of the IL corresponding to the tested attribute. In addition, each neuron of the AL is linked to the neurons of TL corresponding to the test nodes located on the path from the top node towards the terminal node. Finally, each neuron of the output layer is connected to the neurons of AL corresponding to the decision tree terminal nodes. In comparison with the standard MLPs that are fully connected, the entropy network has fewer connections, or equivalently fewer parameters, reducing the time needed for training.

4.7 Hybrid Decision Tree–Neural Network Classifier 211

The entropy network can be used for classification, with some modifications, however, it can also be used for prediction (Wehenkel and Akella 1993). In the latter case, the OL is replaced by a single output neuron, fully connected to all neurons of the AL and the resulting network is retrained. This methodology is called a hybrid decision tree–neural network (HDTNN) approach. After HDTNN convergence, the network is used to predict the test states and after that to classify them accordingly, providing the so-called hybrid decision tree–neural network classifier (HDTNNC).

4.7.1 Example 4.6

Design the entropy network for the decision tree of Example 4.2 shown in Fig. 4.7.

Solution
The number of neurons of the first hidden layer of the entropy network is equal to the number of test nodes of the decision tree of Fig. 4.7. Thus, the first hidden layer of the entropy network has the neurons TL_1 and TL_2 corresponding with the decision tree test nodes 1 and 3, respectively. Because the input I_2 is included at the optimal splitting test of node 1, there is a connection between the input neuron I_2 and the hidden neuron TL_1. For the same reason, the input neuron I_1 is connected with the hidden neuron TL_2.

The number of neurons of the second hidden layer of the entropy network is equal to the number of terminal nodes of the decision tree of Fig. 4.7. Thus, the second hidden layer of the entropy network has the neurons AL_1, AL_2 and AL_3 corresponding with the decision tree terminal nodes 2, 4 and 5, respectively. Because between the root node 1 and the terminal node 4 (neuron AL_2) of the decision tree there are the nodes 1 (neuron TL_1) and 3 (neuron TL_2), the neuron AL_2 of the second hidden layer is connected with neuron TL_1 as well as with neuron TL_2. For the same reason, the neuron AL_3 of the second hidden layer is connected with neuron TL_1 as well as with neuron TL_2.

The number of neurons of the output layer of the entropy network is equal to the number of classes of the decision tree of Fig. 4.7. Thus, there is the neuron A corresponding with the acceptable class, and the neuron NA corresponding with the non-acceptable class. Because the terminal nodes 2 and 5 of the decision tree belong to the non-acceptable class (acceptability index less than 50%), the corresponding neurons AL_1 and AL_3 are connected to output neuron NA. Because the terminal node 4 of the decision tree belongs to the acceptable class, the corresponding neuron AL_2 is connected to output neuron A.

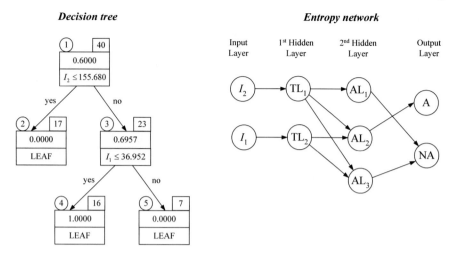

Fig. 4.14 Decision tree and its corresponding entropy network

Following the above analysis, the entropy network is created. Figure 4.14 shows the decision tree, its corresponding entropy network, as well as the correspondence between decision tree nodes and entropy network neurons. As can be seen, the entropy network with architecture 2-2-3-2 (i.e., two input neurons, two neurons in the first hidden layer, three neurons in the second hidden layer, and two output neurons) has 10 connections among its neurons. On the other hand, a fully connected MLP with architecture 2-2-3-2 has 16 connections. Because the entropy network has fewer connections, it requires less time to train, in comparison with its corresponding fully connected MLP.

References

Adly AA (2001) Computation of inrush current forces on transformer windings. IEEE Transactions on Magnetics 37(4):2855–2857

Amjady N, Ehsan M (1999) Evaluation of power systems reliability by an artificial neural network. IEEE Transactions on Power Systems 14(1):287–292

References

Amoiralis EI, Georgilakis PS, Kefalas TD, Tsili MA, Kladas AG (2007) Artificial intelligence combined with hybrid FEM-BE techniques for global transformer optimization. IEEE Transactions on Magnetics 43(4):1633–1636

Battiti R (1992) First and second order methods for learning: between steepest descent and Newton's method. Neural Computation 4(2):141–166

Boyen X, Wehenkel L (1999) Automatic induction of fuzzy decision trees and its applications to power system security assessment. Fuzzy Sets and Systems 102:3–19

Breiman L, Friedman JH, Olshen RA, Stone CJ (1984) Classification and regression trees. Wadsworth International, California

Castro ARG, Miranda V (2005) Knowledge discovery in neural networks with application to transformer failure diagnosis. IEEE Transactions on Power Systems 20(2):717–724

Chen Z, Maun JC (2000) Artificial neural network approach to single-ended fault locator for transmission lines. IEEE Transactions on Power Systems 15(1):370–375

Chen S, Cowan CFN, Grant PM (1991) Orthogonal least squares learning algorithm for radial basis function networks. IEEE Transactions on Neural Networks 2(2):302–309

De A, Chatterjee N (2002) Recognition of impulse fault patterns in transformers using Kohonen's self-organizing feature map. IEEE Transactions on Power Delivery 17(2):489–494

Demuth H, Beale M (2001) Neural network toolbox for use with MATLAB, User's guide, Version 4. MathWorks, MA

Doulamis ND, Doulamis AD, Georgilakis PS, Kollias SD, Hatziargyriou ND (2002) A synergetic neural network-genetic scheme for optimal transformer construction. Integrated Computer-Aided Engineering 9:37–56

Elman JL (1990) Finding structure in time. Cognitive Science 14:179–211

El-Sharkawi MA, Niebur D (1996) Artificial neural networks with applications to power systems. IEEE-PES special publication 96TP-112-0

Farag AS, Mohandes M, Al-Shaikh A (2001) Diagnosing failed distribution transformers using neural networks. IEEE Transactions on Power Delivery 16(4):631–636

Filippeti F, Franceschini G, Tassoni C (1995) Neural networks aided on-line diagnostics of induction motor rotor faults. IEEE Transactions on Industry Applications 31(4):892–899

Foresee FD, Hagan MT (1997) Gauss–Newton approximation to Bayesian regularization. Proc International Joint Conference on Neural Networks

Friedman JH (1977) A recursive partitioning decision rule for nonparametric classification. IEEE Transactions on Computers 26:404–408

Georgilakis PS (2000) Contribution of artificial intelligence techniques in the reduction of distribution transformer iron losses. PhD dissertation. National Technical University of Athens, Athens, Greece

Georgilakis PS (2007) Artificial intelligence solution to electricity price forecasting problem. Applied Artificial Intelligence 21(8):707–727

Georgilakis PS, Amoiralis EI (2007) Spotlight on transformer design. IEEE Power and Energy 5(1):40–50

Georgilakis PS, Hatziargyriou ND, Doulamis ND, Doulamis AD, Kollias SD (1998) Prediction of iron losses of wound core distribution transformers based on artificial neural networks. Neurocomputing 23(1-3):15–29

Georgilakis PS, Hatziargyriou ND, Paparigas D (1999) AI helps reduce transformer iron losses. IEEE Computer Applications in Power 12(4):41–46

Georgilakis PS, Doulamis ND, Doulamis AD, Hatziargyriou ND, Kollias SD (2001) A novel iron loss reduction technique for distribution transformers based on a combined genetic algorithm-neural network approach. IEEE Transactions on Systems, Man, and Cybernetics, Part C 31(1):16–34

Georgilakis PS, Gioulekas AT, Souflaris AT (2007) A decision tree method for the selection of winding material in power transformers. Journal of Materials Processing Technology 181(1-3):281–285

Gerbec D, Gasperic S, Smon I, Gubina F (2005) Allocation of the load profiles to consumers using probabilistic neural networks. IEEE Transactions on Power Systems 20(2):548–555

Gill PE, Murray W, Wright MH (1981) Practical optimization. Academic Press, New York

Guardado JL, Naredo JL, Moreno P, Fuerte CR (2001) A comparative study of neural network efficiency in power transformers diagnosis using dissolved gas analysis. IEEE Transactions on Power Delivery 16(4):643–647

Gubina F, Halilcevic S (1998) Artificial neural network based planning of generation ready reserve capacity. Proc IEEE International Joint Conference on Neural Networks

Guo JJ, Luh PB (2004) Improving market clearing price prediction by using a committee machine of neural networks. IEEE Transactions on Power Systems 19(4):1867–1876

Hagan MT, Menhaj M (1994) Training feedforward networks with the Marquardt algorithm. IEEE Transactions on Neural Networks 5(6):989–993

Hagan MT, Demuth HB, Beale MH (1996) Neural network design. PWS Publishing, MA

Halilcevic S, Gubina F (1999) The generation ready-reserve power determined in the interconnected power system by means of the artificial neural network. Proc International Conference on Electric Power Engineering

Halilcevic SS, Gubina AF, Gubina F (2003) Automatic transmission network capability assessment scheme in daily scheduling using neural networks. Proc Power Tech Conference

Halpin SM, Burch IV RF (1997) Applicability of neural networks to industrial and commercial power systems: a tutorial overview. IEEE Transactions on Industry Applications 33(5):1355–1361

Hammerstrom D (1993) Working with neural networks. IEEE Spectrum 30:46–53

Hao X, Cai-Xin S (2007) Artificial immune network classification algorithm for fault diagnosis of power transformer. IEEE Transactions on Power Delivery 22(2):930–935

Hatziargyriou ND, Contaxis GC, Sireris NC (1994) A decision tree method for on-line steady state security assessment. IEEE Transactions on Power Systems 9(2):1052–1061

Hatziargyriou ND, Papathanassiou SA, Papadopoulos MP (1995) Decision trees for fast security assessment of autonomous power systems with a large penetration from renewables. IEEE Transactions on Energy Conversion 10(2):315–325

Hatziargyriou N, Georgilakis P, Spiliopoulos D, Bakopoulos J (1998) Quality improvement of individual cores of distribution transformers using decision trees. International Journal of Engineering Intelligent Systems for Electrical Engineering and Communications 6(3):141–146

Haykin S (1994) Neural networks: a comprehensive foundation. Macmillan College Publishing, New York

He Q, Si J, Tylavsky DJ (2000) Prediction of top-oil temperature for transformers using neural networks. IEEE Transactions on Power Delivery 15(4):1205–1211

Henrichon EG, Fu KS (1969) A non-parametric partitioning procedure for pattern classification. IEEE Transactions on Computers 7:614–624

Hippert HS, Pedreira CE, Souza RC (2001) Neural networks for short-term load forecasting: a review and evaluation. IEEE Transactions on Power Systems 16(1):44–55

Hoole SRH (1993) Artificial neural networks in the solution of inverse electromagnetic field problems. IEEE Transactions on Magnetics 29(2):1931–1934

Hoole SRH, Haldar MK (1995) Optimization of electromagnetic devices: circuit models, neural networks and gradient methods in concert. IEEE Transactions on Magnetics 31(3):2016–2019

Huang YC (2003) Evolving neural nets for fault diagnosis of power transformers. IEEE Transactions on Power Delivery 18(3):843–848

Huang YC, Huang CM (2002) Evolving wavelet networks for power transformer condition monitoring. IEEE Transactions on Power Delivery 17(2):412–416

Huang CM, Wang FL (2007) An RBF network with OLS and EPSO algorithms for real-time power dispatch. IEEE Transactions on Power Systems 22(1):96–104

Jain T, Srivastava L, Singh SN (2003) Fast voltage contingency screening using radial basis function neural network. IEEE Transactions on Power Systems 18(4):1359–1366

Janikow CZ (1998) Fuzzy decision trees: issues and methods. IEEE Transactions on Systems, Man, and Cybernetics - Part B 28(1):1–14

References

Jensen CA, El-Sharkawi MA, Marks II RJ (2001) Power system security assessment using neural networks: feature selection using Fisher discrimination. IEEE Transactions on Power Systems 16(4):757–763

Karapidakis ES, Hatziargyriou ND (2002) Online preventive dynamic security of isolated power systems using decision trees. IEEE Transactions on Power Systems 17(2):297–304

Khorashadi-Zadeh H, Sanaye-Pasand M (2006) Correction of saturated current transformers secondary current using ANNs. IEEE Transactions on Power Delivery 21(1):73–79

Kiartzis S, Kladas A (2001) Deterministic and artificial intelligence approaches in optimizing permanent magnet generators for wind power applications. Journal of Materials Processing Technology 108(2):232–236

Kohonen T (1987) Self-organization and associative memory, 2nd edn. Springer-Verlag, Berlin

Kononenko I, Bratko I, Roskar E (1984) Experiments in automatic learning of medical diagnosis rules. Technical report, Jozef Stefan Institute

Kosko B (1993) Neural networks and fuzzy systems. Prentice-Hall, Englewood Cliffs, NJ

Leonidaki EA, Georgiadis DP, Hatziargyriou ND (2006) Decision trees for determination of optimal location and rate of series compensation to increase power system loading margin. IEEE Transactions on Power Systems 21(3):1303–1310

Li J, Michel AN, Porod W (1989) Analysis and synthesis of a class of neural networks: linear systems operating on a closed hypercube. IEEE Transactions on Circuits and Systems 36(11):1405–1422

MacKay DJC (1992) Bayesian interpolation. Neural Computation 4(3):415–447

Mao PL, Aggarwal RK (2001) A novel approach to the classification of the transient phenomena in power transformers using combined wavelet transform and neural network. IEEE Transactions on Power Delivery 16(4):654–660

Michie D, Spiegelhalter DJ, Taylor CC (1994) Machine learning, neural and statistical classification. Ellis Horwood. Final report of ESPRIT project 5170 – StatLog

Mingers J (1989) An empirical comparison of pruning methods for decision tree induction. Machine Learning 4:227–243

Miranda V, Castro ARG (2005) Improving the IEC table for transformer failure diagnosis with knowledge extraction from neural networks. IEEE Transactions on Power Delivery 20(4):2509–2516

Mohammed OA, Park DC, Uler FG, Ziqiang C (1992) Design optimization of electromagnetic devices using artificial neural networks. IEEE Transactions on Magnetics 28(5):2805–2807

Moller MF (1993) A scaled conjugate gradient algorithm for fast supervised learning. Neural Networks 6:525–533

Naresh R, Sharma J (2000) Hydro system scheduling using ANN approach. IEEE Transactions on Power Systems 15(1):388–395

Nussbaum C, Pfutzner H, Booth Th, Baumgartinger N, Ilo A, Clabian M (2000) Neural networks for the prediction of magnetic transformer core characteristics. IEEE Transactions on Magnetics 36(1):313–329

Orille-Fernandez AL, Ghonaim NKI, Valencia JA (2001) A FIRANN as a differential relay for three phase power transformer protection. IEEE Transactions on Power Delivery 16(2):215–218

Pao YH (1989) Adaptive pattern recognition and neural networks. Addison-Wesley, Reading, MA

Papalexopoulos AD, Hao S, Peng TM (1994) An implementation of a neural network based load forecasting model for the EMS. IEEE Transactions on Power Systems 9(4):1956–1962

Park DC, El-Sharkawi MA, Marks II RJ, Atlas LE, Damborg MJ (1991) Electric load forecasting using an artificial neural network. IEEE Transactions on Power Systems 6(2):442–449

Peng TM, Hubele NF, Karady GG (1992) Advancement in the application of neural networks for short-term load forecasting. IEEE Transactions on Power Systems 7(1):250–257

Perez LG, Flechsig AJ, Meador JL, Obradovic Z (1994) Training an artificial neural network to discriminate between magnetizing inrush and internal faults. IEEE Transactions on Power Delivery 9(1):434–441

Pihler J, Grcar B, Dolinar D (1997) Improved operation of power transformer protection using artificial neural network. IEEE Transactions on Power Delivery 12(3):1128–1136

Piras A, Germond A, Buchenel B, Imhof K, Jaccard Y (1996) Heterogeneous artificial neural network for short term electrical load forecasting. IEEE Transactions on Power Systems 11(1):397–402

Powell MJD (1977) Restart procedures for the conjugate gradient method. Mathematical Programming 12:241–254

Pylvanainen JK, Nousiainen K, Verho P (2007) Studies to utilize loading guides and ANN for oil-immersed distribution transformer condition monitoring. IEEE Transactions on Power Delivery 22(1):201–207

Quinlan JR (1983) Learning efficient classification procedures and their application to chess endgames. In: Michalski RS, Carbonell J, Mitchell T (eds) Machine learning: an artificial intelligence approach, Morgan Kaufman

Quinlan JR (1986) The effect of noise on concept learning. In: Michalski RS, Carbonell J, Mitchell T (eds) Machine learning II, Morgan Kaufman

Quinlan JR (1987) Simplifying decision trees. International Journal of Man-Machine Studies 27:221–234

Quinlan JR (1993) C4.5 programs for machine learning. Morgan Kaufman

Rebizant W, Bejmert D (2007) Current-transformer saturation detection with genetically optimized neural networks. IEEE Transactions on Power Delivery 22(2):820–827

Reidmiller M, Braun H (1993) A direct adaptive method for faster backpropagation learning: the RPROP algorithm. Proc IEEE International Conference on Neural Networks

Rosenblatt F (1961) Principles of neurodynamics. Spartan Press, Washington D.C.

Rounds EM (1980) A combined nonparametric approach to feature selection and binary decision tree design. Pattern Recognition 12:313–317

Rovnyak S, Kretsinger S, Thorp J, Brown D (1994) Decision trees for real-time transient stability prediction. IEEE Transactions on Power Systems 9(3):1417–1426

Santoso NI, Tan OT (1990) Neural-net based real-time control of capacitors installed on distribution systems. IEEE Transactions on Power Delivery 5:266–272

Scales LE (1985) Introduction to non-linear programming. Springer-Verlag

Segatto EC, Coury DV (2006) A differential relay for power transformers using intelligent tools. IEEE Transactions on Power Systems 21(3):1154–1162

Senroy N, Heydt GT, Vittal V (2006) Decision tree assisted controlled islanding. IEEE Transactions on Power Systems 21(4):1790–1797

Sethi IK (1990) Entropy nets: from decision trees to neural networks. Proceedings of the IEEE 78(10):1605–1613

Sobajic DC, Pao YH (1989) Artificial neural-net based dynamic security assessment for electric power systems. IEEE Transactions on Power Systems 4:220–228

Stankovic AM, Saric AT, Milosevic M (2003) Identification of nonparametric dynamic power system equivalents with artificial neural networks. IEEE Transactions on Power Systems 18(4):1478–1486

Suarez A, Lutsko F (1999) Globally optimal fuzzy decision trees for classification and regression. IEEE Transactions on Pattern Analysis and Machine Intelligence 21(12):1297–1311

Szkuta BR, Sanabria LA, Dillon TS (1999) Electricity price short-term forecasting using artificial neural networks. IEEE Transactions on Power Systems 14(3):851–857

Thalassinakis EJ, Dialynas EN, Agoris D (2006) Method combining ANNs and Monte Carlo simulation for the selection of the load shedding protection strategies in autonomous power systems. IEEE Transactions on Power Systems 21(4):1574–1582

Tsekouras G, Kiartzis S, Kladas AG, Tegopoulos JA (2001) Neural network approach compared to sensitivity analysis based on finite element technique for optimization of permanent magnet generators. IEEE Transactions on Magnetics 37(5):3618–3621

Ugedo A, Lobato E, Peco J, Rouco L (2005) Decision trees applied to the management of voltage constraints in the Spanish market. IEEE Transactions on Power Systems 20(2):963–972

References

Van Cutsem T, Wehenkel L, Pavella M, Heilbronn B, Goubin M (1993) Decision tree approaches for voltage security assessment. IEE Proccedings - Part C 140(3):189–198

Wang MH (2005) Partial discharge pattern recognition of current transformers using an ENN. IEEE Transactions on Power Delivery 20(3):1984–1990

Wang Z, Liu Y, Griffin PJ (1998) A combined ANN and expert system tool for transformer fault diagnosis. IEEE Transactions on Power Delivery 13(4):1224–1229

Wasserman PD (1989) Neural computing. Van Nostrand Reinhold, New York

Wasserman PD (1993) Advanced methods in neural computing. Van Nostrand Reinhold, New York

Wehenkel L (1993) Decision tree pruning using an additive information quality measure. In: Bouchon-Meunier B, Valverde L, Yager RR (eds) Uncertainty in intelligent systems, Elsevier –North Holland

Wehenkel LA (1998) Automatic learning techniques in power systems. Kluwer, Boston

Wehenkel L, Akella VB (1993) A hybrid decision tree – neural network approach for power system dynamic security assessment. Proceedings of the 4th Symposium on Expert Systems Application to Power Systems, Elsevier-North Holland, 397–411

Wehenkel L, Pavella M (1991) Decision trees and transient stability of electric power systems. Automatica 27(1):115–134

Wehenkel L, Pavella M (1993) Decision tree approach to power system security assessment. International Journal of Electrical Power and Energy Systems 15(1):13–36

Wehenkel L, Van Cutsem T, Ribbens-Pavella M (1989a) An artificial intelligence framework for on-line transient stability assessment of electric power systems. IEEE Transactions on Power Systems 4:789–800

Wehenkel L, Van Cutsem T, Ribbens-Pavella M (1989b) Inductive inference applied to on-line transient stability assessment of electric power systems. Automatica 25:445–451

Wehenkel L, Pavella M, Euxibie E, Heilbronn B (1994) Decision tree based transient stability method - a case study. IEEE Transactions on Power Systems 9(2):459–469

Wishart MT, Harley RG (1995) Identification and control of induction machines using artificial neural networks. IEEE Transactions on Industry Applications 31(3):612–619

Yang CC, Hsu YY (1994) Estimation of line flows and bus voltages using decision trees. IEEE Transactions on Power Systems 9(3):1569–1574

Yang HT, Huang YC (1998) Intelligent decision support for diagnosis of incipient transformer faults using self-organizing polynomial networks. IEEE Transactions on Power Systems 13(3):946–952

Yang HT, Liao CC, Chou JH (2001) Fuzzy learning vector quantization networks for power transformer condition assessment. IEEE Transactions on Dielectrics and Electrical Insulation 8(1):143–149

Yu DC, Cummins JC, Wang Z, Yoon HJ, Kojovic LA (2001) Correction of current transformer distorted secondary currents due to saturation using artificial neural networks. IEEE Transactions on Power Delivery 16(2):189–194

Yuan Y, Shaw MJ (1995) Induction of fuzzy decision trees. Fuzzy Sets and Systems 69:125–139

Zaman MR, Rahman MA (1998) Experimental testing of the artificial neural network based protection of power transformers. IEEE Transactions on Power Delivery 13(2):510–517

Zhang Y, Ding X, Liu Y, Griffin PJ (1996) An artificial neural network approach to transformer fault diagnosis. IEEE Transactions on Power Delivery 11(4):1836–1841

5 Optimization

Abstract This chapter is devoted to optimization and is organized in five sections. Section 5.1 is an introduction to optimization. Section 5.2 presents an active set method that effectively solves quadratic programming problems. Section 5.3 describes the sequential quadratic programming method, which is one of the best methods for solving nonlinearly constrained optimization problems. The sequential quadratic programming method iteratively solves a sequence of quadratic programming subproblems. Section 5.4 presents the branch-and-bound method, which, in conjunction with sequential quadratic programming, effectively solves nixed-integer nonlinear programming problems (such as the transformer design optimization problem of Chap. 7). Section 5.5 is devoted to the genetic algorithm method that successfully solves complex optimization problems (such as the transformer no-load loss minimization problem of Chap. 7). The four optimization methods presented in this chapter are accompanied by carefully selected and analytically solved arithmetic examples that make clear the application of the methods to the solution of a variety of optimization problems.

5.1 Introduction

Optimization is the process of making something better. Problems in engineering, economics, and physical and social sciences entail the optimization (minimization or maximization) of an objective function subject to certain constraints (restrictions and tradeoffs).

Optimization problems can be either unconstrained or constrained. Unconstrained optimization deals with the optimization of an objective function that is not subject to constraints. Constrained optimization entails the optimization of an objective function subject to certain constraints.

Optimization problems can be either linear or nonlinear. Linear programming deals only with linear objective functions and linear constraints. When either the set of constraints, the objective function, or both are nonlinear, the optimization problem is called a nonlinear programming problem. Quadratic programming is a special optimization problem in which the objective function is quadratic and the constraints are linear.

A mixed-integer programming problem is an optimization problem in which some of the design variables have to take integer values. A discrete optimization problem is an optimization problem in which some of the design variables have to take discrete values that belong to an ordered set of values.

The best method for solving linear programming problems is the simplex method, which was proposed by Dantzig in 1947 (Dantzig 1963; Castillo et al. 2002). Linear programming can also be solved by the exterior point method, which is equivalent to the dual simplex method (Lemke 1954) and the interior point method (Karmarkar 1984).

Unconstrained nonlinear programming methods can be broadly categorized into direct search methods and indirect search methods. The direct search methods require only the objective function values in finding the optimum. The indirect search methods or descent methods or gradient search methods require, in addition to the objective function values, the first and in some cases the second derivatives of the objective function.

Unconstrained nonlinear programming problems can be solved by various direct search methods, such as:

1. Random search methods, e.g., random jump method, random walk method, and random walk method with direction exploitation (Rao 1996)

2. Univariate method (Rao 1996)

3. Grid search method (Rao 1996)

4. Simplex search method (Spendley et al. 1962; Nelder and Mead 1965)

5. Pattern search methods, i.e., Powell's method (Powell 1964), and Hooke–Jeeves method (Hooke and Jeeves 1961)

6. Rosenbrock's method (Rosenbrock 1960)

Unconstrained nonlinear programming problems can be also solved by several gradient search methods, such as:

1. Steepest descent method or Cauchy method (Cauchy 1847, Rao 1996)

2. Fletcher-Reeves method or conjugate gradient method (Hestenes and Stiefel 1952)

3. Newton's method (Rao 1996)

4. Marquardt method (Marquardt 1963)

5. Quasi-Newton methods, i.e., Davidon–Fletcher–Powell method (Davidon 1959; Fletcher and Powell 1963), and Broyden–Fletcher–Goldfarb–Shanno method (Broyden 1970; Fletcher 1970; Goldfarb 1970; Shanno 1970)

Among the gradient search methods, the most favored are the quasi-Newton methods.

5.1 Introduction

Constrained nonlinear programming problems can be solved using mainly the following methods:

1. Dual methods (Castillo et al. 2002)

2. Penalty methods (Castillo et al. 2002)

3. Augmented Lagrangian or multipliers method (Bazaraa et al. 1993)

4. Feasible direction methods (Bazaraa et al. 1993)

5. Sequential quadratic programming methods (Nocedal and Wright 2006)

Methods for solving discrete optimization problems include dynamic programming (Bellman 1957) and branch-and-bound method (Rao 1996). The branch-and-bound method is also very effective in solving mixed-integer programming problems.

Several metaheuristic techniques have evolved in past decades that facilitate solving optimization problems that were previously difficult or even impossible to solve. Some of the best known and most widely applied metaheuristic optimization techniques are the following:

1. Genetic algorithms (Holland 1975; Goldberg 1989)

2. Evolutionary computation (Fogel 2000)

3. Evolution strategies (Schwefel and Rudolph 1995)

4. Evolutionary programming (Fogel et al. 1966)

5. Differential evolution (Storn and Price 1997; Price et al. 2005)

6. Particle swarm optimization or swarm intelligence (Bonabeau et al. 1999; Kennedy and Eberhart 2001)

7. Ant colony optimization (Dorigo and Stützle 2004)

8. Tabu search (Glover 1989; Glover 1990; Glover and Laguna 1997)

9. Simulated annealing (Kirkpatrick et al. 1983; Cerný 1985)

10. Guided local search (Voudouris and Tsang 1995; Voudouris 1997; Voudouris and Tsang 1999)

11. Greedy randomized adaptive search procedures (Feo and Resende 1989, 1995)

12. Iterated local search (Lourenço et al. 2002)

This chapter presents four optimization methods, namely, quadratic programming (Sect. 5.2), sequential quadratic programming (Sect. 5.3), branch-and-bound (Sect. 5.4), and genetic algorithm method (Sect. 5.5).

5.2 Quadratic Programming

5.2.1 Methodology

Quadratic programming concerns the optimization of a quadratic objective function that is linearly constrained. Quadratic programming is important in its own right, and it also arises as a subproblem in methods for general constrained optimization, such as sequential quadratic programming (Sect. 5.3).

The general quadratic programming problem is stated as:

$$\min_x f(x) = \min_x \left[\frac{1}{2} \cdot x^T \cdot H \cdot x + x^T \cdot c \right], \quad (5.1)$$

subject to:

$$a_i^T \cdot x - b_i = 0, \quad i \in E = \{1, ..., m_e\}, \quad (5.2)$$

$$a_i^T \cdot x - b_i \leq 0, \quad i \in I = \{m_e + 1, ..., m\}, \quad (5.3)$$

where x is the vector of n design parameters, $f(x)$ is the quadratic objective function, m_e is the number of linear equality constraints, $m - m_e$ is the number of linear inequality constraints, H is a symmetric $n \times n$ matrix called the Hessian matrix, and c, a_i, and b_i are vectors in $\Re^n$. If the Hessian matrix H is positive semidefinite, then the optimization problem (5.1) to (5.3) is a convex quadratic program. Moreover, strictly convex quadratic programs are those in which H is positive definite. Nonconvex quadratic programs, in which H is an indefinite matrix, can be difficult to solve because they can have several local minima. It should be noted that a symmetric matrix is positive definite if all its eigenvalues are positive, positive semidefinite if some of its eigenvalues are positive and some are zero, and indefinite if some eigenvalues are positive and some are negative (Rau 2003). In this section, we focus on convex quadratic programs.

The Lagrangian for the quadratic program (5.1) to (5.3) is:

5.2 Quadratic Programming

$$\mathcal{L}(x, \lambda) = \frac{1}{2} \cdot x^T \cdot H \cdot x + x^T \cdot c + \sum_{i=1}^{m} \lambda_i \cdot (a_i^T \cdot x - b_i). \tag{5.4}$$

The *active set* $\mathcal{A}(x^*)$ is defined to consist of the indices of the *active constraints* (5.2) and (5.3), for which equality holds at x^*:

$$\mathcal{A}(x^*) = \left\{ i \mid a_i^T \cdot x^* - b_i = 0 \right\}. \tag{5.5}$$

Any solution x^* of the quadratic program (5.1) to (5.3) satisfies the *Karush–Kuhn–Tucker conditions* (5.6) to (5.9), for some Lagrange multipliers λ_i^*, $i \in \mathcal{A}(x^*)$:

$$H \cdot x^* + c + \sum_{i \in \mathcal{A}(x^*)} \lambda_i^* \cdot a_i = 0, \tag{5.6}$$

$$a_i^T \cdot x^* - b_i = 0 \quad, \quad \forall \, i \in \mathcal{A}(x^*), \tag{5.7}$$

$$a_i^T \cdot x^* - b_i \leq 0 \quad, \quad \forall \, i \in I \setminus \mathcal{A}(x^*), \tag{5.8}$$

$$\lambda_i^* \geq 0 \quad, \quad \forall \, i \in I \cap \mathcal{A}(x^*). \tag{5.9}$$

In (5.8), $I \setminus \mathcal{A}(x^*)$ denotes the subset of inequalities of (5.3) that do not belong to the active set $\mathcal{A}(x^*)$.

It can be proved (Nocedal and Wright 2006) that for convex quadratic programs, i.e., when H is positive semidefinite, the Karush–Kuhn–Tucker conditions (5.6) to (5.9) are sufficient for x^* to be the unique global optimum solution of the quadratic program (5.1) to (5.3).

Active set methods have been widely used and are effective for solving the quadratic program (5.1) to (5.3). The active set method for the solution of convex quadratic programs is composed of the following five steps (Nocedal and Wright 2006):

1. Set the iteration counter k to be zero, i.e., $k = 0$.

2. Compute a feasible starting point x_0.

3. Set working set W_0 to be a subset of the active constraints at x_0.

4. Formulate the equality-constrained quadratic programming subproblem (5.10) to (5.11):

$$\min_{p_k}\left[\frac{1}{2}\cdot p_k^T \cdot H \cdot p_k + g_k^T \cdot p_k\right], \qquad (5.10)$$

subject to:

$$A \cdot p_k = a_i^T \cdot p_k = 0, \quad \forall\, i \in W_k, \qquad (5.11)$$

where:

$$g_k = H \cdot x_k + c. \qquad (5.12)$$

5. Solve the equality-constrained quadratic programming subproblem (5.10) to (5.11) by solving the following Karush–Kuhn–Tucker system of linear equations:

$$\begin{bmatrix} H & A^T \\ A & 0 \end{bmatrix} \cdot \begin{bmatrix} p_k \\ \hat{\lambda}_k \end{bmatrix} = \begin{bmatrix} -g_k \\ 0 \end{bmatrix}. \qquad (5.13)$$

Let p_k and $\hat{\lambda}_k$ be the solution of (5.13), where $\hat{\lambda}_k$ denotes the $\hat{\lambda}_{k,i}$ Lagrange multipliers with $i \in W_k$.

a. If $p_k = 0$ then:

   i. If $\forall\, i \in W_k$ all the Lagrange multipliers $\hat{\lambda}_{k,i}$ are nonnegative, then the algorithm stops because the optimal solution has been found, which is $x^* = x_k$.

   ii. If one or more of the Lagrange multipliers $\hat{\lambda}_{k,i}$ is negative, then we remove an index i corresponding to one of the negative multipliers from the working set W_k and we form the new working set W_{k+1}. We increase the iteration counter by 1, i.e., $k = k+1$, and we proceed to step 4.

b. If $p_k \neq 0$ then we compute the *step length parameter* s_k from the following equation:

$$s_k = \min\left\{1,\ \min_{i \notin W_k,\, a_i^T \cdot p_k > 0}\left[\frac{b_i - a_i^T \cdot x_k}{a_i^T \cdot p_k}\right]\right\}. \qquad (5.14)$$

We call the constraints i for which the minimum in (5.14) is achieved the *blocking constraints*. If $s_k = 1$ and no new constraints are active at the current iteration, then there are no blocking constraints on this iteration k. The new candidate solution is computed as follows:

$$x_{k+1} = x_k + s_k \cdot p_k. \tag{5.15}$$

i. If there are blocking constraints, a new working set W_{k+1} is constructed by adding one of the blocking constraints to W_k.
ii. If there are no blocking constraints on this iteration, the new working set W_{k+1} is the same as W_k.

The iteration counter increases by 1, i.e., $k = k+1$, and we proceed to step 4.

5.2.2 Applications to Power Systems

Quadratic programming has shown its capabilities in solving real-world problems, as reflected by the publications on its applications to power system problems, e.g., Contaxis et al. 1986; Somuah and Schweppe 1987; Rogers and Rolko 1992; Wei et al. 1996; Momoh et al. 1999; Rau 2001; Papageorgiou and Fraga 2007, to name only a few.

5.2.3 Example 5.1

Solve the following problem:

$$\min_x \left[0.5 \cdot x_1^2 + 0.5 \cdot x_2^2 + 4 \cdot x_1 - 8 \cdot x_2 \right], \tag{5.16}$$

subject to:

$$4 \cdot x_1 + 7 \cdot x_2 = 0, \tag{5.17}$$

$$-x_1 \leq 2, \tag{5.18}$$

$$-x_2 \leq 3.5, \tag{5.19}$$

$$x_1 \leq 3, \tag{5.20}$$

$$x_2 \leq 1.5, \tag{5.21}$$

$$28.5 \cdot x_1 + 28 \cdot x_2 \leq 57, \tag{5.22}$$

using the quadratic programming technique with starting point $x_0 = [0\ 0]^T$.

Solution
As can easily be seen, this is a quadratic programming problem of the form (5.1) to (5.3), i.e.:

$$\min_x f(x) = \min_x \left[\frac{1}{2} \cdot x^T \cdot H \cdot x + x^T \cdot c \right],$$

subject to:

$$a_i^T \cdot x - b_i = 0, \quad i \in \{1\},$$

$$a_i^T \cdot x - b_i \leq 0, \quad i \in \{2, ..., 6\},$$

where:

$$x = \begin{bmatrix} x_1 \\ x_2 \end{bmatrix}, \quad x^T = \begin{bmatrix} x_1 & x_2 \end{bmatrix}, \quad H = \begin{bmatrix} 1 & 0 \\ 0 & 1 \end{bmatrix}, \quad c = \begin{bmatrix} 4 \\ -8 \end{bmatrix},$$

$$A = \begin{bmatrix} a_1^T \\ a_2^T \\ a_3^T \\ a_4^T \\ a_5^T \\ a_6^T \end{bmatrix} = \begin{bmatrix} 4 & 7 \\ -1 & 0 \\ 0 & -1 \\ 1 & 0 \\ 0 & 1 \\ 28.5 & 28 \end{bmatrix}, \text{ and } B = \begin{bmatrix} b_1 \\ b_2 \\ b_3 \\ b_4 \\ b_5 \\ b_6 \end{bmatrix} = \begin{bmatrix} 0 \\ 2 \\ 3.5 \\ 3 \\ 1.5 \\ 57 \end{bmatrix}.$$

Iteration 0

We set the iteration counter k to be zero, i.e., $k = 0$.

The starting point $x_0 = [0\ 0]^T$ is a feasible point since it satisfies all the constraints (5.17) to (5.22).

We refer the constraints (5.17) to (5.22), in order, by indices 1 through 6. Constraint 1 is the only constraint that is satisfied as equality at x_0, i.e., constraint 1 is

5.2 Quadratic Programming

an active constraint at x_0, so the initial working set W_0 is composed of the constraint 1, i.e., $W_0 = \{1\}$.

We formulate the equality-constrained quadratic programming subproblem based on (5.10) to (5.11), for $k = 0$:

$$\min_{p_0} \left[\frac{1}{2} \cdot p_0^T \cdot H \cdot p_0 + g_0^T \cdot p_0 \right], \tag{5.23}$$

subject to:

$$a_i^T \cdot p_0 = 0, \quad \forall\, i \in W_0 = \{1\}, \tag{5.24}$$

where:

$$p_0 = \begin{bmatrix} p_1 \\ p_2 \end{bmatrix}, \quad H = \begin{bmatrix} 1 & 0 \\ 0 & 1 \end{bmatrix}, \quad A = a_1^T = \begin{bmatrix} 4 & 7 \end{bmatrix},$$

$$g_0 = H \cdot x_0 + c = \begin{bmatrix} 1 & 0 \\ 0 & 1 \end{bmatrix} \cdot \begin{bmatrix} 0 \\ 0 \end{bmatrix} + \begin{bmatrix} 4 \\ -8 \end{bmatrix} \Rightarrow g_0 = \begin{bmatrix} 4 \\ -8 \end{bmatrix}.$$

Substituting the values of p_0, H, a_1^T, and g_0 into (5.23) and (5.24), we obtain the following equality-constrained quadratic programming subproblem:

$$\min_{p_0} \left[0.5 \cdot p_1^2 + 0.5 \cdot p_2^2 + 4 \cdot p_1 - 8 \cdot p_2 \right], \tag{5.25}$$

subject to:

$$4 \cdot p_1 + 7 \cdot p_2 = 0, \tag{5.26}$$

The solution of the quadratic programming subproblem of (5.25) and (5.26) is obtained by solving the following system of linear equations based on (5.13), for $k = 0$:

$$\begin{bmatrix} H & A^T \\ A & 0 \end{bmatrix} \cdot \begin{bmatrix} p_0 \\ \hat{\lambda}_0 \end{bmatrix} = \begin{bmatrix} -g_0 \\ 0 \end{bmatrix}. \tag{5.27}$$

Substituting the values of p_0, H, A, and g_0 into (5.27), we obtain the following system of linear equations:

$$\begin{bmatrix} 1 & 0 & 4 \\ 0 & 1 & 7 \\ 4 & 7 & 0 \end{bmatrix} \cdot \begin{bmatrix} p_1 \\ p_1 \\ \hat{\lambda}_0 \end{bmatrix} = \begin{bmatrix} -4 \\ 8 \\ 0 \end{bmatrix}. \tag{5.28}$$

The solution of the system of linear equations (5.28) is:

$$\begin{bmatrix} p_1 \\ p_1 \\ \hat{\lambda}_0 \end{bmatrix} = \begin{bmatrix} 1 & 0 & 4 \\ 0 & 1 & 7 \\ 4 & 7 & 0 \end{bmatrix}^{-1} \cdot \begin{bmatrix} -4 \\ 8 \\ 0 \end{bmatrix} \Rightarrow \begin{bmatrix} p_1 \\ p_1 \\ \hat{\lambda}_0 \end{bmatrix} = \begin{bmatrix} -6.4615 \\ 3.6923 \\ 0.6154 \end{bmatrix}.$$

Since $p_0 = \begin{bmatrix} p_1 & p_2 \end{bmatrix}^T = \begin{bmatrix} -6.4615 & 3.6923 \end{bmatrix}^T \neq 0$, we have to compute the step length parameter based on (5.14), for $k = 0$:

$$s_0 = \min\left\{1, \min_{i \notin W_0, a_i^T \cdot p_0 > 0} \left[\frac{b_i - a_i^T \cdot x_0}{a_i^T \cdot p_0}\right]\right\} = \min\left\{1, \min_{i \notin \{1\}, a_i^T \cdot p_k > 0} s_{0,i}\right\}.$$

We do not compute the coefficient $s_{0,1}$ for $i = 1$, since $i = 1 \in W_0 = \{1\}$. We also do not compute the coefficient $s_{0,3}$ for $i = 3$, since $a_3^T \cdot p_0 = -p_2 = -3.6923 < 0$. We do not compute the coefficient $s_{0,4}$ for $i = 4$, since $a_4^T \cdot p_0 = p_1 = -6.4615 < 0$. We also do not compute the coefficient $s_{0,6}$ for $i = 6$, since $a_6^T \cdot p_0 = 28.5 \cdot p_1 + 28 \cdot p_2 = -80.7684 < 0$.

The coefficients $s_{0,2}$ and $s_{0,5}$ are computed as follows:

$$s_{0,2} = \frac{b_2 - a_2^T \cdot x_0}{a_2^T \cdot p_0} = \frac{2 - \begin{bmatrix} -1 & 0 \end{bmatrix} \cdot \begin{bmatrix} 0 & 0 \end{bmatrix}^T}{\begin{bmatrix} -1 & 0 \end{bmatrix} \cdot \begin{bmatrix} -6.4615 & 3.6923 \end{bmatrix}^T} = \frac{2}{6.4615} \Rightarrow s_{0,2} = 0.3095,$$

$$s_{0,5} = \frac{b_5 - a_5^T \cdot x_0}{a_5^T \cdot p_0} = \frac{1.5 - \begin{bmatrix} 0 & 1 \end{bmatrix} \cdot \begin{bmatrix} 0 & 0 \end{bmatrix}^T}{\begin{bmatrix} 0 & 1 \end{bmatrix} \cdot \begin{bmatrix} -6.4615 & 3.6923 \end{bmatrix}^T} = \frac{1.5}{3.6923} \Rightarrow s_{0,5} = 0.4063.$$

The step length parameter is:

$$s_0 = \min\{1, \min\{s_{0,2}, s_{0,5}\}\} = \min\{1, \min\{0.3095, 0.4063\}\} \Rightarrow$$

$$s_0 = 0.3095.$$

5.2 Quadratic Programming

Since $s_0 = s_{0,2}$, the constraint 2 is a blocking constraint that is added to the working set of the next iteration, i.e., the new working set is $W_1 = \{1, 2\}$.

The new candidate solution is computed using (5.15) as follows:

$$x_1 = x_0 + s_0 \cdot p_0 \Rightarrow x_1 = \begin{bmatrix} 0 \\ 0 \end{bmatrix} + 0.3095 \cdot \begin{bmatrix} -6.4615 \\ 3.6923 \end{bmatrix} \Rightarrow x_1 = \begin{bmatrix} -2 \\ 1.1429 \end{bmatrix}.$$

Iteration 1

We increase the iteration counter k by 1, so we have $k = 1$.

At the end of iteration 0 we found that $W_1 = \{1, 2\}$ and $x_1 = \begin{bmatrix} -2 & 1.1429 \end{bmatrix}^T$.

We formulate the equality-constrained quadratic programming subproblem based on (5.10) and (5.11), for $k = 1$:

$$\min_{P_1} \left[\frac{1}{2} \cdot P_1^T \cdot H \cdot P_1 + g_1^T \cdot P_1 \right], \tag{5.29}$$

subject to:

$$a_i^T \cdot P_1 = 0, \quad \forall i \in W_1 = \{1, 2\}, \tag{5.30}$$

where:

$$P_1 = \begin{bmatrix} p_1 \\ p_2 \end{bmatrix}, \quad H = \begin{bmatrix} 1 & 0 \\ 0 & 1 \end{bmatrix},$$

$$a_1^T = \begin{bmatrix} 4 & 7 \end{bmatrix}, \quad a_2^T = \begin{bmatrix} -1 & 0 \end{bmatrix}, \quad A = \begin{bmatrix} a_1^T \\ a_2^T \end{bmatrix} = \begin{bmatrix} 4 & 7 \\ -1 & 0 \end{bmatrix},$$

$$g_1 = H \cdot x_1 + c = \begin{bmatrix} 1 & 0 \\ 0 & 1 \end{bmatrix} \cdot \begin{bmatrix} -2 \\ 1.1429 \end{bmatrix} + \begin{bmatrix} 4 \\ -8 \end{bmatrix} \Rightarrow g_1 = \begin{bmatrix} 2 \\ -6.8571 \end{bmatrix}.$$

Substituting the values of P_1, H, a_1^T, a_2^T, and g_1 into (5.29) and (5.30), we obtain the following equality-constrained quadratic programming subproblem:

$$\min_{P_1} \left[0.5 \cdot P_1^2 + 0.5 \cdot P_2^2 + 2 \cdot P_1 - 6.8571 \cdot P_2 \right], \tag{5.31}$$

subject to:

$$4 \cdot P_1 + 7 \cdot P_2 = 0, \tag{5.32}$$

$$-P_1 = 0. \tag{5.33}$$

The solution of the quadratic programming subproblem (5.31) to (5.33) is obtained by solving the following system of linear equations based on (5.13), for $k=1$:

$$\begin{bmatrix} H & A^T \\ A & 0 \end{bmatrix} \cdot \begin{bmatrix} P_1 \\ \hat{\lambda}_1 \end{bmatrix} = \begin{bmatrix} -g_1 \\ 0 \end{bmatrix}. \tag{5.34}$$

Substituting the values of P_1, H, A, and g_1 into (5.34), we obtain the following system of linear equations:

$$\begin{bmatrix} 1 & 0 & 4 & -1 \\ 0 & 1 & 7 & 0 \\ 4 & 7 & 0 & 0 \\ -1 & 0 & 0 & 0 \end{bmatrix} \cdot \begin{bmatrix} p_1 \\ p_2 \\ \hat{\lambda}_1 \\ \hat{\lambda}_2 \end{bmatrix} = \begin{bmatrix} -2 \\ 6.8571 \\ 0 \\ 0 \end{bmatrix}. \tag{5.35}$$

The solution of the system of linear (5.35) is:

$$\begin{bmatrix} p_1 \\ p_2 \\ \hat{\lambda}_1 \\ \hat{\lambda}_2 \end{bmatrix} = \begin{bmatrix} 1 & 0 & 4 & -1 \\ 0 & 1 & 7 & 0 \\ 4 & 7 & 0 & 0 \\ -1 & 0 & 0 & 0 \end{bmatrix}^{-1} \cdot \begin{bmatrix} -2 \\ 6.8571 \\ 0 \\ 0 \end{bmatrix} \Rightarrow \begin{bmatrix} p_1 \\ p_2 \\ \hat{\lambda}_1 \\ \hat{\lambda}_2 \end{bmatrix} = \begin{bmatrix} 0 \\ 0 \\ 0.9796 \\ 5.9184 \end{bmatrix}.$$

Since $P_1 = \begin{bmatrix} p_1 & p_2 \end{bmatrix}^T = \begin{bmatrix} 0 & 0 \end{bmatrix}^T = 0$, and because all the Lagrange multipliers for the constraints belonging to the working set W_1 are nonnegative ($\hat{\lambda}_1 = 0.9796 > 0$ and $\hat{\lambda}_2 = 5.9184 > 0$), the algorithm stops because the optimal solution has been found, this being $x^* = x_1 = \begin{bmatrix} -2 & 1.1429 \end{bmatrix}^T$.

Since constraints 3 to 6 do not belong to the working set $W_1 = \{1, 2\}$, their Lagrange multipliers are equal to zero.

Consequently, the optimum solution of the quadratic programming problem (5.16) to (5.22) is given by:

$$x_1^* = -2, \ x_2^* = 1.1429, \ \lambda_1^* = 0.9796, \ \lambda_2^* = 5.9184, \ \lambda_3^* = 0, \ \lambda_4^* = 0,$$
$$\lambda_5^* = 0, \ \lambda_6^* = 0.$$

5.3 Sequential Quadratic Programming

5.3.1 Methodology

Sequential quadratic programming is one of the best methods for solving nonlinearly constrained optimization problems (Boggs and Tolle 1995; Rao 1996; Nocedal and Wright 2006). Sequential quadratic programming is also known as iterative quadratic programming, recursive quadratic programming, and constrained variable metric method. The basic form of sequential quadratic programming method dates back to Wilson (Wilson 1963) and was popularized by Han (Han 1977) and Powell (Powell 1978).

The nonlinear programming problem is stated as:

$$\min_{\mathbf{x}} f(\mathbf{x}), \tag{5.36}$$

subject to:

$$g_i(\mathbf{x}) = 0, \quad i = 1, ..., m_e, \tag{5.37}$$

$$g_i(\mathbf{x}) \leq 0, \quad i = m_e + 1, ..., m, \tag{5.38}$$

where $\mathbf{x}$ is the vector of n design parameters, $f(\mathbf{x})$ is the objective function, m_e is the number of equality constraints, and $m - m_e$ is the number of inequality constraints. The great advantage of the sequential quadratic programming method is its ability to solve problems with nonlinear constraints. For this reason it is assumed that the nonlinear programming problem (5.36)–(5.38) contains at least one nonlinear constraint function.

The Lagrangian for the nonlinear programming problem (5.36)–(5.38) is:

$$\mathcal{L}(\mathbf{x}, \boldsymbol{\lambda}) = f(\mathbf{x}) + \sum_{i=1}^{m} \lambda_i \cdot g_i(\mathbf{x}). \tag{5.39}$$

The basic idea of sequential quadratic programming is to model the nonlinear programming problem (5.36)–(5.38) at a given approximate solution, say $\mathbf{x}^{(k)}$, by a quadratic programming subproblem, and then to use the solution to this subproblem to construct a better approximation $\mathbf{x}^{(k+1)}$. This process is iterated to create a sequence of approximations that will converge to the optimal solution $\mathbf{x}^*$.

The principal idea of sequential quadratic programming is the formulation of a quadratic programming subproblem based on a quadratic approximation of the Lagrangian function $\mathcal{L}(\mathbf{x}, \boldsymbol{\lambda})$ of (5.39) and by linearizing the nonlinear constraints of (5.37) and (5.38). It can be proved (Boggs and Tolle 1995) that at a current approximation $\mathbf{x}^{(k)}$ the quadratic programming subproblem (5.40)–(5.42) has to be solved:

$$\min_{\mathbf{s}} f(\mathbf{s}) = \min_{\mathbf{s}} \left[\frac{1}{2} \cdot \mathbf{s}^T \cdot \mathbf{H}_k \cdot \mathbf{s} + \nabla f(\mathbf{x}^{(k)})^T \cdot \mathbf{s} \right], \tag{5.40}$$

subject to:

$$\nabla g_i(\mathbf{x}^{(k)})^T \cdot \mathbf{s} + g_i(\mathbf{x}^{(k)}) = 0, \quad i = 1, \ldots, m_e, \tag{5.41}$$

$$\nabla g_i(\mathbf{x}^{(k)})^T \cdot \mathbf{s} + g_i(\mathbf{x}^{(k)}) \leq 0, \quad i = m_e + 1, \ldots, m, \tag{5.42}$$

where $\mathbf{H}_k$ is a positive definite approximation of the Hessian matrix of the Lagrangian function $\mathcal{L}(\mathbf{x}, \lambda)$ of (5.39) at $\mathbf{x}^{(k)}$. Initially, the matrix $\mathbf{H}_k$ is taken as the identity matrix and is updated in subsequent iterations by the BFGS method (Nocedal and Wright 2006).

The quadratic programming subproblem (5.40)–(5.42) is solved using the active set method of Sect. 5.2.1. The solution $\mathbf{s}$ of the quadratic programming subproblem of (5.40) to (5.42) is then used to form a new approximation $\mathbf{x}^{(k+1)}$ as follows:

$$\mathbf{x}^{(k+1)} = \mathbf{x}^{(k)} + \alpha^{(k)} \cdot \mathbf{s}, \tag{5.43}$$

where $\alpha^{(k)}$ is the *step length parameter*, which is determined by the following *line search procedure*:

1. Initially it is assumed that $\alpha^{(k)} = 1$.

2. The new approximation $\mathbf{x}^{(k+1)}$ is computed using (5.43).

3. If the condition $f(\mathbf{x}^{(k+1)}) < f(\mathbf{x}^{(k)})$ is satisfied then go to step 5 else go to step 4.

4. The step length is halved, i.e., $\alpha^{(k)} = 0.5 \cdot \alpha^{(k)}$. Next, go to step 2.

5. The step length $\alpha^{(k)}$ has been found.

5.3.2 Applications to Power Systems

Sequential quadratic programming has shown its capabilities in solving real-world problems, as reflected by the number of publications on its applications to power system problems, e.g., Lu et al. 1988; Grudinin 1998; Nejdawi et al. 2000; Abril and Quintero 2003; Coelho and Mariani 2006; Finardi and da Silva 2006, to name only a few.

5.3.3 Example 5.2

Solve the following problem:

$$\min_{\mathbf{x}} f(\mathbf{x}) = \min_{\mathbf{x}} \left[x_1^4 - 2 \cdot x_1^2 \cdot x_2 + 7 \right], \quad (5.44)$$

subject to:

$$g_1(\mathbf{x}) = x_1^2 + x_2^2 - 16.25 = 0, \quad (5.45)$$

$$g_2(\mathbf{x}) = -x_1 \leq 0, \quad (5.46)$$

$$g_3(\mathbf{x}) = -x_2 \leq 0, \quad (5.47)$$

$$g_4(\mathbf{x}) = x_1 - 5 \leq 0, \quad (5.48)$$

$$g_5(\mathbf{x}) = x_2 - 5 \leq 0, \quad (5.49)$$

$$g_6(\mathbf{x}) = x_1^2 + 2 \cdot x_1 \cdot x_2^2 - 110 \leq 0, \quad (5.50)$$

using the sequential quadratic programming technique with starting point $\mathbf{x}^{(0)} = [2 \ \ 3.5]^T$.

Solution
The gradients of the objective and the constraint functions are:

$$\nabla f(\mathbf{x}) = \begin{bmatrix} \dfrac{\partial f}{\partial x_1} \\ \dfrac{\partial f}{\partial x_2} \end{bmatrix} = \begin{bmatrix} 4 \cdot x_1^3 - 4 \cdot x_1 \cdot x_2 \\ -2 \cdot x_1^2 \end{bmatrix}, \quad \nabla g_1(\mathbf{x}) = \begin{bmatrix} \dfrac{\partial g_1}{\partial x_1} \\ \dfrac{\partial g_1}{\partial x_2} \end{bmatrix} = \begin{bmatrix} 2 \cdot x_1 \\ 2 \cdot x_2 \end{bmatrix},$$

$$\nabla g_2(\mathbf{x}) = \begin{bmatrix} -1 \\ 0 \end{bmatrix}, \ \nabla g_3(\mathbf{x}) = \begin{bmatrix} 0 \\ -1 \end{bmatrix}, \ \nabla g_4(\mathbf{x}) = \begin{bmatrix} 1 \\ 0 \end{bmatrix}, \ \nabla g_5(\mathbf{x}) = \begin{bmatrix} 0 \\ 1 \end{bmatrix},$$

$$\nabla g_6(\mathbf{x}) = \begin{bmatrix} 2 \cdot x_1 + 2 \cdot x_2^2 \\ 4 \cdot x_1 \cdot x_2 \end{bmatrix}.$$

Iteration 0

The starting point $\mathbf{x}^{(0)} = [2\ 3.5]^T$ is feasible, since it satisfies all the constraints (5.45)–(5.50):

$$g_1(\mathbf{x}^{(0)}) = 2^2 + 3.5^2 - 16.25 \Rightarrow g_1(\mathbf{x}^{(0)}) = 0,$$

$$g_2(\mathbf{x}^{(0)}) = -2 \Rightarrow g_2(\mathbf{x}^{(0)}) < 0,$$

$$g_3(\mathbf{x}^{(0)}) = -3.5 \Rightarrow g_3(\mathbf{x}^{(0)}) < 0,$$

$$g_4(\mathbf{x}^{(0)}) = 2 - 5 \Rightarrow g_4(\mathbf{x}^{(0)}) = -3 \Rightarrow g_4(\mathbf{x}^{(0)}) < 0,$$

$$g_5(\mathbf{x}^{(0)}) = 3.5 - 5 \Rightarrow g_5(\mathbf{x}^{(0)}) = -1.5 \Rightarrow g_5(\mathbf{x}^{(0)}) < 0,$$

$$g_6(\mathbf{x}^{(0)}) = 2^2 + 2 \cdot 2 \cdot 3.5^2 - 110 = -57 \Rightarrow g_6(\mathbf{x}^{(0)}) < 0.$$

The value of the objective function at the starting point $\mathbf{x}^{(0)} = [2\ 3.5]^T$ is:

$$f(\mathbf{x}^{(0)}) = 2^4 - 2 \cdot 2^2 \cdot 3.5 + 7 \Rightarrow f(\mathbf{x}^{(0)}) = 5.$$

The gradient of the objective function and the gradient of the constraint functions at the initial point $\mathbf{x}^{(0)} = [2\ 3.5]^T$ are:

$$\nabla f(\mathbf{x}^{(0)}) = \begin{bmatrix} 4 \cdot 2^3 - 4 \cdot 2 \cdot 3.5 \\ -2 \cdot 2^2 \end{bmatrix} = \begin{bmatrix} 4 \\ -8 \end{bmatrix}, \ \nabla g_1(\mathbf{x}^{(0)}) = \begin{bmatrix} 2 \cdot 2 \\ 2 \cdot 3.5 \end{bmatrix} = \begin{bmatrix} 4 \\ 7 \end{bmatrix},$$

$$\nabla g_2(\mathbf{x}^{(0)}) = \begin{bmatrix} -1 \\ 0 \end{bmatrix}, \ \nabla g_3(\mathbf{x}^{(0)}) = \begin{bmatrix} 0 \\ -1 \end{bmatrix}, \ \nabla g_4(\mathbf{x}^{(0)}) = \begin{bmatrix} 1 \\ 0 \end{bmatrix},$$

$$\nabla g_5(\mathbf{x}^{(0)}) = \begin{bmatrix} 0 \\ 1 \end{bmatrix}, \ \nabla g_6(\mathbf{x}^{(0)}) = \begin{bmatrix} 2 \cdot 2 + 2 \cdot 3.5^2 \\ 4 \cdot 2 \cdot 3.5 \end{bmatrix} = \begin{bmatrix} 28.5 \\ 28 \end{bmatrix}.$$

5.3 Sequential Quadratic Programming

Iteration 1

The following quadratic programming subproblem has to be formulated and solved:

$$\min_{\mathbf{s}} f(\mathbf{s}) = \min_{\mathbf{s}} \left[\frac{1}{2} \cdot \mathbf{s}^T \cdot \mathbf{H}_0 \cdot \mathbf{s} + \nabla f(\mathbf{x}^{(0)})^T \cdot \mathbf{s} \right], \qquad (5.51)$$

subject to:

$$\nabla g_1(\mathbf{x}^{(0)})^T \cdot \mathbf{s} + g_1(\mathbf{x}^{(0)}) = 0, \qquad (5.52)$$

$$\nabla g_i(\mathbf{x}^{(0)})^T \cdot \mathbf{s} + g_i(\mathbf{x}^{(0)}) \leq 0, \quad \forall\, i = 2, ..., 6, \qquad (5.53)$$

where $\mathbf{H}_0$ is initially taken as the identity matrix.

By substituted the values of the objective function, the constraint functions, the gradient of the objective function, and the gradients of the constraints functions at the initial point $\mathbf{x}^{(0)}$ into (5.51) to (5.53) we have:

$$\min_{\mathbf{s}} f(\mathbf{s}) = \min_{\mathbf{s}} \left\{ \frac{1}{2} \cdot \mathbf{s}^T \cdot \mathbf{H}_0 \cdot \mathbf{s} + \nabla f(\mathbf{x}^{(0)})^T \cdot \mathbf{s} \right\} =$$

$$= \min_{\mathbf{s}} \left\{ \frac{1}{2} \cdot \begin{bmatrix} s_1 \\ s_2 \end{bmatrix}^T \cdot \begin{bmatrix} 1 & 0 \\ 0 & 1 \end{bmatrix} \cdot \begin{bmatrix} s_1 \\ s_2 \end{bmatrix} + \begin{bmatrix} 4 \\ -8 \end{bmatrix}^T \cdot \begin{bmatrix} s_1 \\ s_2 \end{bmatrix} \right\} \Rightarrow$$

$$\min_{\mathbf{s}} f(\mathbf{s}) = \min_{\mathbf{s}} \left[0.5 \cdot s_1^2 + 0.5 \cdot s_2^2 + 4 \cdot s_1 - 8 \cdot s_2 \right], \qquad (5.54)$$

subject to:

$$\nabla g_1(\mathbf{x}^{(0)})^T \cdot \mathbf{s} + g_1(\mathbf{x}^{(0)}) = 0 \Rightarrow \begin{bmatrix} 4 \\ 7 \end{bmatrix}^T \cdot \begin{bmatrix} s_1 \\ s_2 \end{bmatrix} + 0 = 0 \Rightarrow$$

$$g_1(\mathbf{s}) = 4 \cdot s_1 + 7 \cdot s_2 = 0, \qquad (5.55)$$

$$\nabla g_2(\mathbf{x}^{(0)})^T \cdot \mathbf{s} + g_2(\mathbf{x}^{(0)}) \leq 0 \Rightarrow \begin{bmatrix} -1 \\ 0 \end{bmatrix}^T \cdot \begin{bmatrix} s_1 \\ s_2 \end{bmatrix} + (-2) \leq 0 \Rightarrow$$

$$g_2(\mathbf{s}) = -s_1 \leq 2, \qquad (5.56)$$

$$g_3(\mathbf{s}) = -s_2 \leq 3.5, \qquad (5.57)$$

$$g_4(\mathbf{s}) = s_1 \leq 3, \qquad (5.58)$$

$$g_5(\mathbf{s}) = s_2 \leq 1.5, \tag{5.59}$$

$$g_6(\mathbf{s}) = 28.5 \cdot s_1 + 28 \cdot s_2 \leq 57. \tag{5.60}$$

The quadratic programming subproblem (5.54)–(5.60) has been solved in Example 5.1 and the solution is:

$$s_1^{(1)} = -2, \quad s_2^{(1)} = 1.1429, \quad \lambda_1^{(1)} = 0.9796, \quad \lambda_2^{(1)} = 5.9184, \quad \lambda_3^{(1)} = 0,$$
$$\lambda_4^{(1)} = 0, \quad \lambda_5^{(1)} = 0, \quad \lambda_6^{(1)} = 0.$$

The new candidate point is computed as follows:

$$\mathbf{x}_k^{(1)} = \mathbf{x}^{(0)} + \alpha_k^{(1)} \cdot \begin{bmatrix} s_1^{(1)} \\ s_2^{(1)} \end{bmatrix}, \tag{5.61}$$

so that:

$$f(\mathbf{x}_k^{(1)}) < f(\mathbf{x}^{(0)}), \tag{5.62}$$

where $\alpha^{(1)}$ is the line search step length.

The process of finding the appropriate $\alpha^{(1)}$ starts with unitary line search step length, i.e., $\alpha_k^{(1)} = \alpha_1^{(1)} = 1$. Then the new point is computed from (5.61). If (5.62) is satisfied, then the current $\alpha_k^{(1)}$ is the desired line search step length, otherwise the line search step length is halved and the process is repeated until (5.62) is satisfied.

The search starts with $k = 1$ and $\alpha_1^{(1)} = 1$. The new candidate point is computed by (5.61):

$$\mathbf{x}_1^{(1)} = \mathbf{x}^{(0)} + \alpha_1^{(1)} \cdot \begin{bmatrix} s_1^{(1)} \\ s_2^{(1)} \end{bmatrix} = \begin{bmatrix} 2 \\ 3.5 \end{bmatrix} + 1 \cdot \begin{bmatrix} -2 \\ 1.1429 \end{bmatrix} \Rightarrow \mathbf{x}_1^{(1)} = \begin{bmatrix} 0 \\ 4.6429 \end{bmatrix},$$

while the value of the objective function at the new candidate point $\mathbf{x}_1^{(1)}$ is:

$$f(\mathbf{x}_1^{(1)}) = 0^4 - 2 \cdot 0^2 \cdot 4.6429 + 7 \Rightarrow f(\mathbf{x}_1^{(1)}) = 7.$$

Since $f(\mathbf{x}^{(0)}) = -5$, the new candidate point $\mathbf{x}_1^{(1)}$ does not satisfy (5.62), so we proceed with $k = 2$ and $\alpha_2^{(1)} = 0.5 \cdot \alpha_1^{(1)} = 0.5$. The new candidate point $\mathbf{x}_2^{(1)}$ is computed by (5.61):

$$\mathbf{x}_2^{(1)} = \mathbf{x}^{(0)} + \alpha_2^{(1)} \cdot \begin{bmatrix} s_1^{(1)} \\ s_2^{(1)} \end{bmatrix} = \begin{bmatrix} 2 \\ 3.5 \end{bmatrix} + 0.5 \cdot \begin{bmatrix} -2 \\ 1.1429 \end{bmatrix} \Rightarrow \mathbf{x}_2^{(1)} = \begin{bmatrix} 1 \\ 4.0714 \end{bmatrix},$$

while the value of the objective function at the new candidate point $\mathbf{x}_2^{(1)}$ is:

$$f(\mathbf{x}_2^{(1)}) = 1^4 - 2 \cdot 1^2 \cdot 4.0714 + 7 \Rightarrow f(\mathbf{x}_2^{(1)}) = -0.1429.$$

5.3 Sequential Quadratic Programming

Table 5.1 Calculation of line search step length $\alpha^{(1)}$ for the Example 5.2

k	$\alpha_k^{(1)}$	$\mathbf{x}_k^{(1)}$	$f(\mathbf{x}_k^{(1)})$	$f(\mathbf{x}^{(0)})$	$f(\mathbf{x}_k^{(1)}) < f(\mathbf{x}^{(0)})$
1	1	$[0 \quad 4.6429]^T$	7	-5	No
2	0.5	$[1 \quad 4.0714]^T$	-0.1429	-5	No
3	0.25	$[1.5 \quad 3.7857]^T$	-4.9732	-5	No
4	0.125	$[1.75 \quad 3.6429]^T$	-5.9336	-5	Yes

Since $f(\mathbf{x}^{(0)}) = -5$, the new candidate point $\mathbf{x}_2^{(1)}$ does not satisfy (5.62), so we proceed with $k = 3$ and $\alpha_3^{(1)} = 0.5 \cdot \alpha_2^{(1)} = 0.25$. The required computations for the calculation of line search step length are shown in Table 5.1, from which we can see that for $k = 1, 2, 3$, (5.62) is not satisfied, while for $k = 4$, (5.62) is satisfied, so the desired line search step length is $\alpha^{(1)} = 0.125$, the new point is $\mathbf{x}^{(1)} = [1.75 \quad 3.6429]^T$ for which $f(\mathbf{x}^{(1)}) = -5.9336$.

The values of the constraint functions at the new point $\mathbf{x}^{(1)} = [1.75 \quad 3.6429]^T$ are:

$$g_1(\mathbf{x}^{(1)}) = 1.75^2 + 3.6429^2 - 16.25 \Rightarrow g_1(\mathbf{x}^{(1)}) = 0.0829,$$

$$g_2(\mathbf{x}^{(1)}) = -1.75, \quad g_3(\mathbf{x}^{(1)}) = -3.6429,$$

$$g_4(\mathbf{x}^{(1)}) = 1.75 - 5 \Rightarrow g_4(\mathbf{x}^{(1)}) = -3.25,$$

$$g_5(\mathbf{x}^{(1)}) = 3.6429 - 5 \Rightarrow g_5(\mathbf{x}^{(1)}) = -1.3571,$$

$$g_6(\mathbf{x}^{(1)}) = 1.75^2 + 2 \cdot 1.75 \cdot 3.6429^2 - 110 \Rightarrow g_6(\mathbf{x}^{(1)}) < -60.4911.$$

The gradient of the objective function and the gradients of the constraint functions at the new point $\mathbf{x}^{(1)} = [1.75 \quad 3.6429]^T$ are:

$$\nabla f(\mathbf{x}^{(1)}) = \begin{bmatrix} 4 \cdot 1.75^3 - 4 \cdot 1.75 \cdot 3.6429 \\ -2 \cdot 1.75^2 \end{bmatrix} = \begin{bmatrix} -4.0625 \\ -6.125 \end{bmatrix},$$

$$\nabla g_1(\mathbf{x}^{(1)}) = \begin{bmatrix} 2 \cdot 1.75 \\ 2 \cdot 3.6429 \end{bmatrix} = \begin{bmatrix} 3.5 \\ 7.2857 \end{bmatrix}, \quad \nabla g_2(\mathbf{x}^{(1)}) = \begin{bmatrix} -1 \\ 0 \end{bmatrix},$$

$$\nabla g_3(\mathbf{x}^{(1)}) = \begin{bmatrix} 0 \\ -1 \end{bmatrix}, \quad \nabla g_4(\mathbf{x}^{(1)}) = \begin{bmatrix} 1 \\ 0 \end{bmatrix}, \quad \nabla g_5(\mathbf{x}^{(1)}) = \begin{bmatrix} 0 \\ 1 \end{bmatrix},$$

$$\nabla g_6(\mathbf{x}^{(1)}) = \begin{bmatrix} 2 \cdot 1.75 + 2 \cdot 3.6429^2 \\ 4 \cdot 1.75 \cdot 3.6429 \end{bmatrix} = \begin{bmatrix} 30.0408 \\ 25.5 \end{bmatrix}.$$

In order to calculate the optimization error, the parameters $NGL^{(1)}$ and $NC^{(1)}$ are needed. The parameter $NGL^{(1)}$ is computed as follows:

$$NGL^{(1)} = \max \left\{ \begin{array}{l} \left| \nabla f(x_1^{(1)}) + \sum_{i=1}^{6} \nabla g_i(x_1^{(1)}) \cdot \lambda_i^{(1)} \right|, \\ \left| \nabla f(x_2^{(1)}) + \sum_{i=1}^{6} \nabla g_i(x_2^{(1)}) \cdot \lambda_i^{(1)} \right| \end{array} \right\} \Rightarrow$$

$$NGL^{(1)} = \max \left\{ \begin{array}{l} |-4.0625 + 3.5 \cdot 0.9796 + (-1) \cdot 5.9184|, \\ |-6.125 + 7.2857 \cdot 0.9796 + 0 \cdot 5.9184| \end{array} \right\} \Rightarrow$$

$$NGL^{(1)} = \max \{|-6.5523|, |1.012|\} \Rightarrow NGL^{(1)} = 6.5523.$$

The parameter $NC^{(1)}$ is computed as follows:

$$NC^{(1)} = \max \left\{ \begin{array}{l} \left| \lambda_2^{(1)} \cdot g_2(\mathbf{x}^{(1)}) \right|, \left| \lambda_3^{(1)} \cdot g_3(\mathbf{x}^{(1)}) \right|, \left| \lambda_4^{(1)} \cdot g_4(\mathbf{x}^{(1)}) \right|, \\ \left| \lambda_5^{(1)} \cdot g_5(\mathbf{x}^{(1)}) \right|, \left| \lambda_6^{(1)} \cdot g_6(\mathbf{x}^{(1)}) \right| \end{array} \right\} \Rightarrow$$

$$NC^{(1)} = \max \{|5.9184 \cdot (-1.75)|, |0|, |0|, |0|, |0|\} \Rightarrow NC^{(1)} = 10.3571.$$

The *optimization error* at the first iteration is:

$$OE^{(1)} = \max \{NGL^{(1)}, NC^{(1)}\} = \max \{6.5523, 10.3571\} \Rightarrow$$

$$OE^{(1)} = 10.3571.$$

The *feasibility error* at the first iteration is:

$$FE^{(1)} = \max \{g_1(\mathbf{x}^{(1)}), g_2(\mathbf{x}^{(1)}), g_3(\mathbf{x}^{(1)}), g_4(\mathbf{x}^{(1)}), g_5(\mathbf{x}^{(1)}), g_6(\mathbf{x}^{(1)})\} \Rightarrow$$

$$FE^{(1)} = 0.0829.$$

The iterations continue, because the following two inequalities are not satisfied:

$$OE^{(1)} < 10^{-6} \text{ and } FE^{(1)} < 10^{-6}.$$

Table 5.2 Convergence of sequential quadratic programming for the Example 5.2

Iteration j	Objective function value $f(\mathbf{x}^{(j)})$	Feasibility error $FE^{(j)}$	Optimization error $OE^{(j)}$
0	-5	0	
1	-5.93359	0.08291	10.4
2	-5.88173	0.001497	0.106
3	-5.88059	$1.306 \cdot 10^{-5}$	0.00175
4	-5.88058	$5.929 \cdot 10^{-9}$	$6.45 \cdot 10^{-6}$
5	-5.88058	$2.842 \cdot 10^{-14}$	$4.65 \cdot 10^{-9}$

Optimum solution
As can be seen from Table 5.2, after five iterations the procedure converges, since:

$$OE^{(5)} < 10^{-6} \text{ and } FE^{(5)} < 10^{-6}.$$

The optimum solution after five iterations is $\mathbf{x} = \mathbf{x}^{(5)} = \begin{bmatrix} 1.7823 & 3.6157 \end{bmatrix}^T$ and the value of the objective function is $f(\mathbf{x}) = f(\mathbf{x}^{(5)}) = -5.8806$.

5.4 Branch-and-Bound

5.4.1 Methodology

The branch-and-bound method is very effective in solving mixed-integer linear and nonlinear programming problems (Rao 1996). The branch-and-bound method was originally developed by Land and Doig to solve integer linear programming problems (Land and Doig 1960) and was later modified by Dakin to solve mixed-integer programming problems (Dakin 1965). Subsequently, the branch-and-bound method was extended to solve nonlinear mixed-integer programming problems (Borchers and Mitchell 1994; Leyffer 2001).

The mixed-integer nonlinear programming problem is stated as:

$$\min_{\mathbf{x}} f(\mathbf{x}), \qquad (5.63)$$

subject to:

$$g_i(\mathbf{x}) = 0, \quad i = 1, \ldots, m_e, \qquad (5.64)$$

$$g_i(\mathbf{x}) \leq 0, \quad i = m_e + 1, \ldots, m, \tag{5.65}$$

$$x_j \text{ integer}, \quad j = 1, \ldots, n_0 \ (n_0 \leq n), \tag{5.66}$$

where $\mathbf{x} = [x_1 \ x_2 \ \ldots \ x_n]^T$ is the vector of n design variables among which the first n_0 variables are identified as the integer variables.

According to the branch-and-bound method, initially, all integer restrictions of (5.66) are relaxed and the resulting nonlinear programming problem (5.63)–(5.65) is solved using an appropriate method, e.g., the sequential quadratic programming method of Sect. 5.3. If all integer variables take an integer value at the solution then this is also the solution of the mixed-integer nonlinear programming problem (5.63)–(5.66). However, usually, some integer variables take a non-integer value. The branch-and-bound method then selects one of those integer variables that take a non-integer value, say x_i, with value $\hat{x}_i$, and branches on it. Branching generates two new nonlinear programming problems:

1. The first nonlinear programming problem is composed of (5.63)–(5.65) plus the simple bound constraint $x_i \leq [\hat{x}_i]$, where $[\hat{x}_i]$ is the largest integer not greater than $\hat{x}_i$.

2. The second nonlinear programming problem is composed of (5.63)–(5.65) plus the simple bound constraint $x_i \geq [\hat{x}_i] + 1$.

One of the above two nonlinear programming problems is selected and solved next using an appropriate method, e.g., the sequential quadratic programming method of Sect. 5.3. If the integer variables take non-integer values then branching is repeated, thus generating a branch-and-bound tree whose nodes correspond to nonlinear programming problems and where an edge indicates the addition of a branching bound. If one of the following three fathoming rules is satisfied, then no branching is required; the corresponding node has been fully fathomed (explored) and can be abandoned. The *fathoming rules* are the following:

1. An infeasible node is detected. In this case the whole sub-tree starting at this node is infeasible and the node has been fathomed.

2. An integer feasible node is detected. This provides an upper bound on the optimum of the mixed-integer nonlinear programming problem; no branching is possible and the node has been fathomed.

3. A lower bound on the nonlinear programming solution of a node is greater than or equal to the current upper bound. In this case the node is fathomed, since this nonlinear programming solution provides a lower bound for all problems in the corresponding sub-tree.

5.4.2 Applications to Power Systems

Branch-and-bound has been shown capable of solving real-world problems, as reflected by the growing number of publications on its application to power system problems, e.g., Thompson and Wall 1981; Lauer et al. 1982; Cohen and Yoshimura 1983; Boardman and Meckiff 1985; Haffner et al. 2001; Choi et al. 2005; Amoiralis et al. 2008; Rider et al. 2008; Amoiralis et al. 2009, to name only a few.

5.4.3 Example 5.3

Solve the following mixed-integer nonlinear programming problem:

$$\min_{\mathbf{x}} f(\mathbf{x}) = \min_{\mathbf{x}} \left[x_1^4 - 2 \cdot x_1^2 \cdot x_2 + 7 \right], \tag{5.67}$$

subject to:

$$g_1(\mathbf{x}) = x_1^2 + x_2^2 - 16.25 = 0, \tag{5.68}$$

$$g_2(\mathbf{x}) = -x_1 \leq 0, \tag{5.69}$$

$$g_3(\mathbf{x}) = -x_2 \leq 0, \tag{5.70}$$

$$g_4(\mathbf{x}) = x_1 - 5 \leq 0, \tag{5.71}$$

$$g_5(\mathbf{x}) = x_2 - 5 \leq 0, \tag{5.72}$$

$$g_6(\mathbf{x}) = x_1^2 + 2 \cdot x_1 \cdot x_2^2 - 110 \leq 0, \tag{5.73}$$

$$x_2 \text{ integer}, \tag{5.74}$$

using the branch-and-bound technique with starting point $\mathbf{x}^{(0)} = [2 \ \ 3.5]^T$.

Solution
The various steps of the branch-and-bound method are presented in the following.

Step 1
First the problem is solved as a continuous variable problem, i.e., without (5.74). This problem has been solved in Example 5.2 and its optimum solution is:

$$x_1 = 1.7823, \ x_2 = 3.6157, \ f(\mathbf{x}) = -5.8806.$$

Step 2

It was found in Step 1 that the optimum value of x_2 is $x_2 = 3.6157$. However, x_2 has to be an integer, as can be seen from (5.74). Based on branch-and-bound method, two nonlinear programming problems have to be solved:

1. The first nonlinear programming problem is composed of (5.67)–(5.73) plus the simple bound constraint $x_2 \leq \lfloor 3.6157 \rfloor \Rightarrow x_2 \leq 3$. This first problem is solved in Step 3.

2. The second nonlinear programming problem is composed of (5.67)–(5.73) plus the simple bound constraint $x_2 \geq \lfloor 3.6157 \rfloor + 1 \Rightarrow x_2 \geq 4$. This second problem is solved in Step 4.

Step 3

The following problem has to be solved:

$$\min_{\mathbf{x}} f(\mathbf{x}) = \min_{\mathbf{x}} \left[x_1^4 - 2 \cdot x_1^2 \cdot x_2 + 7 \right], \tag{5.75}$$

subject to:

$$g_1(\mathbf{x}) = x_1^2 + x_2^2 - 16.25 = 0, \tag{5.76}$$

$$g_2(\mathbf{x}) = -x_1 \leq 0, \tag{5.77}$$

$$g_3(\mathbf{x}) = -x_2 \leq 0, \tag{5.78}$$

$$g_4(\mathbf{x}) = x_1 - 5 \leq 0, \tag{5.79}$$

$$g_5(\mathbf{x}) = x_2 - 5 \leq 0, \tag{5.80}$$

$$g_6(\mathbf{x}) = x_1^2 + 2 \cdot x_1 \cdot x_2^2 - 110 \leq 0, \tag{5.81}$$

$$g_7(\mathbf{x}) = x_2 - 3 \leq 0. \tag{5.82}$$

The optimization problem (5.75)–(5.82) is solved using the sequential quadratic programming method (Sect. 5.3) and its optimum solution is:

$$x_1^{(L)} = 2.6926, \ x_2^{(L)} = 3, \ f(\mathbf{x}^{(L)}) = 16.0625.$$

Step 4

The following problem has to be solved:

5.4 Branch-and-Bound

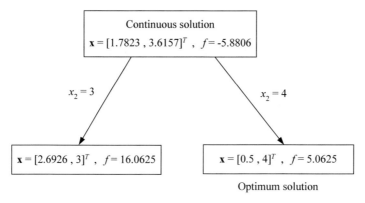

Fig. 5.1 Solution of Example 5.3 using the branch-and-bound method

$$\min_{\mathbf{x}} f(\mathbf{x}) = \min_{\mathbf{x}} \left[x_1^4 - 2 \cdot x_1^2 \cdot x_2 + 7 \right], \tag{5.83}$$

subject to:

$$g_1(\mathbf{x}) = x_1^2 + x_2^2 - 16.25 = 0, \tag{5.84}$$

$$g_2(\mathbf{x}) = -x_1 \leq 0, \tag{5.85}$$

$$g_3(\mathbf{x}) = -x_2 \leq 0, \tag{5.86}$$

$$g_4(\mathbf{x}) = x_1 - 5 \leq 0, \tag{5.87}$$

$$g_5(\mathbf{x}) = x_2 - 5 \leq 0, \tag{5.88}$$

$$g_6(\mathbf{x}) = x_1^2 + 2 \cdot x_1 \cdot x_2^2 - 110 \leq 0, \tag{5.89}$$

$$g_7(\mathbf{x}) = -x_2 + 4 \leq 0. \tag{5.90}$$

The optimization problem (5.83)–(5.90) is solved using the sequential quadratic programming method and its optimum solution is:

$$x_1^{(R)} = 0.5, \ x_2^{(R)} = 4, \ f(\mathbf{x}^{(R)}) = 5.0625.$$

Step 5
The left branch problem, solved in Step 3, and the right branch problem, solved in Step 4, give feasible solutions. However, since the problem under consideration is a minimization problem, and because:

$$5.0625 < 16.0625 \Rightarrow f(\mathbf{x}^{(R)}) < f(\mathbf{x}^{(L)}),$$

the optimum solution to the initial mixed-integer nonlinear programming problem (5.67)–(5.74) is the following:

$$x_1 = 0.5, \; x_2 = 4, \; f(\mathbf{x}) = 5.0625.$$

The results are shown in Fig. 5.1.

5.5 Genetic Algorithms

5.5.1 Methodology

Genetic algorithms are general-purpose optimization algorithms based on the mechanics of natural selection and genetics. The genetic algorithm method was developed by Holland (Holland 1975) and popularized by Goldberg (Goldberg 1989).

Genetic algorithms operate on string structures (*chromosomes*), typically a concatenated list of binary digits representing a coding of the control parameters (phenotype) of a given problem. Chromosomes themselves are composed of genes. The real value of a control parameter, encoded in a gene, is called an allele.

Genetic algorithms are an attractive alternative to other optimization methods because of their robustness. The three major advantages of genetic algorithms in comparison with conventional optimization algorithms are the following:

1. Genetic algorithms operate on the encoded string of the problem parameters rather than the actual parameters of the problem. Each string can be thought of as a chromosome that completely describes one candidate solution to the problem.

2. Genetic algorithms use a population of points rather than a single point in their search. This allows the genetic algorithm to explore several areas of the search space simultaneously, reducing the probability of finding local optima.

3. Genetic algorithms do not require any prior knowledge, space limitations, or special properties of the function to be optimized, such as smoothness, convexity, unimodality, or existence of derivatives. They only require the evaluation of the so-called *fitness function* to assign a quality value to every solution produced.

5.5 Genetic Algorithms

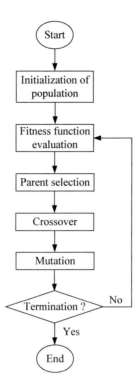

Fig. 5.2 Simple genetic algorithm

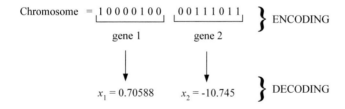

Fig. 5.3 Example of a chromosome representation with two genes and 8 bits per gene

A simple genetic algorithm process is illustrated in Fig. 5.2. After an initial population is randomly or heuristically produced, the fitness function of the population is evaluated and the genetic algorithm evolves the population through sequential and iterative application of three genetic operators: *parent selection*, *crossover*, and *mutation*. A new generation is formed at the end of each iteration.

5.5.1.1 Encoding and Decoding

The problem to be solved by a genetic algorithm is encoded as two distinct parts: the genotype called the chromosome and the phenotype called the fitness function. In computing terms the fitness function is a subroutine representing the given problem or the problem domain knowledge, while the chromosome refers to the parameters of this fitness function.

Traditionally the genotype is coded using a programming language vector, array, or record-like chromosome consisting of the problem parameters. Binary, integer, and real (floating point) codings are the most frequently used basic data types to represent genes in this immediate coding approach. In the case of binary representation of genes, the algorithm is called a *binary genetic algorithm*. On the other hand, a *continuous genetic algorithm* uses real coding of genes within chromosomes.

As an example, let us assume that an optimization problem has two variables, x_1 and x_2, and these variables are bounded as follows: $-20 \leq x_1 \leq 20$ and $-20 \leq x_2 \leq 20$. Let us also assume that binary representation is selected with 8 bits per variable. Figure 5.3 shows an example of such a chromosome with two genes: the first gene corresponds to variable x_1 and the second gene corresponds to variable x_2. The binary representation of the first gene is 10000100 (encoding), which corresponds to $x_1 = 0.70588$ (decoding), as can be seen in Example 5.4, solved in Sect. 5.5.3. The binary representation of the second gene is 00111011 (encoding), which corresponds to $x_2 = -10.745$ (decoding), as can be seen in Example 5.4.

5.5.1.2 Fitness Function

Each chromosome is evaluated and assigned a fitness value after the creation of a population of chromosomes. It is useful to distinguish between the objective function and the fitness function used by a genetic algorithm. The *objective function* provides a performance measure with respect to a particular set of gene values, independently of any other chromosome. The fitness function transforms that performance measure into an allocation of reproductive opportunities, i.e., the fitness of a chromosome is defined with respect to other members of the current population. In brief, the fitness function is a measure of the quality of each chromosome (candidate solution) of the current population. The fitness function may be a mathematical function, a simulator program, an experiment, or a human expert that decides the quality of a chromosome.

The determination of an appropriate fitness function is crucial for the correct operation of a genetic algorithm. As an optimization tool, genetic algorithms face the task of dealing with problem constraints. Crossover and mutation, i.e., the variation mechanism of genetic algorithms, are operators that do not take into ac-

5.5 Genetic Algorithms

count the feasibility region. Consequently, infeasible offspring appear quite frequently. One technique that is commonly used for handling constraints when using genetic algorithms is to incorporate a penalty function within the fitness function.

5.5.1.3 Natural Selection

Natural selection, i.e., survival of the fittest translates into discarding the chromosomes with the lowest fitness values. The number of chromosomes that are kept for mating, N_k, is computed as follows:

$$N_k = fr \cdot N_p, \qquad (5.91)$$

where N_p is the number of chromosomes in the population and fr is the fraction of N_p that survives for mating. At the end of natural selection, N_k chromosomes are in the mating pool.

For example, if the fraction of chromosomes that survive for mating is 50%, then for a population that is composed of 12 chromosomes, only the six chromosomes with the highest fitness values survive for mating.

5.5.1.4 Parent Selection

At the end of natural selection, N_k chromosomes are in the mating pool. Two chromosomes are selected from the mating pool of N_k chromosomes to produce two new offspring. Pairing takes place in the mating pool until $N_p - N_k$ offspring are born to replace the chromosomes that were discarded during natural selection.

Many parent selection schemes are currently in use for pairing chromosomes in a genetic algorithm. They can be classified in two groups: proportionate selection and ordinal selection. Proportionate methods select chromosomes on their fitness values relative to the fitness of the other chromosomes in the population. Ordinal methods select chromosomes not based on their fitness but based on their rank within the population. The *rank weighting method* that is used for the solution of Example 5.4 in Sect. 5.5.3 is an ordinal selection method.

5.5.1.5 Crossover

Crossover is an extremely important operator for the genetic algorithm. It is responsible for the information exchange between mating chromosomes and the convergence speed of the genetic algorithm. The crossover is usually applied with high crossover probability (0.6 to 0.9). The chromosomes of the two parents selected are combined to form new chromosomes that inherit segments of information stored in parent chromosomes. The most common crossover types are one-point, two-point, and uniform crossovers.

An example of one-point crossover is shown in Fig. 5.4. The parent (mother and father) chromosomes are composed of 16 bits each. A crossover point is randomly selected between the first and the last (16th) bit of the parent chromosomes. Let us suppose that the 12th crossover point has been selected. Then, offspring 1 is composed of the first 12 bits of mother chromosome plus the last 4 bits of father chromosome. Similarly, offspring 2 is composed of the first 12 bits of father chromosome plus the last 4 bits of mother chromosome, as Fig. 5.4 shows.

5.5.1.6 Mutation

Mutation is the operator responsible for the injection of new information. With a small mutation probability, random bits of the offspring chromosomes flip from 0 to 1 and vice versa and give new characteristics that do not exist in the parent population.

5.5.2 Applications to Power Systems

Genetic algorithms have been shown capable of solving real-world problems, as reflected by the growing number of publications on their applications to the following power system problems:

1. *Power system planning*: Iba 1994; Lee and Park 1995; Fukuyama and Chiang 1996; Gallego et al. 1998; Lee 1998; Urdaneta el al. 1999; da Silva et al. 2000; Delfanti et al. 2000; Park et al. 2000.

2. *Power system operation*: Dasgupta and McGregor 1994; Chen and Chang 1995; Sheblé and Brittig 1995; Chen and Chang 1996; Kazarlis et al. 1996; Maifeld and Sheblé 1996; Orero and Irving 1996; Chang and Chen 1998; Orero and Irving 1998; Rudolf and Bayrleithner 1999; Richter and Sheblé 2000; Sinha et al. 2003.

3. *Power system control*: Taranto and Falcão 1998; Abdel-Magid et al. 1999; Abido and Abdel-Magid 1999; do Bomfim et al. 2000; Zhang and Coonick 2000; Rerkpreedapong et al. 2003; Malachi and Singer 2006.

Genetic algorithms have been also applied to solve transformer design problems (Bai et al. 1995; Nims et al. 1996; Galdi et al. 2001; Georgilakis et al. 2001; Doulamis et al. 2002; Tutkun and Moses 2004a, 2004b; Pan et al. 2008; Georgilakis 2009).

5.5.3 Example 5.4

Solve the following problem:

$$\min_{x_1, x_2} f(x_1, x_2) = \min_{x_1, x_2} \left\{ \begin{array}{l} x_1 \cdot \sin\left[\sqrt{|x_1 - (x_2 + 9)|}\right] \\ -(x_2 + 9) \cdot \sin\left[\sqrt{|x_2 + 0.5 \cdot x_1 + 9|}\right] \end{array} \right\}, \quad (5.92)$$

subject to:

$$-20 \leq x_1 \leq 20, \quad (5.93)$$

$$-20 \leq x_2 \leq 20, \quad (5.94)$$

using the binary genetic algorithm technique.

Solution
The various steps of the binary genetic algorithm method are presented in the following.

Step 1 Initialization
First the genetic algorithm parameters are selected as follows:

1. Population size = 12

2. Mutation probability = 0.15

3. Number of bits per design variable = 8

4. Maximum number of generations = 100

5. Fraction of chromosomes that survive for mating = 50%

The optimization problem has two variables (x_1 and x_2) and each variable is coded using 8 bits, so the total number of bits per chromosome is $2 \cdot 8 = 16$, where the first 8 bits correspond to variable x_1 and the next 8 bits correspond to variable x_2.

The maximum value that can de coded with 8 bits is:

$$p^{max} = (11111111)_{(2)} \Rightarrow$$

$$p^{max} = 1 \cdot 2^0 + 1 \cdot 2^1 + 1 \cdot 2^2 + 1 \cdot 2^3 + 1 \cdot 2^4 + 1 \cdot 2^5 + 1 \cdot 2^6 + 1 \cdot 2^7 \Rightarrow$$

$$p^{max} = 255.$$

The minimum value that can de coded with 8 bits is:

$$p^{min} = (00000000)_{(2)} \Rightarrow p^{min} = 0.$$

The design variables are constrained according to (5.93) and (5.94), from where we can see that the minimum and maximum parameter values are:

$$x^{min} = -20 \text{ and } x^{max} = 20.$$

The initial population of 12 chromosomes is created randomly. These binary chromosomes are shown in the second column of Table 5.3. For example, the first chromosome of Table 5.3 is the binary string $(1000010000111011)_{(2)}$, which corresponds to:

$$x_1^d = (10000100)_{(2)} \Rightarrow$$

$$x_1^d = 0 \cdot 2^0 + 0 \cdot 2^1 + 1 \cdot 2^2 + 0 \cdot 2^3 + 0 \cdot 2^4 + 0 \cdot 2^5 + 0 \cdot 2^6 + 1 \cdot 2^7 \Rightarrow$$

$$x_1^d = 132,$$

and

$$x_2^d = (00111011)_{(2)} \Rightarrow$$

$$x_2^d = 1 \cdot 2^0 + 1 \cdot 2^1 + 0 \cdot 2^2 + 1 \cdot 2^3 + 1 \cdot 2^4 + 1 \cdot 2^5 + 0 \cdot 2^6 + 0 \cdot 2^7 \Rightarrow$$

$$x_2^d = 59.$$

Next, the binary chromosomes are decoded to continuous values corresponding to the design variables x_1 and x_2. For example, for the first chromosome of Table 5.3, the x_1^d is decoded to continuous value x_1 as follows:

$$x_1 = x^{min} + (x_1^d - p^{min}) \cdot \frac{(x^{max} - x^{min})}{(p^{max} - p^{min})} \Rightarrow$$

$$x_1 = -20 + (132 - 0) \cdot \frac{(20 + 20)}{(255 - 0)} \Rightarrow x_1 = 0.70588.$$

Similarly, for the first chromosome of Table 5.3, the x_2^d is decoded to continuous value x_2 as follows:

5.5 Genetic Algorithms

Table 5.3 Initial population of the binary genetic algorithm for the Example 5.4

Code	Chromosome	x_1	x_2	$f(x_1, x_2)$
1	1000010000111011	0.70588	-10.745	2.3193
2	1110011001010111	16.078	-6.3529	-7.6986
3	0001111000100101	-15.294	-14.196	-1.6705
4	0010010100100110	-14.196	-14.039	-3.3281
5	0110001000111000	-4.6275	-11.216	-2.7465
6	1101100111001010	14.039	11.686	25.12
7	0100100010010010	-8.7059	2.902	4.0067
8	1110001000111101	15.451	-10.431	-11.874
9	0010110001101000	-13.098	-3.6863	7.1888
10	1111011100011011	18.745	-15.765	-10.924
11	1111111010000110	19.843	1.0196	9.8616
12	1100000001101111	10.118	-2.5882	11.046

Table 5.4 Initial population sorted by cost function $f(x_1, x_2)$

Code	Chromosome	x_1	x_2	$f(x_1, x_2)$
8	1110001000111101	15.451	-10.431	-11.874
10	1111011100011011	18.745	-15.765	-10.924
2	1110011001010111	16.078	-6.3529	-7.6986
4	0010010100100110	-14.196	-14.039	-3.3281
5	0110001000111000	-4.6275	-11.216	-2.7465
3	0001111000100101	-15.294	-14.196	-1.6705
1	1000010000111011	0.70588	-10.745	2.3193
7	0100100010010010	-8.7059	2.902	4.0067
9	0010110001101000	-13.098	-3.6863	7.1888
11	1111111010000110	19.843	1.0196	9.8616
12	1100000001101111	10.118	-2.5882	11.046
6	1101100111001010	14.039	11.686	25.12

$$x_2 = x^{min} + (x_2^d - p^{min}) \cdot \frac{(x^{max} - x^{min})}{(p^{max} - p^{min})} \Rightarrow$$

$$x_2 = -20 + (59 - 0) \cdot \frac{(20 + 20)}{(255 - 0)} \Rightarrow x_2 = -10.745.$$

The fitness value is equal to the value of the cost function $f(x_1, x_2)$. For example, for the first chromosome of Table 5.3, the fitness value is:

Table 5.5 Rank weighting of the chromosomes of the mating pool

Code	Rank r	Chromosome	P_r	$\sum_{i=1}^{r} P_r$
8	1	1110001000111101	0.28571	0.28571
10	2	1111011100011011	0.2381	0.52381
2	3	1110011001010111	0.19048	0.71429
4	4	0010010100100110	0.14286	0.85714
5	5	0110001000111000	0.095238	0.95238
3	6	0001111000100101	0.047619	1.00000

$$f(x_1, x_2) = 0.70588 \cdot \sin\left[\sqrt{|0.70588 - (-10.745 + 9)|}\right]$$
$$- (-10.745 + 9) \cdot \sin\left[\sqrt{|-10.745 + 0.5 \cdot 0.70588 + 9|}\right] \Rightarrow$$
$$f(x_1, x_2) = 2.3193.$$

Next the chromosomes of Table 5.3 are sorted according to their fitness value and the results are shown in Table 5.4.

Step 2 Generation 1

The fraction of chromosomes that survive for mating is 50%, which means that six out of 12 chromosomes survive. In particular:

1. The chromosomes of Table 5.4 with codes 8, 10, 2, 4, 5, 3 survive for mating, since they have the lowest cost values, i.e., the highest fitness values. These chromosomes (a) are copied to the population of the next generation, and (b) are also copied to the mating pool.

2. The chromosomes of Table 5.4 with codes 1, 7, 9, 11, 12, 6 are deleted, since they have the highest cost values, i.e., the lowest fitness values.

The chromosomes that have been selected for mating are shown in the third column of Table 5.5 and their rank is shown in the second column of Table 5.5. The probability P_r of chromosome with rank r being selected for mating is:

$$P_r = \frac{N_k - r + 1}{\sum_{r=1}^{N_k} r},$$

where $N_k = 6$ is the number of chromosomes that survive for mating.

For example, the probability P_1 of chromosome with rank 1 being selected for mating is:

5.5 Genetic Algorithms

$$P_1 = \frac{N_k - r + 1}{\sum_{r=1}^{N_k} r} = \frac{6-1+1}{1+2+3+4+5+6} = \frac{6}{21} \Rightarrow P_1 = 0.28571.$$

The cumulative probabilities, shown in the last column of Table 5.5, are used to select the pairs of chromosomes from the mating pool to produce the new offspring according to roulette wheel weighting as follows:

1. A random number between zero and one is generated.

2. Starting from the chromosome with rank 1, the first chromosome with a cumulative probability that is greater than the random number is selected for the mating pool.

Applying the above algorithm we have:

1. The random numbers $\begin{bmatrix} 0.87131 & 0.60613 & 0.16366 \end{bmatrix}$ were generated to select the mother chromosomes. The chromosome of Table 5.5 with rank 5 is the first chromosome that has cumulative probability that is greater than the random number 0.87131, so the chromosome of Table 5.5 with rank 5 is selected as the first mother chromosome of the mating pool. Similarly, the rest mother chromosomes are the chromosomes of Table 5.5 with ranks 3 and 1. Consequently:

    Ranks of mother chromosomes $= \begin{bmatrix} 5 & 3 & 1 \end{bmatrix}$.

2. The random numbers $\begin{bmatrix} 0.86312 & 0.30487 & 0.028166 \end{bmatrix}$ were generated for selecting the father chromosomes. Consequently:

    Ranks of father chromosomes $= \begin{bmatrix} 5 & 2 & 1 \end{bmatrix}$.

One-point crossover is selected. The crossover points, that are produced randomly, are the following: 13, 12, 4. This means that:

1. Mother chromosome with rank 5 mates with father chromosome with rank 5 and the crossover is at point 13. Since mother and father chromosomes are the same (rank 5), they produce two identical offspring.

2. Mother chromosome with rank 3 mates with father chromosome with rank 2 and the crossover is at point 12 and they produce the two offspring of Fig. 5.4.

3. Mother chromosome with rank 1 mates with father chromosome with rank 1 and the crossover is at point 4. Since mother and father chromosomes are the same (rank 1), they produce two identical offspring.

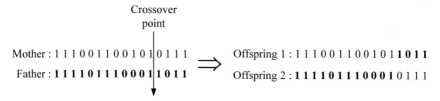

Fig. 5.4 Mother and father mate to produce two offspring using one-point crossover

Table 5.6 Population after mating

New code	Chromosome	Remark
1	1110001000111101	A copy of chromosome with rank 1 of Table 5.5
2	1111011100011011	A copy of chromosome with rank 2 of Table 5.5
3	1110011001010111	A copy of chromosome with rank 3 of Table 5.5
4	0010010100100110	A copy of chromosome with rank 4 of Table 5.5
5	0110001000111000	A copy of chromosome with rank 5 of Table 5.5
6	0001111000100101	A copy of chromosome with rank 6 of Table 5.5
7	0110001000111000	Offspring 1 by mating chromosomes with rank 5 & 5
8	0110001000111000	Offspring 2 by mating chromosomes with rank 5 & 5
9	1110011001011011	Offspring 1 by mating chromosomes with rank 3 & 2
10	1111011100010111	Offspring 2 by mating chromosomes with rank 3 & 2
11	1110001000111101	Offspring 1 by mating chromosomes with rank 1 & 1
12	1110001000111101	Offspring 2 by mating chromosomes with rank 1 & 1

Table 5.7 Mutation pairs

Mutation pair		Mutation pair		Mutation pair	
Row	Column	Row	Column	Row	Column
11	3	4	9	12	3
12	10	7	6	10	1
4	8	11	4	2	15
8	10	6	9	5	10
2	10	2	12	7	2
5	11	4	4	10	5
6	15	2	16	12	9
11	10	12	12	4	10
3	15	8	3	11	4

The new population after crossover is shown in Table 5.6.

Next, mutation is applied with 15% mutation probability, which means that the total number of mutations, nm, is:

5.5 Genetic Algorithms

Table 5.8 Population at the end of the first generation

New code	Chromosome after mutation	x_1	x_2	$f(x_1, x_2)$
1	1110001000111101	15.451	-10.431	-11.874
2	1111011101001000	18.745	-8.7059	-17.149
3	1110011001010101	16.078	-6.6667	-8.4359
4	0011010011100110	-11.843	16.078	26.113
5	0110001001011000	-4.6275	-6.1961	-3.6747
6	0001111010100111	-15.294	6.1961	4.7177
7	0010011000111000	-14.039	-11.216	4.3349
8	0100001001111000	-9.6471	-1.176	0.58878
9	1110011001011011	16.078	-5.7255	-6.0789
10	0111111100010111	-0.078431	-16.392	2.9509
11	1100001001111101	10.431	-0.39216	14.87
12	1100001011101101	10.431	17.176	8.7936

Table 5.9 Population at the end of the first generation sorted by cost function $f(x_1, x_2)$

New code	Chromosome after mutation	x_1	x_2	$f(x_1, x_2)$
2	1111011101001000	18.745	-8.7059	-17.149
1	1110001000111101	15.451	-10.431	-11.874
3	1110011001010101	16.078	-6.6667	-8.4359
9	1110011001011011	16.078	-5.7255	-6.0789
5	0110001001011000	-4.6275	-6.1961	-3.6747
8	0100001001111000	-9.6471	-1.176	0.58878
10	0111111100010111	-0.078431	-16.392	2.9509
7	0010011000111000	-14.039	-11.216	4.3349
6	0001111010100111	-15.294	6.1961	4.7177
12	1100001011101101	10.431	17.176	8.7936
11	1100001001111101	10.431	-0.39216	14.87
4	0011010011100110	-11.843	16.078	26.113

$$nm = (N_p - 1) \cdot NT \cdot m = (12-1) \cdot 16 \cdot 0.15 = 26.4 \Rightarrow nm \approx 27,$$

where N_p is the population size, NT is the total number of bits in a chromosome, and m is the mutation probability. It should be noted that the best chromosome, i.e., the chromosome with code 1 in Table 5.6, is not mutated.

The bits that will be mutated are selected randomly and the selected mating pairs are shown in Table 5.7. For example, the first random pair is [11, 3], which means that the bit in the 11th row and the 3rd column of the chromosomes appearing in the second column of Table 5.6 has to be mutated, which is the third bit of the chromosome with code 11. This particular bit is 1, so it has to be mutated to 0. The chromosomes after mutation are shown in the second column of Table 5.8.

Next the chromosomes of Table 5.8 are sorted according to their fitness value and the results are shown in Table 5.9. At the initialization of the genetic algorithm, the best solution had a cost of −11.874, as Table 5.4 shows. After the end of generation 1, the best solution has a cost of −17.149, as Table 5.9 shows. This means that after only one generation (iteration) the genetic algorithm improved its best solution by:

$$Impr = \left[\frac{-17.149-(-11.874)}{-11.874}\right] \cdot 100\% = 44.4\%.$$

Optimum solution
After 71 generations (iterations), the optimum solution is found, which is:

$$x_1 = -14.5098, \ x_2 = -20, \ f(x_1, x_2) = -23.8035.$$

References

Abdel-Magid YL, Abido MA, Mantawy AH (1999) Simultaneous stabilization of multimachine power systems via genetic algorithms. IEEE Transactions on Power Systems 14(4):1428–1439

Abido MA and Abdel-Magid YL (1999) Hybridizing rule-based power system stabilizers with genetic algorithms. IEEE Transactions on Power Systems 14(2):600–607

Abril IP and Quintero JAG (2003) VAr compensation by sequential quadratic programming. IEEE Transactions on Power Systems 18(1):36–41

Amoiralis EI, Tsili MA, Georgilakis PS, Kladas AG, Souflaris AT (2008) A parallel mixed integer programing-finite element method technique for global design optimization of power transformers. IEEE Transactions on Magnetics 44(6):1022–1025

Amoiralis EI, Georgilakis PS, Tsili MA, Kladas AG (2009) Global transformer optimization method using evolutionary design and numerical field computation. IEEE Transactions on Magnetics 45(3):1720–1723

Bai B, Xie D, Cui J, Fei ZY, Mohammed OA (1995) Optimal transposition design of transformer windings by genetic algorithms. IEEE Transactions on Magnetics 31(6):3572–3574

Bazaraa MS, Sherali HD, Shetty CM (1993) Nonlinear programming: theory and algorithms, 2nd edn. Wiley, New York

Bellman R (1957) Dynamic programming. Princeton University Press, Princeton, NJ

Boardman JT and Meckiff CC (1985) A branch and bound formulation to an electricity distribution planning problem. IEEE Transactions on Power Apparatus and Systems 104(8):2112–2118

Boggs PT and Tolle JW (1995) Sequential quadratic programming. Acta Numerica 4:1–51

Bonabeau E, Dorigo M, Theraulaz G (1999) Swarm intelligence: from natural to artificial systems. Oxford University Press, Oxford

Borchers B and Mitchell JE (1994) An improved branch and bound algorithm for mixed integer nonlinear programs. Computers and Operations Research 21(4):359–367

Broyden CG (1970) The convergence of a class of double-rank minimization algorithms – 2: The new algorithm. IMA Journal of Applied Mathematics 6(3):222–231

Castillo E, Conejo AJ, Pedregal P, García R, Alguacil N (2002) Building and solving mathematical programming models in engineering and science. John Wiley & Sons, New York

Cauchy AL (1847) Méthode générale pour la resolution des systèmes d'équations simultanées. Comptes Rendus de l'Academie des Sciences 25:536–538

Cerný V (1985) Thermodynamical approach to the traveling salesman problem: An efficient simulation algorithm. Journal of Optimization Theory and Applications 45(1):41–51

Chang H-C and Chen P-H (1998) Hydrothermal generation scheduling package: a genetic based approach. IEE Proceedings Generation, Transmission and Distribution 145(4):451–457

Chen P-H and Chang H-C (1995) Large-scale economic dispatch by genetic algorithm. IEEE Transactions on Power Systems 10(4):1919–1926

Chen P-H and Chang H-C (1996) Genetic aided scheduling of hydraulically coupled plants in hydro-thermal coordination. IEEE Transactions on Power Systems 11(2):975–981

Choi J, El-Keib AA, Tran T (2005) A fuzzy branch and bound-based transmission system expansion planning for the highest satisfaction level of the decision maker. IEEE Transactions on Power Systems 20(1):476–484

Coelho LdS and Mariani VC (2006) Combining of chaotic differential evolution and quadratic programming for economic dispatch optimization with valve-point effect. IEEE Transactions on Power Systems 21(2):989–996

Cohen AI and Yoshimura M (1983) A branch and bound algorithm for unit commitment. IEEE Transactions on Power Apparatus and Systems 102(2):444–451

Contaxis GC, Delkis C, Korres G (1986) Decoupled optimal load flow using linear or quadratic programming. IEEE Transactions on Power Systems 1(2):1–7

da Silva EL, Gil HA, Areiza JM (2000) Transmission network expansion planning under an improved genetic algorithm. IEEE Transactions on Power Systems 15(3):1168–1174

Dakin RJ (1965) A tree-search algorithm for mixed integer programming problems. The Computer Journal 8(3):250–255

Dantzig GB (1963) Linear programming and extensions. Princeton University Press, Princeton, NJ

Dasgupta D and McGregor DR (1994) Thermal unit commitment using genetic algorithms. IEE Proceedings Generation, Transmission and Distribution 141(5):459–465

Davidon WC (1959) Variable metric method for minimization. Report ANL-5990, Argonne National Laboratory, Argonne, Ill

Delfanti M, Granelli GP, Marannino P, Montagna M (2000) Optimal capacitor placement using deterministic and genetic algorithms. IEEE Transactions on Power Systems 15(3):1041–1046

Do Bomfim ALB, Taranto GN, Falcão DM (2000) Simultaneous tuning of power damping controllers using genetic algorithms. IEEE Transactions on Power Systems 15(1):163–169

Dorigo M and Stützle T (2004) Ant colony optimization. MIT Press, Cambridge, MA

Doulamis ND, Doulamis AD, Georgilakis PS, Kollias SD, Hatziargyriou ND (2002) A synergetic neural network-genetic scheme for optimal transformer construction. Integrated Computer-Aided Engineering 9(1):37–56

Feo TA and Resende MGC (1989) A probabilistic heuristic for a computationally difficult set covering problem. Operations Research Letters 8(2):67–71

Feo TA and Resende MGC (1995) Greedy randomized adaptive search procedures. Journal of Global Optimization 6(2):109–133

Finardi EC and da Silva EL (2006) Solving the hydro unit commitment problem via dual decomposition and sequential quadratic programming. IEEE Transactions on Power Systems 21(2):835–844

Fletcher R (1970) A new approach to variable metric algorithms. The Computer Journal 13(3):317–322

Fletcher R and Powell MJD (1963) A rapidly convergent descent method for minimization. The Computer Journal 6(2):163–168

Fogel DB (2000) Evolutionary computation: toward a new philosophy of machine intelligence, 2nd edn. IEEE Press, Piscataway, NJ

Fogel LJ, Owens AJ, Walsh MJ (1966) Artificial intelligence through simulated evolution. John Wiley & Sons, New York

Fukuyama Y and Chiang H-D (1996) A parallel genetic algorithm for generation expansion planning. IEEE Transactions on Power Systems 11(2):955961

Galdi V, Ippolito L, Piccolo A, Vaccaro A (2001) Parameter identification of power transformer thermal model via genetic algorithms. Electric Power Systems Research 60(2):107–113

Gallego RA, Monticelli AJ, Romero R (1998) Comparative studies on non-convex optimization methods for transmission expansion planning. IEEE Transactions on Power Systems 13(3):822–828

Georgilakis PS (2009) A recursive genetic algorithm–finite element method technique for the solution of transformer manufacturing cost minimization problem. IET Electric Power Applications, accepted for publication

Georgilakis PS, Doulamis ND, Doulamis AD, Hatziargyriou ND, Kollias SD (2001) A novel iron loss reduction technique for distribution transformers based on a genetic algorithm-neural network approach. IEEE Transactions on Systems, Man, and Cybernetics, Part C 31(1):16–34

Glover F (1989) Tabu search – Part I. ORSA Journal on Computing 1(3):190–206

Glover F (1990) Tabu search – Part II. ORSA Journal on Computing 2(1):4–32

Glover F and Laguna M (1997) Tabu search. Kluwer Academic Publishers, Boston

Goldberg DE (1989) Genetic algorithms in search, optimization, and machine learning. Addison-Wesley, Reading, MA

Goldfarb D (1970) A family of variable metric methods derived by variational means. Mathematics of Computation 24:23–26

Grudinin N (1998) Reactive power optimization using successive quadratic programming method. IEEE Transactions on Power Systems 13(4):1219–1225

Haffner S, Monticelli A, Garcia A, Romero R (2001) Specialised branch-and-bound algorithm for transmission network expansion planning. IEE Proceedings Generation, Transmission and Distribution 148(5):482–488

Han SP (1977) A globally convergent method for nonlinear programming. Journal of Optimization Theory and Applications 22(3):297–309

Hestenes MR and Stiefel E (1952) Methods of conjugate gradients for solving linear systems. Report 1659, National Bureau of Standards, Washington D.C.

Holland JH (1975) Adaptation in natural and artificial systems. University of Michigan Press, Ann Arbor, MI

Hooke R and Jeeves TA (1961) Direct search solution of numerical and statistical problems. Journal of the ACM 8(2):212–229

Iba K (1994) Reactive power optimization by genetic algorithm. IEEE Transactions on Power Systems 9(2):685–692

Karmarkar N (1984) A new polynomial-time algorithm for linear programming. Combinatorica 4(4):373–395

Kazarlis SA, Bakirtzis AG, Petridis V (1996) A genetic algorithm solution to the unit commitment problem. IEEE Transactions on Power Systems 11(1):83–92

Kennedy J and Eberhart R (2001) Swarm intelligence. Morgan Kaufmann, San Mateo, CA

Kirkpatrick S, Gelatt CD Jr, Vecchi MP (1983) Optimization by simulated annealing. Science 220:671–680

Land AH, Doig AG (1960) An automatic method for solving discrete programming problems. Econometrica 28(3):497–520

Lauer GS, Sandell NR Jr., Bertsekas DP, Posbergh TA (1982) Solution of large-scale optimal unit commitment problems. IEEE Transactions on Power Apparatus and Systems 101(1):79–86

References

Lee KY (1998) Optimal reactive power planning using evolutionary algorithms: a comparative study for evolutionary programming, evolutionary strategy, genetic algorithm, and linear programming. IEEE Transactions on Power Systems 13(1):101–108

Lee KY and Park Y-M (1995) Optimization method for reactive power planning by using a modified simple genetic algorithm. IEEE Transactions on Power Systems 10(4):1843–1850

Lemke CE (1954) The dual method of solving the linear programming problem. Naval Research Logistics Quarterly 1(1):36–47

Leyffer S (2001) Integrating SQP and branch-and-bound for mixed integer nonlinear programming. Computational Optimization and Applications 18:295–309

Lourenço HR, Martin O, Stützle T (2002) Iterated local search. In: Glover F and Kochenberger G (eds) Handbook of metaheuristics. Kluwer Academic Publishers, Norwell, MA

Lu CN, Chen SS, Ing CM (1988) The incorporation of HVDC equations in optimal power flow methods using sequential quadratic programming techniques. IEEE Transactions on Power Systems 3(3):1005–1011

Maifeld TT and Sheblé GB (1996) Genetic-based unit commitment algorithm. IEEE Transactions on Power Systems 11(3):1359–1370

Malachi Y and Singer S (2006) A genetic algorithm for the corrective control of voltage and reactive power. IEEE Transactions on Power Systems 21(1):295–300

Marquardt D (1963) An algorithm for least-squares estimation of nonlinear parameters. SIAM Journal on Applied Mathematics 11(2):431–441

Momoh JA, Adapa R, El-Hawari ME (1999) A review of selected optimal power flow literature to 1993. I. Nonlinear and quadratic programming approaches. IEEE Transactions on Power Systems 14(1):96–104

Nejdawi IM, Clements KA, Davis PW (2000) An efficient interior point method for sequential quadratic programming based optimal power flow. IEEE Transactions on Power Systems 15(4):1179–1183

Nelder JA and Mead R (1965) A simplex method for function minimization. The Computer Journal 7(4):308–313

Nims JW III, Smith RE, El-Keib AA (1996) Application of genetic algorithm to power transformer design. Electric Machines and Power Systems 24(6):669–680

Nocedal J and Wright SJ (2006) Numerical optimization, 2nd edn. Springer, New York

Orero SO and Irving MR (1996) Economic dispatch of generators with prohibited operating zones: a genetic algorithm approach. IEE Proceedings Generation, Transmission and Distribution 143(6):529–534

Orero SO and Irving MR (1998) A genetic algorithm modelling framework and solution technique for short term optimal hydrothermal scheduling. IEEE Transactions on Power Systems 13(2):501–518

Pan C, Chen W, Yun Y (2008) Fault diagnostic method of power transformers based on hybrid genetic algorithm evolving wavelet neural network. IET Electric Power Applications 2(1):71–76

Papageorgiou LG and Fraga ES (2007) A mixed integer quadratic programming formulation for the economic dispatch of generators with prohibited operating zones. Electric Power Systems Research 77(10):1292–1296

Park J-B, Park Y-M, Won J-R, Lee KY (2000) An improved genetic algorithm for generation expansion planning. IEEE Transactions on Power Systems 15(3):916–922

Powell MJD (1964) An efficient method for finding the minimum of a function of several variables without calculating derivatives. The Computer Journal 7(2):155–162

Powell MJD (1978) A fast algorithm for nonlinearly constrained optimization calculations. In: Watson GA (ed) Numerical analysis. Springer Verlag, Berlin

Price KV, Storn RM, Lampinen JA (2005) Differential evolution: a practical approach to global optimization. Springer, Berlin

Rao SS (1996) Engineering optimization: theory and practice, 3rd edn. John Wiley & Sons, New York

Rau NS (2001) Radial equivalents to map networks to market formats–an approach using quadratic programming. IEEE Transactions on Power Systems 16(4):856–861

Rau NS (2003) Optimization principles: practical applications to the operation and markets of the electric power industry. IEEE Press, New Jersey

Rerkpreedapong D, Hasanovic A, Feliachi A (2003) Robust load frequency control using genetic algorithms and linear matrix inequalities. IEEE Transactions on Power Systems 18(2):855–861

Richter CW and Sheblé GB (2000) A profit-based unit commitment GA for the competitive environment. IEEE Transactions on Power Systems 15(2):715–721

Rider MJ, Garcia AV, Romero R (2008) Transmission system expansion planning by a branch-and-bound algorithm. IET Generation, Transmission and Distribution 2(1):90–99

Rogers JS and Rolko M (1992) A quadratic programming model for planning generation and inter-utility transmission. International Journal of Electrical Power & Energy Systems 14(1):18–22

Rosenbrock HH (1960) An automatic method for finding the greatest or least value of a function. The Computer Journal 3(3):175–184

Rudolf A and Bayrleithner R (1999) A genetic algorithm for solving the unit commitment problem of a hydro-thermal power system. IEEE Transactions on Power Systems 14(4):1460–1468

Schwefel H-P and Rudolph G (1995) Contemporary evolution strategies. In: Morán F, Moreno A, Merelo JJ, Chacón P (eds) Advances in artificial life. Springer, Berlin

Shanno DF (1970) Conditioning of quasi-Newton methods for function minimization. Mathematics of Computation 24:647–656

Sheblé GB and Brittig K (1995) Refined genetic algorithms–economic dispatch example. IEEE Transactions on Power Systems 10(1):117–123

Sinha N, Chakrabarti R, Chattopadhyay PK (2003) Evolutionary programming techniques for economic load dispatch. IEEE Transactions on Evolutionary Computation 7(1):83–94

Somuah CB and Schweppe FC (1987) Minimum frequency constrained generation margin allocation using quadratic programming. International Journal of Electrical Power & Energy Systems 9(2):105–112

Spendley W, Hext GR, Himsworth FR (1962) Sequential application of simplex designs in optimisation and evolutionary operation. Technometrics 4:441–461

Storn R and Price KV (1997) Differential evolution – a simple and efficient heuristic for global optimization over continuous spaces. Journal of Global Optimization 11:341–359

Taranto GM and Falcão DM (1998) Robust decentralised control design using genetic algorithms in power system damping control. IEE Proceedings Generation, Transmission and Distribution 145(1):1–6

Thompson GL and Wall DL (1981) A branch and bound model for choosing optimal substation locations. IEEE Transactions on Power Apparatus and Systems 100(5):2683–2688

Tutkun N and Moses AJ (2004a) Design optimisation of a typical strip-wound toroidal core using genetic algorithms. Journal of Magnetism and Magnetic Materials 277(1-2):216–220

Tutkun N and Moses AJ (2004b) Estimates of simplified equivalent circuit parameters of a typical wound toroidal core using genetic algorithms. Journal of Magnetism and Magnetic Materials 284:201–205

Urdaneta AJ, Gómez JF, Sorrentino E, Flores L, Diaz R (1999) A hybrid genetic algorithm for optimal reactive power planning based upon successive linear programming. IEEE Transactions on Power Systems 14(4):1292–1298

Voudouris C (1997) Guided local search for combinatorial optimization problems. Ph.D. Thesis, University of Essex, Colchester, UK

Voudouris C and Tsang E (1995) Guided local search. Technical Report CSM-247, University of Essex, Colchester, UK

Voudouris C and Tsang EPK (1999) Guided local search. European Journal of Operational Research 113(2):469–499

References

Wei H, Sasaki H, Yokoyama R (1996) An application of interior point quadratic programming algorithm to power system optimization problems. IEEE Transactions on Power Systems 11(1):260–266

Wilson RB (1963) A simplified algorithm for concave programming. Ph.D. Thesis, Harvard University, USA

Zhang P and Coonick AH (2000) Coordinated synthesis of PSS parameters in multi-machine power systems using the method of inequalities applied to genetic algorithms. IEEE Transactions on Power Systems 15(2):811–816

Part III
Modern Transformer Design

6 Evaluation of Transformer Technical Characteristics

Abstract This chapter is devoted to the evaluation of transformer technical characteristics. Decision trees and artificial neural networks are used to solve the no-load loss classification problem. Artificial neural networks are used to solve the no-load loss prediction problem. Impedance voltage evaluation is implemented using a particular finite element model with detailed representation of winding geometry.

6.1 Introduction

In the light of the twenty-first century energy market, where competition continues to accelerate in the electricity supply industry, utilities will try to further improve system reliability and quality, while simultaneously being cost effective. The transformer manufacturing industry must improve transformer efficiency and reliability while reducing cost, since high quality, low cost products have become the key to survival (Georgilakis et al. 1999a, 2001a). Transformer efficiency is improved by reducing load and no-load losses. Transformer reliability is improved mainly by the accurate evaluation of the leakage field, the short-circuit impedance and the resulting forces on transformer windings under short-circuit, since these enable the avoidance of mechanical damage and failures during short-circuit tests and power system faults.

Variability in the design process, material, and production process cause significant variability in measured transformer performance parameters, especially transformer no-load losses (teNyenhuis and Girgis 2006). That is why design margins are considered during transformer design. Moreover, there is an ever-increasing need in today's transformer market to minimize design margins without increasing the risk of violating guaranteed values of transformer performance parameters. Consequently, it should not come as a surprise that the evaluation of transformer technical characteristics is a subject of permanent research effort (Olivares et al. 2002, 2003a, 2003b, 2004a, 2004b; Kulkarni et al. 2004; Escarela-Perez et al. 2007; Amoiralis et al. 2008; Olivares-Galván et al. 2009).

This chapter is focused on the evaluation of transformer no-load losses and impedance voltage (short-circuit impedance). In particular:

1. Artificial intelligence based methods, namely decision trees and artificial neural networks, for no-load loss classification and prediction are presented and compared. The no-load loss classification and prediction is implemented for in-

dividual cores as well as for assembled transformers. The classification and prediction is based on actual industrial measurements on individual cores taken at the early stages of their construction. Decision trees are used to select the most relevant attributes among a large set of candidates and to produce if-then-else decision rules. These rules are applicable at the early stages of core production and allow possible corrective actions during the manufacturing process. Artificial neural networks are used to predict no-load losses at the early stages of transformer manufacturing. The attributes selected by the decision trees are used as inputs to the neural networks. Results from the application of these methods to the transformer manufacturing industry demonstrate their effectiveness and practicality.

2. An efficient three-dimensional (3D) finite element method (FEM) model of power transformers for the leakage field and short-circuit impedance evaluation, suitable for design office use, has been developed and applied in the transformer manufacturing industry. Detailed representation of the transformer (focusing on the winding geometry and cooling ducts) and the particular reduced scalar potential technique adopted are the main advantages of the model, with respect to standard FEM codes. It has been validated through local field measurements and short-circuit impedance calculations for several three-phase, wound core, single and dual voltage, power transformers. The computed results compared favorable with measured values and the mean deviation in the impedance value was less than 3%. The method is very cost effective, as high accuracy is obtained for low mesh densities, requiring little computational time. This ability, along with the development of an automated, user-oriented, transformer short-circuit impedance calculation program based on the FEM model overcome the main deficiencies of the method and enable its use during transformer design.

6.2 No-Load Loss Classification with Decision Trees and Artificial Neural Networks

6.2.1 Introduction

No-load loss classification aims at classifying the no-load losses into two classes: acceptable and non-acceptable. In this chapter, artificial intelligence based no-load loss classification techniques will be presented and applied for both the individual core and the transformer (Hatziargyriou et al. 1998a; Georgilakis et al. 1998b; Georgilakis 2000).

It is very important to have an efficient no-load loss classification tool, i.e., tool with high classification accuracy, since it offers the ability to know in advance if an individual core or transformer will be of acceptable quality or not. It is neces-

6.2 No-Load Loss Classification with Decision Trees and Artificial Neural Networks

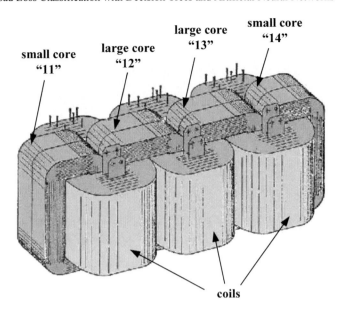

Fig. 6.1 Assembled active part of wound core distribution transformer

sary to determine the values of the attributes (input parameters) that are known at the early stages of individual core or transformer manufacturing and afterwards to use the classification tool. Thus, the existence of an efficient no-load loss classification tool aims at predicting the quality of individual cores and transformers as well as at investigating the impact that variation of the attribute values will have on no-load loss quality (Georgilakis 2000).

6.2.2 Individual Core

6.2.2.1 Production Process

The wound core shell type distribution transformer is composed of two small individual cores and two large individual cores as shown in Fig. 6.1. We denote with "11" the left small individual core, with "12" the left large individual core, and with "13" and "14" the other two individual cores, so the arrangement of individual cores from left to right is "11"–"12"–"13"–"14", as Fig. 6.1 shows.

The production of individual cores includes, at the first stage, the slitting of magnetic material into bands of standard width. Then, the slit sheets are cut to predetermined lengths and are wound on a circular mandrel. After that, a suitable press gives a rectangular shape to the circular core. However, the previously described process significantly deteriorates core characteristics and especially its physical and electrical properties. To restore these properties, annealing follows at temperatures in the range 760–860°C in a protective environment containing pure dry nitrogen mixed with up to 2% hydrogen (Georgilakis et al. 1998b).

The *annealing cycle* adopted is divided into four phases:

1. Starting and heating up phase, to avoid oxidation and to normally achieve a temperature of 825°C.

2. Soaking phase, to achieve homogeneous temperature distribution for all cores.

3. Slow cooling phase, to slowly cool the load to avoid the development of internal stresses in the cores.

4. Fast cooling phase, with reduction of the temperature to 380°C to avoid oxidation of cores when they are exposed to the natural environment.

In contrast to production of stacked cores, wound cores present the following additional difficulties:

1. Air gaps may diverge due to tolerances of the machine performing the cutting and winding of sheets and due to difficulties in processing the magnetic material (slide).

2. The desirable dimensions of wound cores cannot be obtained accurately as in stacked cores.

3. Core formation may deteriorate the magnetic material insulation.

4. Homogeneous temperature distribution is hard to obtain during the annealing procedure.

6.2.2.2 Parameters Affecting No-load Losses of Individual Cores

The no-load losses of individual cores are affected by magnetic material properties (Moses 1992; Georgilakis et al. 2001b; Kalokiris et al. 2007), and by design and production factors (Godec 1977; Valkovic 1982; Moses 1984; Valkovic 1984; Nakata et al. 1984; Fecich and Balmer 1985; Ling et al. 1992; Basak and Bonyar 1992; Valkovic and Rezic 1992; Ilo et al. 1996; Girgis et al. 1998; Mechler and Girgis 2000; Soda and Enokizono 2000; teNyenhuis et al. 2001).

In the case of wound core shell type distribution transformers, the following three categories of parameters influence the no-load losses of the individual core (Hatziargyriou et al. 1998a):

1. Magnetic material parameters, e.g.:
    – Supplier of magnetic material

6.2 No-Load Loss Classification with Decision Trees and Artificial Neural Networks   269

- Thickness of magnetic material (e.g., 0.23, 0.27, 0.30 mm)
- Type of magnetic material (e.g., M3, M4, M5, Hi-B)
- Specific no-load loss (W/kg) of magnetic material
- Resistivity of surface insulation of magnetic material
- Hardness of magnetic material

2. Design parameters, e.g.:
   - Rated magnetic induction
   - Thickness of core leg
   - Width of core leg
   - Height of core window
   - Width of core window

3. Production parameters, e.g.:
   - Parameters of the annealing process
   - Grade of destruction of magnetic material insulation
   - Mechanical stresses during the formation of core
   - Actual weight of core
   - Cutting quality of magnetic material

6.2.2.3 Database

The first step in the application of artificial intelligence to the classification of individual core no-load losses is to create a database with all the attributes (input parameters) that affect the no-load losses of individual cores.

In the case of individual cores, eight attributes have been selected and used as the input vector for the artificial intelligence techniques. The selection of these attributes was based on extensive research and transformer designers' experience. These attributes correspond to parameters that actually affect the no-load losses of individual cores. In particular, the impact of the annealing cycle, the divergence of the actual core weight from its theoretical value, and the quality of core magnetic material are taken into consideration as elements of the network input vector. Six attributes (ATTR1 to ATTR6) have been investigated corresponding to the annealing process, depicted in Table 6.1. The other two attributes are:

Table 6.1 Attributes for the annealing process

Symbol	Attribute name	Low value (L)	High value (H)
ATTR1	Annealing final temperature	825 ^{0}C	855 ^{0}C
ATTR2	Temperature rise time	3 hours	4 hours
ATTR3	Furnace opening temperature	250 ^{0}C	350 ^{0}C
ATTR4	Duration of constant temperature	2 hours	3 hours
ATTR5	Position of core in the furnace	Down	Up
ATTR6	Protective atmosphere	100% N_2	98% N_2 and 2% H_2

Table 6.2 Conditions of the eight annealing experiments according to OA_8 orthogonal array

Experiment	ATTR1	ATTR2	ATTR3	ATTR4	ATTR5	ATTR6
1	L	L	L	L	H	H
2	H	L	H	L	L	H
3	H	H	H	H	H	H
4	L	H	L	H	L	H
5	L	H	H	L	H	L
6	H	H	L	L	L	L
7	H	L	L	H	H	L
8	L	L	H	H	L	L

Table 6.3 Percentage of acceptable individual cores per annealing experiment

	Annealing experiment							
	1	2	3	4	5	6	7	8
Acceptable cores (%)	94	95	93	69	94	98	98	93

1. The ratio of actual core weight to theoretical core weight (ATTR7). This attribute represents the divergence between the actual and the theoretical weight of the individual core.

2. The specific no-load losses (W/kg at 15000 Gauss) of the magnetic material used for the construction of the individual core (ATTR8). This attribute reflects the quality of the magnetic material of the individual core as it is expressed from the specific no-load losses of the magnetic material measured at a magnetic induction of 15000 Gauss.

6.2 No-Load Loss Classification with Decision Trees and Artificial Neural Networks   271

In order to take into account all combinations of the six attributes of the annealing process (Table 6.1) with two values (Low and High value), 64 experiments are required (i.e., 2^6 experiments). However, these experiments are time consuming and costly, therefore reduction of the implemented experiments is achieved through the *statistical design of experiments* method (SDE) proposed by Taguchi (Taguchi and Konishi 1987). According to SDE (Taguchi and Konishi 1987, Logothetis 1992), the parameters are varied at the same time in a systematic way, using orthogonal arrays, assuring the reliable and independent study of the impact and interaction of all main parameters in the production procedure. This means that a small number of representative experiments can characterize the process. Among the various orthogonal arrays, the orthogonal array OA_8 is selected, which requires only eight experiments instead of 64 experiments. The parameters characterizing each of the eight experiments of the OA_8 array are shown in Table 6.2. It can be seen that, due to the symmetric property, four experiments are carried out with low value (L) of each attribute, and the other four with high value (H).

All annealing experiments of Table 6.2 were done in a real industrial environment using the same 160 kVA transformer design and the same supplier of cores magnetic material. The magnetic steel was of grade M3, according to USA AISI, 1983, with thickness 0.23 mm. For every one of the eight annealing experiments, 96 individual cores (48 small and 48 large) were constructed. It should be noticed that all cores were annealed in the same furnace. 768 measurement sets (i.e., $8 \cdot 96$) were collected in a database for the creation of the learning and test sets. 576 measurement sets were used as the learning set and the rest 192 as the test set.

6.2.2.4 Classification Criterion

It is required to classify the no-load losses of each individual core into two classes: acceptable and non-acceptable no-load losses. The classification criterion is the following: one individual core belongs to the non-acceptable class if its actual specific no-load losses are greater than 10% of its theoretical specific no-load losses. Otherwise, the individual core belongs to the acceptable class.

Table 6.3 presents the percentage of acceptable individual cores per annealing experiment using the above described classification criterion. It can be seen from Table 6.3 that the individual cores annealed under the conditions of experiments 6 and 7 have the best quality (98% acceptable cores), while the worst results are obtained for the annealing experiment 4 (only 69% acceptable cores).

6.2.2.5 Decision Trees

Many decision trees were developed using the learning set of the 576 measurement sets and the eight candidate attributes (ATTR1 to ATTR8) of Sect. 6.2.2.3 and the χ^2 test of Sect. 4.5.6 with risk levels from 0.001% to 10%. The minimum

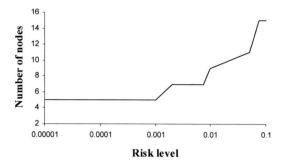

Fig. 6.2 Variation of decision tree size (number of nodes) with risk level

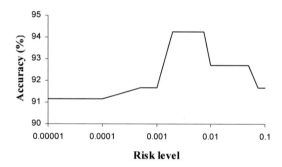

Fig. 6.3 Variation of decision tree accuracy with risk level

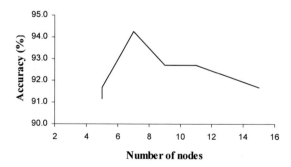

Fig. 6.4 Variation of decision tree accuracy with decision tree size

6.2 No-Load Loss Classification with Decision Trees and Artificial Neural Networks 273

classification entropy (i.e., the entropy below which the node is declared a leaf), H_{min}, was set to 0.1. All decision trees were tested using the independent test set of 192 measurement sets of Sect. 6.2.2.3. The results are presented in Figs. 6.2 to 6.4.

In Fig. 6.2, the decision tree size (total number of nodes) is plotted against the risk level. A first conclusion readily derived is that reduced risk levels result in decision trees of smaller size. This is expected, since higher risks permit easier expansion of the lower nodes of the decision trees (the question of the statistical significance of the test outcome is treated more leniently).

Figures 6.3 and 6.4 present the classification accuracy results of the decision trees when tested with the independent test set of 192 measurement sets of Sect. 6.2.2.3. The classification accuracy or classification success rate on the test set, expressed as a percentage (%), is defined as the ratio of correctly classified measurement sets (MS) of the test set to the total number of MS of the test set. In particular, Fig. 6.3 plots decision tree accuracy against risk level, while Fig. 6.4 presents decision tree accuracy versus decision tree size.

Figure 6.3 shows that as the risk level is increased from 0.001% to 0.75% the classification accuracy is increased, while when the risk level is increased from 0.75% to 10% the classification accuracy is decreased. Figure 6.3 shows that the minimum classification success rate (CSR) on the test set is 91.1%, while the maximum CSR is 94.3%, corresponding to 181 correct classifications out of the 192 measurement sets of the test set. This result (i.e., 3.2% difference between the minimum and maximum CSR) is very important, since it shows that the CSR on the test set can be increased by up to 3.2% in this particular example if many decision trees are constructed with different risk levels. The maximum CSR is valid for all risk levels in the range 0.2% to 0.75%, as Fig. 6.3 shows. In general, the value (or the range of values) of the risk level that corresponds to the maximum CSR on the test set is not known in advance and it requires many decision trees to be built.

Figure 6.4 shows that as the number of nodes is increased from 5 to 7, the CSR is increased, reaching its maximum value (94.3%) for seven nodes, while when the number of nodes becomes greater than seven, the CSR is decreased. The reason that the CSR is reduced for decision trees having 11 or 15 nodes is that in these decision trees the lower nodes are composed of few measurement sets (on average 15 to 20), so the lower nodes are not representative.

Figure 6.5 shows a decision tree that corresponds to 0.5% risk level, composed of seven nodes, and having 94.3% classification success rate on the test set (Hatziargyriou et al. 1998a). Based on the previous analysis regarding the variation of classification accuracy in relation to the risk level and the number of nodes, the decision tree of Fig. 6.5 is considered as an optimum decision tree since it has the maximum classification accuracy while its complexity (number of nodes) is low.

Table 6.4 presents calculations of the CSR for each of the four terminal nodes of the decision tree of Fig. 6.5 as well as the total CSR of the whole decision tree. Table 6.4 shows that the most successful terminal nodes are nodes 4 and 7, which

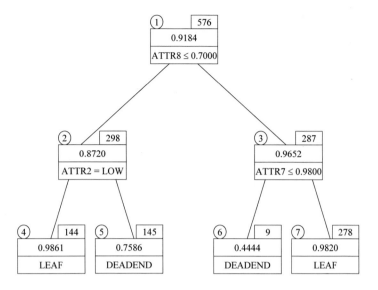

Fig. 6.5 Optimum decision tree with 0.5% risk level for the classification of no-load losses of individual cores

Table 6.4 Calculation of the classification success rate of the decision tree of Fig. 6.5

Node number	Node type	Acceptability index	Tested MS	Correctly classified MS	Classification success rate (%)
4	LEAF	0.9861	48	48	100.0
5	DEADEND	0.7586	48	39	81.3
6	DEADEND	0.4444	3	1	33.3
7	LEAF	0.9820	93	93	100.0
Total			192	181	94.3

Table 6.5 Calculation of the information of attributes of the decision tree of Fig. 6.5

Node	Attribute	H_C	H_C^T	I_C^T	N	$N \cdot I_C^T$	$N \cdot I_C^T$ (%)
1	ATTR8	0.408	0.386	0.022	576	12.8	21.7
2	ATTR2	0.552	0.453	0.099	289	28.7	48.6
3	ATTR7	0.218	0.157	0.061	287	17.5	29.7
					Total	59.0	100.0

6.2 No-Load Loss Classification with Decision Trees and Artificial Neural Networks

have CSR equal to 100%, while it should also be noted that both nodes 4 and 7 are of leaf type. As can be seen from Table 6.4, the total classification success rate of the whole decision tree is 94.3%, since it correctly classifies the 181 of the 192 measurement sets of the test set.

Table 6.5 presents calculations of the information of the attributes of the decision tree of Fig. 6.5. The notation used for the columns of Table 6.5 is as follows: H_C is the prior classification entropy, H_C^T is the posterior classification entropy, I_C^T is the information, N is the number of measurement sets of the learning set that fall into the node under consideration, and $N \cdot I_C^T$ is the total information of each test node. Table 6.5 shows that the total information of the decision tree is 59.0. Among the three test attributes, ATTR2 has the highest information, i.e., 28.7, corresponding to 48.6% of the total information of the decision tree.

The decision tree of Fig. 6.5 is composed of three test nodes and four terminal nodes and has automatically selected only three of the eight candidate attributes of Sect. 6.2.2.3. These three attributes, appearing at the test nodes of Fig. 6.5, are ATTR8, ATTR2, and ATTR7. Parameter ATTR8 reflects the quality of the material, as it is equal to the specific no-load losses (W/kg at 15000 Gauss) of core magnetic material. Parameter ATTR2 represents the temperature rise time of the annealing cycle, while parameter ATTR7 expresses the ratio of actual core weight to theoretical core weight.

The selection of these attributes by the decision tree of Fig. 6.5 is reasonable and expected, since they are all related to the quality of the individual core. It is notable that the only variable relevant to the annealing cycle that appears in the node splitting tests of the decision tree is ATTR2. This is due to the fact that ATTR2, ATTR4, and also the duration of the slow and fast cooling stages are strongly correlated, since the total annealing time is considered constant. On the other hand, ATTR5, which denotes the position of core in the furnace, is not important.

Based on the decision tree of Fig. 6.5, four rules are derived, one per decision tree terminal node. These rules are useful, since they will help the transformer production engineers to maximize the number of individual cores with no-load losses belonging to the acceptable class. In particular, it is desirable to construct individual cores leading to nodes 4 and 7 of the decision tree of Fig. 6.5, if it is technically and economically feasible, since these nodes have acceptability index greater than 98%.

Figure 6.5 shows that the measurement sets following the rule ATTR8>0.7 and ATTR7≤0.98 lead to node 6 and are characterized as non-acceptable. In order to avoid this, the transformer production engineers must increase ATTR7. This is equivalent to increasing the actual weight of the individual core by adding more magnetic material, so that the ratio of actual core weight to theoretical core weight (ATTR7) is greater than 0.98. This is also equivalent to monitoring the weight of each individual core and taking care that the actual weight of each individual core is at least equal to 98% of its theoretical weight (Hatziargyriou et al. 1998a).

Given the quality of the magnetic material (ATTR8), the most suitable annealing experiment can be selected as follows (Georgilakis et al. 1998a):

1. If ATTR8≤0.7, the annealing experiment 7 of Table 6.2 must be selected. The reason is that, since ATTR8≤0.7, it is desirable to lead to node 4, since it has higher acceptability index than terminal node 5. From the splitting rule of node 2 it can be derived that ATTR2 must be set to its Low value. Additionally, from Table 6.3 it can be seen that the best annealing cycles are those of experiments 6 and 7, which lead to 98% acceptable cores. From these two experiments, only experiment 7 has the ATTR2 equal to Low (see Table 6.2). At first sight, this result seems unexpected. However, this can easily be explained given that the duration of the total annealing cycle is considered constant for all the annealing experiments. The total annealing cycle includes not only the temperature rise time, but also the duration of constant temperature and the slow and fast cooling stages.

2. If ATTR8>0.7, the standard annealing cycle must be selected. The reason is that the splitting rule of node 3 does not depend on annealing variables (ATTR1 to ATTR6).

Taking into account that the transformer manufacturer follows only one annealing cycle, the above analysis leads to the conclusion that the best annealing cycle is the one of experiment 7.

6.2.2.6 Multi-layer Perceptrons

This section studies the classification of no-load losses of individual cores using a fully-connected multi-layer perceptron (MLP) with one input layer, one or two hidden layers, and one output layer. The input layer is composed of eight neurons corresponding to the eight attributes (ATTR1 to ATTR8) of Sect. 6.2.2.3. The output layer has two neurons: one corresponds to the acceptable and the other to the non-acceptable class.

The determination of the optimum number of hidden layers as well as the optimum number of hidden neurons was done by trial and error. In particular, we started with a small number of hidden neurons and this number was progressively increased (Georgilakis et al. 1999b). For each one of the different neural network architectures, the neural network was trained and afterwards was tested by measuring its classification success rate on the unknown test set of 192 measurement sets. Table 6.6 shows the CSR on the test set for 12 trials. It is concluded from Table 6.6 that the optimum MLP architecture is 8-6-2, i.e., eight input neurons, six hidden neurons and two output neurons, because this architecture presents the highest CSR on the test set, i.e., 96.9%.

As activation function for the hidden and output neurons, the following form of the sigmoid function was used:

6.2 No-Load Loss Classification with Decision Trees and Artificial Neural Networks

Table 6.6 CSR on test set versus MLP architecture

Trial	MLP architecture	CSR (%)
1	8-3-2	94.8
2	8-4-2	95.8
3	8-6-2	96.9
4	8-7-2	95.3
5	8-8-2	93.2
6	8-9-2	91.7
7	8-10-2	90.6
8	8-4-2-2	95.3
9	8-6-3-2	94.8
10	8-8-4-2	93.8
11	8-8-6-2	91.1
12	8-10-5-2	90.1

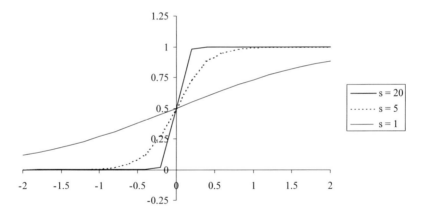

Fig. 6.6 Sigmoid functions for different values of the sigmoid slope s

$$f(x) = \frac{1}{1+e^{-s \cdot x}}, \quad (6.1)$$

where s is the *sigmoid slope*, i.e., the slope of the sigmoid function. It should be noted that if we set $s=1$ in (6.1), we obtain the sigmoid function of (4.37). Sigmoid functions for different values of the sigmoid slope s are plotted in Fig. 6.6.

Table 6.7 CSR on test set for 8-6-2 MLP architecture versus sigmoid slope

Sigmoid slope s	CSR (%)
20	93.8
10	95.3
5	95.8
2	96.4
1	96.9
0.1	95.8

Table 6.8 Configuration and classification results of the optimum MLP for the classification of no-load losses of individual cores

Parameter	Value
Input neurons	8
Hidden neurons	6
Output neurons	2
Slope of sigmoid function	1
Learning rate	0.3
Momentum	0.5

Result	Value
Classification success rate on test set (%)	96.9

Table 6.7 presents the impact of the value of the sigmoid slope on test set classification success rate provided by a trained MLP with architecture 8-6-2, i.e., the optimum architecture of Table 6.6. It can be seen from Table 6.7 that the best CSR corresponds to $s = 1$.

Finally, the impact of the learning rate (lr) and momentum (m) on the CSR was studied, while keeping the remaining MLP parameters at their optimal values, i.e., network architecture 8-6-2 and $s = 1$. The conclusion was that the optimum CSR, i.e., 96.9%, was obtained for $lr = 0.3$ and $m = 0.5$.

Table 6.8 presents the configuration and the classification results of the optimal MLP for the classification of no-load losses of individual cores.

6.2.2.7 Entropy Networks

Starting from the decision tree of Fig. 6.5 designed for solution of the individual core no-load loss classification problem, and using the rules presented in Sect. 4.7, we obtain the entropy network of Fig. 6.7, designed for the solution of the same classification problem (Georgilakis and Hatziargyriou 2002).

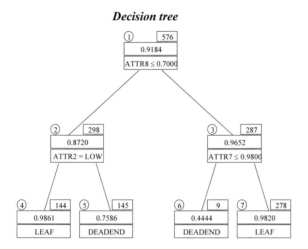

Correspondence between decision tree nodes and entropy network neurons

Decision tree node	1	2	3	4	5	6	7
Entropy network neuron	TL1	TL2	TL3	AL1	AL2	AL3	AL4

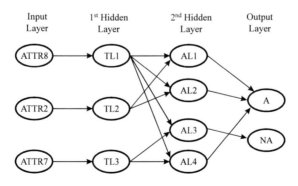

Fig. 6.7 Entropy network for the solution of individual core no-load loss classification problem

Table 6.9 Comparison of methods for the classification of no-load losses of individual cores based on the classification success rate on the test set

Method	Architecture	CSR on test set (%)
Decision tree		94.3
Multi-layer perceptron 1	8-6-2	96.9
Entropy network	3-3-4-2	94.8
Hybrid decision tree-neural network classifier	3-3-4-1	95.8
Multi-layer perceptron 2	3-6-2	94.8

It can be seen from Fig. 6.7 that the entropy network has three input neurons, three neurons in the first hidden layer, four neurons in the second hidden layer and two output neurons. The entropy network of Fig. 6.7 has only 15 connections, while the corresponding fully connected MLP with architecture 3-3-4-2 has 29 connections $(3 \cdot 3 + 3 \cdot 4 + 4 \cdot 2 = 29)$. Moreover, the fully connected MLP of Sect. 6.2.2.6 with architecture 8-6-2 has 60 connections, in comparison with only 15 connections of the entropy network.

It is interesting to train the entropy network with different values of the sigmoid slope s. If a high value of s is selected, e.g., $s = 20$, the entropy network copies the discrete information of the decision tree, since for $s = 20$ the majority of output values are either near to zero (for negative input values) or near to one (for positive input values), as Fig. 6.6 shows. Similarly, an output value close to one is interpreted as acceptable by the decision tree, while an output value close to zero is of non-acceptable class. Having selected $s = 20$, we train the entropy network and next we test the trained entropy network on the unknown test set and find that the CSR is 94.3%. If we repeat the same procedure for $s = 5$, we find that the CSR is 94.8%.

We also test the hybrid decision tree-neural network classifier (HDTNNC), which is created from the entropy network if its two output neurons are substituted by one output neuron that is fully connected with all the neurons of the second hidden layer. The trained HDTNNC with architecture 3-3-4-1 provides 95.8% CSR.

6.2.2.8 Synthesis

Table 6.9 compares all methods tested for the classification of no-load losses of individual cores based on the CSR on the test set. The last row of Table 6.9 presents the CSR obtained from a fully connected MLP with three input neurons corresponding to the three attributes selected by the decision tree of Fig. 6.5 and with architecture 3-6-2 computed by trial and error, according to the methodology presented in Sect. 6.2.2.6. Table 6.9 shows that the best performing method is the fully connected MLP with architecture 8-6-2, which provided 96.9% CSR on the unknown test set. It is also important to mention that the decision tree is the only method that provides if-then-else rules, which are very useful for transformer design engineers. Moreover, the decision tree automatically selected the three most important attributes among the eight candidates.

6.2.3 Transformer

6.2.3.1 Database

The no-load losses of the assembled transformer are affected by magnetic material properties (Moses 1992; Georgilakis et al. 2001b; Kalokiris et al. 2007), as well as by design and production factors (Godec 1977; Valkovic 1982; Moses 1984; Valkovic 1984; Nakata et al. 1984; Fecich and Balmer 1985; Ling et al. 1992; Basak and Bonyar 1992; Valkovic and Rezic 1992; Ilo et al. 1996; Girgis et al. 1998; Mechler and Girgis 2000; Soda and Enokizono 2000; teNyenhuis et al. 2001).

After sufficient experimentation on the selection of candidate attributes, construction of many decision trees for different risk levels, and selection of the optimum decision trees, the following conclusions can be drawn (Georgilakis 2000):

1. If the supplier, grade and thickness of magnetic material are considered as candidate attributes, then the resulting decision trees have very low classification success rate. On the other hand, if we construct one decision tree per different supplier, grade and thickness of magnetic material, then the CSR results are significantly improved.

2. If, except for the above grouping per supplier, grade and thickness of magnetic material, the measurement sets are also grouped per transformer rated power (kVA), then the CSR of the decision tree is reduced.

Based on the above, it was decided that the best solution is the construction of one decision tree per different supplier, grade and thickness of magnetic material that define a different environment (Georgilakis 2000). Table 6.10 presents the three different environments considered (Georgilakis 2000). For example, environment 1 is characterized by magnetic material of grade M3, according to USA AISI 1983, thickness 0.23 mm, while the supplier of material was A. Decision trees were constructed using many different lists of candidate attributes. Two of the most representative attribute lists are presented in Tables 6.11 and 6.12 (Hatziargyriou et al. 1998b; Georgilakis and Hatziargyriou 1999, 2002; Georgilakis et al. 1999b, 1999c, 1999d).

Table 6.10 Environments considered

	Environment		
Characteristic	1	2	3
Supplier	A	B	A
Steel grade	M3	M4	Hi-B
Thickness (mm)	0.23	0.27	0.23

Table 6.11 List of nine candidate attributes

Parameter	Description
ATTR1	$ASFL_{TF}/DSFL_{TF}$
ATTR2	AKg_{TF}/DKg_{TF}
ATTR3	$(WPK_{"11",mat,a}+WPK_{"12",mat,a}+WPK_{"13",mat,a}+WPK_{"14",mat,a})/4$
ATTR4	Rated magnetic induction
ATTR5	Thickness of core leg
ATTR6	Width of core leg
ATTR7	Height of core window
ATTR8	Width of core window
ATTR9	Transformer volts per turn

In Table 6.11, the attribute ATTR1 represents the ratio of actual ($ASFL_{TF}$) to theoretical ($DSFL_{TF}$) total no-load losses of the four individual cores. The attribute ATTR2 represents the ratio of actual (AKg_{TF}) to theoretical (DKg_{TF}) total weight of the four individual cores. Finally, the attribute ATTR3 represents the average specific no-load losses of the magnetic material of the four individual cores, where $WPK_{"11",mat,a}$ denotes the specific no-load losses (W/kg) at 15000 Gauss of the magnetic material of the individual core that is placed at position "11" shown in Fig. 6.1.

In addition to the variables used in Table 6.11 and defined previously, Table 6.12 uses some more variables. In particular, the parameter $WPK_{"11",mat,b}$ denotes the specific no-load losses (W/kg) at 17000 Gauss of the magnetic material of the individual core that is placed at position "11" shown in Fig. 6.1. The variable $AWPK_{"11"}$ represents the actual specific no-load losses (W/kg) of the individual core at place "11", i.e., the ratio of actual no-load losses of the individual core at place "11" to its actual weight. The variable $DWPK_{"11"}$ represents the theoretical specific no-load losses of the individual core at place "11" calculated from the no-load loss curve of the individual core. The parameter $AKg_{"11"}$ represents the actual weight of the individual core at place "11", while the variable $DKg_{"11"}$ represents the theoretical weight of the individual core at place "11".

6.2 No-Load Loss Classification with Decision Trees and Artificial Neural Networks 283

Table 6.12 List of 19 candidate attributes

Parameter	Description
ATTR1	Rated magnetic induction
ATTR2	$(WPK_{``11",mat,a}+WPK_{``12",mat,a}+WPK_{``13",mat,a}+WPK_{``14",mat,a})/4$
ATTR3	$(WPK_{``11",mat,b}+WPK_{``12",mat,b}+WPK_{``13",mat,b}+WPK_{``14",mat,b})/4$
ATTR4	AKg_{TF}/DKg_{TF}
ATTR5	$ASFL_{TF}/DSFL_{TF}$
ATTR6	$AWPK_{``11"}/DWPK_{``11"}$
ATTR7	$AWPK_{``12"}/DWPK_{``12"}$
ATTR8	$AWPK_{``13"}/DWPK_{``13"}$
ATTR9	$AWPK_{``14"}/DWPK_{``14"}$
ATTR10	$AKg_{``11"}/DKg_{``11"}$
ATTR11	$AKg_{``12"}/DKg_{``12"}$
ATTR12	$AKg_{``13"}/DKg_{``13"}$
ATTR13	$AKg_{``14"}/DKg_{``14"}$
ATTR14	$(AWPK_{``11"}+AWPK_{``12"})/(DWPK_{``11"}+DWPK_{``12"})$
ATTR15	$(AWPK_{``12"}+AWPK_{``13"})/(DWPK_{``12"}+DWPK_{``13"})$
ATTR16	$(AWPK_{``13"}+AWPK_{``14"})/(DWPK_{``13"}+DWPK_{``14"})$
ATTR17	$(AWPK_{``11"}+AWPK_{``12"})/(AWPK_{``12"}+AWPK_{``13"})$
ATTR18	$(AWPK_{``12"}+AWPK_{``13"})/(AWPK_{``13"}+AWPK_{``14"})$
ATTR19	$(AWPK_{``12"}+AWPK_{``13"})/(AWPK_{``11"}+AWPK_{``14"})$

In total, 2595, 2225, and 2385 measurement sets were collected for the environments 1, 2, 3, respectively (the environments are defined in Table 6.10). The measurement sets of each environment were split into two independent sets: the learning set and the test set. More specifically, the learning sets for the environments 1, 2, 3 are composed of 1730, 1485, and 1590 measurement sets, respectively, and the test sets are composed of the remaining 865, 740, and 795 measurement sets, respectively.

6.2.3.2 Classification Criterion

It is required to classify the no-load losses of the assembled transformer into two classes: acceptable and non-acceptable no-load losses. The classification criterion is the following: one assembled transformer belongs to the non-acceptable class if its actual specific no-load losses are greater than 10% of its theoretical specific no-load losses. Otherwise, the assembled transformer belongs to the acceptable class.

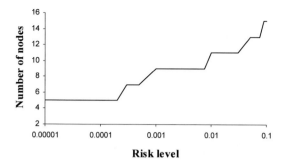

Fig. 6.8 Variation of decision tree size (number of nodes) with risk level

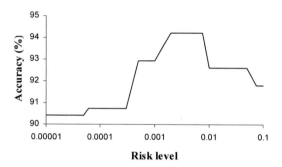

Fig. 6.9 Variation of decision tree accuracy with risk level

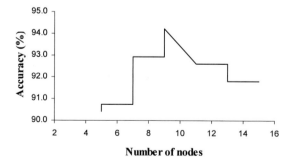

Fig. 6.10 Variation of decision tree accuracy with number of nodes

6.2 No-Load Loss Classification with Decision Trees and Artificial Neural Networks

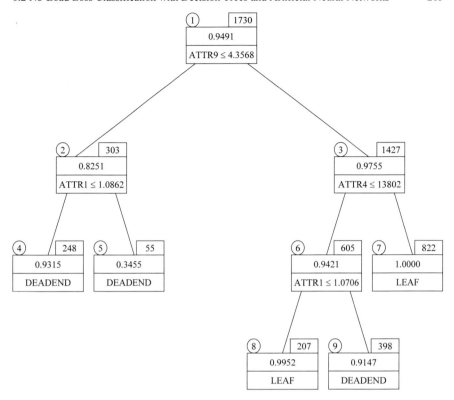

Fig. 6.11 Optimim decision tree with 0.25% risk level for the classification of no-load losses of assembled transformers for environment 1 using the nine-attribute list of Table 6.11

6.2.3.3 Decision Trees

Based on the learning set of 1730 measurement sets for the environment 1 (Table 6.10) and the list of nine candidate attributes (Table 6.11) many decision trees were constructed for risk levels from 0.001% to 10%. The accuracy of these decision trees was evaluated using the independent test set of 865 measurement sets. Figure 6.8 shows the variation of the number of decision tree nodes as a function of the risk level. Figures 6.9 and 6.10 present the classification accuracy results of the decision trees as a function of the risk level and the number of nodes, respectively. Applying the methodology presented in Sect. 6.2.2.5 for the analysis of Figs. 6.8 to 6.10 we conclude that the optimal decision trees for environment 1 have nine nodes and are built with risk levels from 0.2% to 0.75%. Such a decision tree is shown in Fig. 6.11, built with 0.25% risk level (Georgilakis et al. 1997; Georgilakis and Hatziargyriou 1999).

Table 6.13 Calculation of the classification success rate of the decision tree of Fig. 6.11

Node number	Node type	Acceptability index	Tested MS	Correctly classified MS	Classification success rate (%)
4	DEADEND	0.9315	123	113	91.9
5	DEADEND	0.3455	14	7	50.0
7	LEAF	1.0000	414	405	97.8
8	LEAF	0.9952	106	103	97.2
9	DEADEND	0.9147	208	187	89.9
Total			865	815	94.2

The decision tree of Fig. 6.11 consists of four test and five terminal nodes, and has automatically selected only three attributes among the nine candidates of Table 6.11. These attributes are ATTR9, ATTR1 and ATTR4. ATTR9 corresponds to transformer volts per turn, ATTR1 is the ratio of actual to theoretical total no-load losses of the four individual cores, and ATTR4 represents the rated magnetic induction. The selection of these attributes is reasonable and expected, since they are all related to transformer no-load losses.

Each terminal node of the decision tree of Fig. 6.11 produces one decision rule, on the basis of its acceptability index. For example, from terminal node 7 the following rule is derived: if ATTR9>4.3568 and ATTR4>13802, then transformer specific no-load losses are of acceptable quality. Consequently, based on the decision tree of Fig. 6.11, rules useful for the design (parameters ATTR4 and ATTR9) and also for core production (parameter ATTR1) can be derived. In particular (Georgilakis and Hatziargyriou 1999):

1. It is desirable to construct transformers leading to nodes 7, 8, and 4, if it technically and economically feasible. These nodes have acceptability indices greater than 93%.

2. The measurement sets following the rule ATTR9≤4.3568 and ATTR1>1.0862 lead to node 5, and are characterized as non-acceptable. In order to avoid this, ATTR1 must be reduced during transformer construction. The method is to reduce the actual total no-load losses of the four individual cores of the transformer by removing from the transformer cores set one or more individual cores with high no-load losses and adding individual cores with lower no-load losses.

3. The measurement sets following the rule ATTR9>4.3568 and ATTR4>13802 lead to node 7, and are characterized as acceptable. This is equivalent to increasing the volts per turn (ATTR9), and also increasing the rated magnetic induction (ATTR4). Transformer design engineers determine both these parame-

6.2 No-Load Loss Classification with Decision Trees and Artificial Neural Networks

ters. In fact, the rated magnetic induction offers enough flexibility, therefore it is desirable to design transformers leading to this node, if it is technically and economically feasible.

Table 6.13 presents the classification success rate of each terminal node of the decision tree of Fig. 6.11. It can be seen from Table 6.13 that the total CSR of the decision tree of Fig. 6.11 is 94.2%. In order to increase the CSR of the decision tree methodology, alternative attributes were selected and tested, and finally the 19-attribute list of Table 6.12 was selected.

Many decision trees were built for environment 1 using the 19-attribute list of Table 6.12 and various risk levels and the optimum decision tree is shown in Fig. 6.12. Table 6.14 shows that the decision tree of Fig. 6.12 provides 95.3% CSR on the test set, i.e., for the classification of no-load losses of assembled transformers for the environment 1, the decision tree of Fig. 6.12 that was built with the 19-attribute list provides 1.1% higher accuracy (CSR) in comparison with the accuracy of the decision tree of Fig. 6.11 that was built with the nine-attribute list.

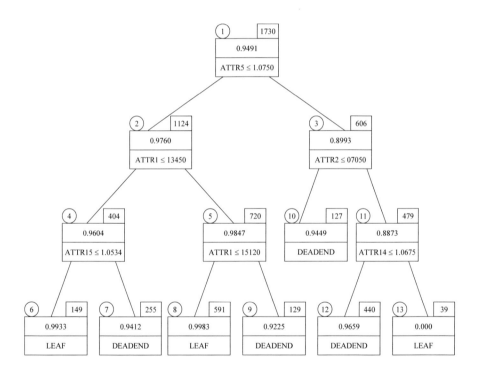

Fig. 6.12 Optimum decision tree for the classification of no-load losses of assembled transformers for environment 1 using the 19-attribute list of Table 6.12

Table 6.14 Calculation of the classification success rate of the decision tree of Fig. 6.12

Node number	Node type	Acceptability index	Tested MS	Correctly classified MS	Classification success rate (%)
6	LEAF	0.9933	74	70	94.6
7	DEADEND	0.9412	127	118	92.9
8	LEAF	0.9983	294	287	97.6
9	DEADEND	0.9225	68	64	94.1
10	DEADEND	0.9449	64	61	95.3
12	DEADEND	0.9659	215	208	96.7
13	LEAF	0.0000	23	16	69.6
Total			865	824	95.3

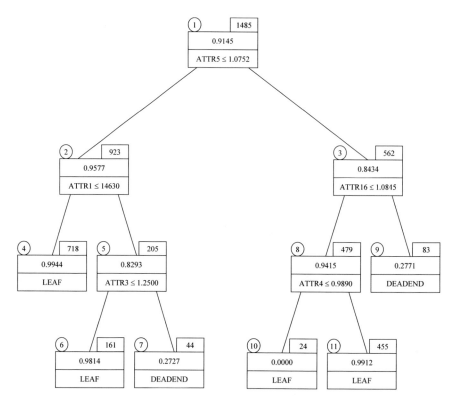

Fig. 6.13 Optimum decision tree with 95.7% CSR on the test set designed for the classification of no-load losses of assembled transformers for the environment 2 using the 19-attribute list of Table 6.12

6.2 No-Load Loss Classification with Decision Trees and Artificial Neural Networks   289

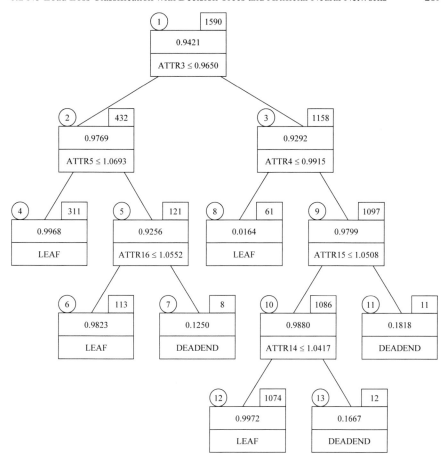

Fig. 6.14 Optimum decision tree with 96.1% CSR on the test set designed for the classification of no-load losses of assembled transformers for environment 3 using the 19-attribute list of Table 6.12

Using the same method, decision trees were also built for environments 2 and 3 of Table 6.10 using the nine-attribute list of Table 6.11 as well as the 19-attribute list of Table 6.12. In both environments, the decision trees with the best accuracy were built on the basis of the 19-attribute list. Figures 6.13 and 6.14 show the optimum decision trees for environments 2 and 3, respectively. The decision tree of Fig. 6.13 provides 95.7% CSR on the test set. The decision tree of Fig. 6.14 provides 96.1% CSR on the test set.

6.2.3.4 Multi-layer Perceptrons

Applying the methodology presented in Sect. 6.2.2.6, several MLP architectures (one and two hidden layers, and various neurons per hidden layer) as well as various configuration parameters (sigmoid slope, learning rate, momentum) were tested using the nine-attribute and the 19-attribute lists so as to determine the optimal MLP configuration for each of the three environments of Table 6.10. Table 6.15 presents the results of this investigation, i.e., the optimal MLP configuration for the classification of no-load losses of assembled transformers for all the environments (Georgilakis et al. 1999c, 1999d). Table 6.15 shows that in all cases the 19-attribute list gave the highest CSR on the test set. In all cases, one hidden layer was enough.

6.2.3.5 Entropy Networks

Applying the methodology presented in Sect. 6.2.2.7, entropy networks were trained for the classification of no-load losses of assembled transformers for all the environments (Georgilakis et al. 1999c; Georgilakis and Hatziargyriou 2002). Table 6.16 presents the architecture and the classification results of the optimum entropy networks for the three environments.

Table 6.15 Configuration and classification results of the optimum MLP for the classification of no-load losses of assembled transformers

Environment	1	2	3
Parameter			
Input neurons	19	19	19
Hidden neurons	5	6	4
Output neurons	2	2	2
Slope of the sigmoid function	1	1	1
Learning rate	0.1	0.3	0.3
Momentum	0.5	0.5	0.4
Result			
Classification success rate on test set (%)	97.5	97.6	97.9

6.2 No-Load Loss Classification with Decision Trees and Artificial Neural Networks

Table 6.16 Architecture and classification results of the optimum entropy networks for the classification of no-load losses of assembled transformers

Environment	1	2	3
Architecture			
Entropy network	5-6-7-2	5-5-6-2	6-6-7-2
Hybrid decision tree-neural network classifier	5-6-7-1	5-5-6-1	6-6-7-1
Classification success rate on test set (%)			
Entropy network	95.8	96.1	96.6
Hybrid decision tree-neural network classifier	96.6	96.8	97.4

6.2.3.6 Synthesis

Table 6.17 compares all methods tested for the classification of no-load losses of assembled transformers based on the CSR on the test set.

The last row of Table 6.17 presents the CSR obtained from a fully connected MLP with number of inputs equal to the number of attributes selected by the decision tree and with number of hidden neurons computed by trial and error, according to the methodology presented in Sect. 6.2.2.6. In particular, this MLP (denoted multi-layer perceptron 2 in Table 6.17) has:

1. Five input neurons for environment 1, corresponding to the five attributes automatically selected by the decision tree of Fig. 6.12

2. Five input neurons for environment 2, corresponding to the five attributes automatically selected by the decision tree of Fig. 6.13

3. Six input neurons for environment 3, corresponding to the six attributes automatically selected by the decision tree of Fig. 6.14

Table 6.17 shows that the best performing method is the fully connected MLP (denoted multi-layer perceptron 1 in Table 6.17) with 19 input neurons corresponding to the 19 attributes of Table 6.12. It is also important to mention that the decision tree is the only method that provides if-then-else rules, which are very useful for transformer design engineers. Moreover, the decision tree automatically selected the most important attributes among the candidate ones.

Table 6.17 Comparison of methods for the classification of no-load losses of assembled transformers based on the classification success rate on the test set

Environment	1	2	3
Architecture			
Decision tree			
Multi-layer perceptron 1	19-5-2	19-6-2	19-4-2
Entropy network	5-6-7-2	5-5-6-2	6-6-7-2
Hybrid decision tree-neural network classifier	5-6-7-1	5-5-6-1	6-6-7-1
Multi-layer perceptron 2	5-5-2	5-6-2	6-4-2
Classification success rate on test set (%)			
Decision tree	95.3	95.7	96.1
Multi-layer perceptron 1	97.5	97.6	97.9
Entropy network	95.8	96.1	96.6
Hybrid decision tree-neural network classifier	96.6	96.8	97.4
Multi-layer perceptron 2	96.0	96.2	96.9

6.3 No-Load Loss Forecasting with Artificial Neural Networks

6.3.1 Introduction

No-load loss forecasting aims at predicting the no-load losses. In this chapter, artificial neural network based no-load loss forecasting techniques will be presented and applied for both the individual core and the transformer.

It is very important to have an efficient no-load loss forecasting tool, i.e., tool with high forecasting accuracy, since it offers the ability to know in advance the no-load loss value of an individual core or transformer. It is necessary to determine the values of the attributes (input parameters) that are known at the early stages of individual core or transformer manufacturing and then to use the forecasting tool. Thus, the existence of an efficient no-load loss forecasting tool aims at predicting the no-load losses of individual cores and transformers as well as at investigating the impact that variation of the attribute values will have on no-load losses. Moreover, if an efficient no-load loss forecasting tool is combined with an optimization tool, transformers with actual no-load losses very close to the desired no-load losses can be constructed (Georgilakis et al. 1999a, 2001a). Thus, in a production batch with, say, 50 transformers of the same design, the application of an efficient no-load loss forecasting tool minimizes the possibility of paying no-load loss penalties. Moreover, if the forecasting tool is combined with an optimization tool, then the no-load losses of the 50 transformers will be minimum (Georgilakis et al. 1999a, 2001a).

Several methods have been proposed in the literature for the estimation of transformer no-load losses during the design phase. These approaches can be grouped into four main categories:

1. *Empirical methods* (Moses 2003). These methods are based on experimental observations. Experimental curves are usually extracted using a large number of measurements to investigate the effect of several transformer parameters on no-load losses. However, due to the continuous evolution both of technical characteristics of the magnetic materials and the design of cores, the experimental curves should be systematically reconstructed when data change.

2. *Analytical methods* (Jiles and Atherton 1986; Bertotti 1988; Semlyen and Rajakovic 1989; Rajakovic and Semlyen 1989a, 1989b; Sato and Sakaki 1990; Fiorello and Novikov 1990; Hatziargyriou et al. 1993; Elleuch and Poloujadoff 1996; Mork 1999; Annakkage et al. 2000; Proussalidis et al. 2001; Elleuch and Poloujadoff 2003). These methods model three-phase transformers based on the equivalent magnetic circuit of the transformer cores. These methods are based on semiempirical description of various components of no-load losses (hysteresis losses, classical eddy-current losses, and excess losses) that are functions of frequency and maximum flux density. No-load losses result in the definition of a resistance that is added to the general model of the transformer.

3. *Numerical methods* (Basak et al. 1994; Enokizono et al. 1995; Kanada et al. 1996; Enokizono and Soda 1997; Mechler and Girgis 1998; Moses 1998; Enokizono and Soda 1998; Enokizono and Soda 1999; teNyenhuis et al. 2000; Im et al. 2001; Tatis et al. 2004; Kaimori et al. 2007; Rovolis et al. 2007; Kefalas et al. 2008). These methods are based on the arithmetic analysis of the electromagnetic field of the transformer cores. Finite elements and finite difference methods are mainly used. The potentials of the electromagnetic fields are calculated by creating mesh models of the transformer geometry and using several field parameters, such as the magnetic flux distribution. This analysis is very important during the transformer design phase, when the manufacturer needs to check the correctness of the transformer drawings.

4. *Artificial intelligence methods* (Nussbaum et al. 1996; Georgilakis et al. 1998b, 1999a, 2001a; Nussbaum et al. 2000; Miti et al. 2003). Artificial neural networks are used to predict no-load losses as a function of core design parameters. These models can be successful only in case that representative and accurate data are used for training artificial neural networks.

6.3.2 Forecasting Accuracy

6.3.2.1 Individual Core

In the individual core problem, various neural networks are trained to predict the specific no-load losses of the individual core. After training the neural network, its forecasting accuracy is evaluated based on the *average absolute relative error* (AARE) on the test set, which is defined as follows:

$$AARE = \sum_{i=1}^{N} \frac{\left|S_i^a - S_i^p\right|}{S_i^a} \cdot 100\%, \qquad (6.2)$$

where S_i^a are the actual specific no-load losses of the ith individual core, S_i^p are the forecasted specific no-load losses of the ith individual core, and N is the number of measurement sets of the test set.

6.3.2.2 Transformer

In the transformer problem, various neural networks are trained to predict the specific no-load losses of the transformer. After training the neural network, its forecasting accuracy is evaluated based on the average absolute relative error on the test set defined by (6.2), where S_i^a are the actual specific no-load losses of the ith transformer, S_i^p are the forecasted specific no-load losses of the ith transformer, and N is the number of measurement sets of the test set.

6.3.3 Individual Core

6.3.3.1 Multi-layer Perceptrons

Applying the methodology presented in Sect. 6.2.2.6, several MLP architectures (one and two hidden layers, and various neurons per hidden layer) as well as various configuration parameters (sigmoid slope, learning rate, momentum) were tested using the eight-attribute list of Sect. 6.2.2.3 to determine the optimal MLP configuration, i.e., the MLP with the minimum average absolute relative error on the test set. Table 6.18 presents the results of this investigation, i.e., the optimal MLP configuration for the prediction of no-load losses of individual cores (Georgilakis et al. 1999b). The optimal MLP of Table 6.18 provides 2.32% average absolute relative error on the unknown test set.

6.3 No-Load Loss Forecasting with Artificial Neural Networks

Table 6.18 Configuration and forecasting accuracy of the optimum MLP for the prediction of no-load losses of individual cores

Parameter	Value
Input neurons	8
Hidden neurons	7
Output neurons	1
Slope of the sigmoid function	1
Learning rate	0.1
Momentum	0.5

Result	Value
AARE on test set (%)	2.32

Figure 6.15 presents the fractile diagram or the Q–Q plot (quantile–quantile) (Kobayashi 1981) of the specific no-load losses. According to this method, the data of actual specific no-load losses are plotted against the predicted values. Perfect prediction lies on a line of 45° slope. It is observed that prediction of the individual core specific no-load losses, based only on rated magnetic induction and ignoring all other parameters (current practice), provides a constant (equal to 0.78 W/kg) estimate for all measurement sets belonging to the test set. This occurs since, as mentioned in Sect. 6.2.2.3, all the individual cores have been constructed from the same design (i.e., same magnetic induction) and the same magnetic material (i.e., same no-load loss curve), so the specific no-load losses obtained by the no-load loss curve are the same for all the individual cores of the database (learning set and test set). Therefore, the estimate of the current practice significantly diverges from the optimal line of 45°, providing an erroneous prediction, especially at large or small actual W/kg values in the range 0.5925 to 0.9433 W/kg. On the contrary, the MLP method is able to accurately estimate the no-load losses of individual cores for all the test samples, due to the neural network learning capabilities. The maximum absolute relative error is 23.7% for the current practice (no-load loss curve), while the respective error for the MLP method is 4.9%. The average absolute relative error is 7.8% for the current practice and 2.32% for the MLP method. It is observed that the MLP architecture gives much better results as far as the average absolute relative error and the maximum absolute relative error (worst case error) are concerned.

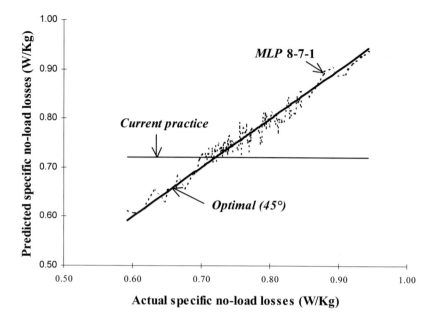

Fig. 6.15 Prediction of specific no-load losses of the individual core using the typical no-load loss curve (current practice) and the MLP of Table 6.18

6.3.3.2 Hybrid Multi-layer Perceptrons

Two different hybrid multi-layer perceptrons will be used to predict the no-load losses of the individual core:

1. The *hybrid entropy network*. This network results from the entropy network of Fig. 6.7 with architecture 3-3-4-2 by substituting the two output neurons with one single neuron that will predict the specific no-load losses of the individual core. Moreover, the single output neuron of the hybrid entropy network is fully connected with all neurons of the second hidden layer. It is concluded from the above that the architecture of the hybrid entropy network is 3-3-4-1. When trained, this network presents 2.36% AARE on the test set.

2. The *hybrid decision tree network*. This network has as inputs only the three attributes selected by the decision tree of Fig. 6.5. The number of hidden layers and the number of neurons per hidden layer are calculated by trial and error to minimize the AARE on the test set. When trained, this network has 2.39% AARE on the test set.

6.3 No-Load Loss Forecasting with Artificial Neural Networks   297

Table 6.19 Comparison of methods for the prediction of no-load losses of individual cores based on the AARE on the test set

Method	Architecture	AARE on test set (%)
Multi-layer perceptron	8-7-1	2.32
Hybrid entropy network	3-3-4-1	2.36
Hybrid decision tree network	3-8-1	2.39

Table 6.20 Configuration and forecasting accuracy results of the optimum MLP for the prediction of no-load losses of assembled transformers

Environment	1	2	3
Parameter			
Input neurons	19	19	19
Hidden neurons	5	6	4
Output neurons	1	1	1
Slope of the sigmoid function	1	1	1
Learning rate	0.1	0.3	0.3
Momentum	0.5	0.5	0.4
Result			
AARE on test set (%)	1.63	1.52	1.74

6.3.3.3 Synthesis

Table 6.19 compares all methods tested for the prediction of no-load losses of individual cores based on the AARE on the test set. Table 6.19 shows that the best performing method is the fully connected MLP with architecture 8-7-1, which provided 2.32% AARE on the unknown test set. This is a significant improvement in comparison with the current practice (no-load loss curve), which presents 7.8% AARE on the test set.

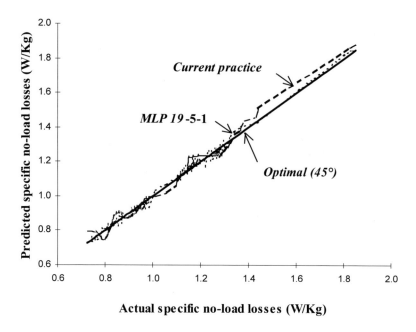

Fig. 6.16 Prediction of specific no-load losses of the assembled transformer for environment 1 using the typical no-load loss curve (current practice) and the 19-5-1 MLP of Table 6.20

6.3.4 Transformer

6.3.4.1 Multi-layer Perceptrons

Applying the methodology presented in Sect. 6.2.2.6, several MLP architectures (one and two hidden layers, and various neurons per hidden layer) as well as various configuration parameters (sigmoid slope, learning rate, momentum) were tested using the nine-attribute (Table 6.11) and the 19-attribute (Table 6.12) lists to determine the optimal MLP configuration for each of the three environments of Table 6.10. Table 6.20 presents the results of this investigation, i.e., the optimal MLP configuration for the prediction of no-load losses of assembled transformers for all the environments (Georgilakis et al. 1999e; Doulamis et al. 2002). Table 6.20 shows that in all cases the 19-attribute list gave the lowest AARE on the test set. In all cases, one hidden layer was enough.

Figure 6.16 presents prediction of the specific no-load losses of the assembled transformer for environment 1 using the typical no-load loss curve (current prac-

6.3 No-Load Loss Forecasting with Artificial Neural Networks

tice) and the 19-5-1 MLP of Table 6.20. The maximum absolute relative error is 11.4% for the current practice (no-load loss curve), while the respective error for the MLP method is 4.5%. The average absolute relative error is 3.1% for the current practice and 1.6% for the MLP method. It is observed that the MLP architecture gives much better results as far as the average absolute relative error and the maximum absolute relative error (worst case error) are concerned.

6.3.4.2 Hybrid Multi-layer Perceptrons

Similar to Sect. 6.3.3.2, two hybrid multi-layer perceptrons will be used to predict the no-load losses of the assembled transformer: (1) the hybrid entropy network; and (2) the hybrid decision tree network. The results are shown in Table 6.21.

6.3.4.3 Synthesis

Table 6.22 compares all methods tested for the prediction of no-load losses of assembled transformers based on the AARE on the test set. Table 6.22 shows that the best performing method (lowest AARE on the test set) is the fully connected MLP for all the environments.

Table 6.21 Architecture and forecasting accuracy results of the optimum hybrid multi-layer perceptrons for the prediction of no-load losses of assembled transformers

Environment	1	2	3
Architecture			
Hybrid entropy network	5-6-7-1	5-5-6-1	6-6-7-1
Hybrid decision tree network	5-5-1	5-6-1	6-4-1
AARE on test set (%)			
Hybrid entropy network	1.66	1.56	1.78
Hybrid decision tree network	1.68	1.59	1.80

Table 6.22 Comparison of methods for the prediction of no-load losses of assembled transformers based on the AARE on the test set

Environment	1	2	3
Architecture			
Multi-layer perceptron	19-5-1	19-6-1	19-4-1
Hybrid entropy network	5-6-7-1	5-5-6-1	6-6-7-1
Hybrid decision tree network	5-5-1	5-6-1	6-4-1
AARE on test set (%)			
Multi-layer perceptron	1.63	1.52	1.74
Hybrid entropy network	1.66	1.56	1.78
Hybrid decision tree network	1.68	1.59	1.80

6.3.4.4 Retraining Neural Networks

Despite the good performance of the neural network in predicting no-load losses of assembled transformers, there are some cases after the completion of the transformer construction, where the prediction error is not acceptable. It should be mentioned that the error is calculated based on the difference between the predicted value, provided by the neural network output, and the actual value of the specific no-load losses, which is available after the completion of the transformer construction. In the case of not acceptable performance, the neural network is retrained using additional measurement sets. The retrained neural network will have improved performance (lower AARE) on future measurement sets.

In the experiments considered, we suppose that an AARE 10% above the average is the upper tolerance limit. That is, for environment 2 where the AARE for the initial test set (TS) is equal to 1.52% (Table 6.20), the acceptance limit (upper limit, UL) is 1.67%.

Figure 6.17 illustrates the AARE for 19 production batches during transformer construction. It is observed that at the 19th production batch, the AARE violates the defined threshold UL and the neural network is retrained. After retraining the neural network, the AARE on the TS is 1.14% and the new upper limit is set to 1.25% (i.e., 10% above the AARE). Figure 6.18 depicts the neural network prediction results (denoted NN-New on Fig. 6.18) obtained after retraining the neural network for the following 11 production batches. In all cases, the AARE was within the tolerance interval (Georgilakis et al. 1999e).

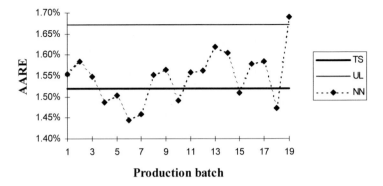

Fig. 6.17 Prediction error (AARE) provided by the 19-6-1 MLP of Table 6.20 for 19 production batches of environment 2

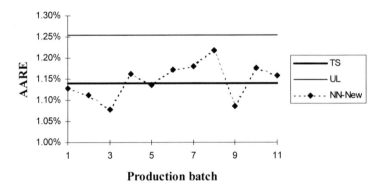

Fig. 6.18 Prediction error (AARE) provided by the retrained 19-6-1 MLP of Table 6.20 for 11 production batches of environment 2

6.4 Impedance Voltage Evaluation with Numerical Models

6.4.1 Introduction

The transformer impedance voltage or short-circuit impedance or leakage impedance is one of the most important technical characteristics of transformer. The impedance voltage consists of resistive component and reactive component (*leakage reactance*). Accurate estimation of transformer impedance voltage (short-circuit impedance) during the transformer design phase is crucial, since (1) it increases transformer reliability and manufacturer credibility, and (2) it reduces the material cost, since smaller impedance voltage design margin is used.

However, some of the existing design methodologies used by manufacturers still rely on leakage field and impedance voltage calculations that include gross approximations and assumptions and incorporate empirical factors in the transformer magnetic field simulation (Dymkov 1975; Mittle and Mittal 1996; Raitsios 2001). This approach is likely to result in significant deviations from the measured impedance voltage values, augmenting the risk of transformer failure and overstepping of the respective guaranteed values. It is therefore necessary to develop improved methods of leakage field evaluation, incorporable to the transformer design process. For this purpose, research efforts in this field focus on the use of advanced power transformer modeling techniques that take into account the constructional details of these devices. That is why numerical methods, mainly finite element models, have attracted the attention of researchers (Andersen 1973; Djurovic and Carpenter 1975; Tomczuk 1988; Zakrzewski and Kukaniszyn 1992; Kladas et al. 1994; Zakrzewski and Tomczuk 1996; Xiang et al. 1997; Kulkarni and Khaparde 2004; Tsili et al. 2004; Tsili 2005; Tsili et al. 2005, 2006). Other techniques that have been developed for evaluation of the impedance voltage include the reluctance network method (Turowski et al. 1990) and the surface magnetic charge method (Lu et al. 1998).

6.4.2 Finite Element Model

6.4.2.1 Field Equations

The finite element method is a numerical technique for the solution of problems described by partial differential equations. The governing equation in the case of a magnetostatic field is the Laplace equation:

$$\nabla^2 \Phi_m = 0, \tag{6.3}$$

where Φ_m is the scalar magnetic potential. The field considered is represented by a group of finite elements. The space discretization is realized by triangles or tetrahedra if the problem is two- or three-dimensional, respectively. Therefore, a continuous physical problem is converted into a discrete problem of finite elements with unknown field values in their vertices nodes. The solution of such a problem reduces to a system of algebraic equations and the field values inside the elements can be retrieved with the use of calculated values in their indices.

6.4 Impedance Voltage Evaluation with Numerical Models

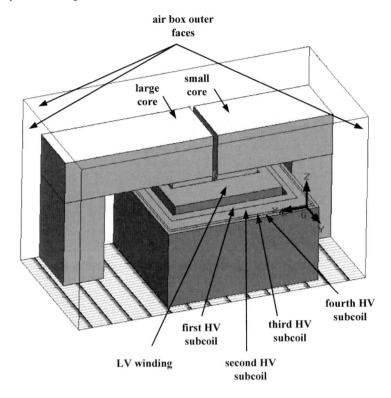

Fig. 6.19 Perspective view of the transformer one phase part modeled (simplified winding geometry with orthogonal approximation of winding corners)

Many scalar potential formulations have been developed for 3D magnetostatics, but they usually necessitate a prior source field calculation using Biot-Savart's law. This has the drawback of considerable computational effort.

In the case of impedance voltage evaluation, a particular scalar potential formulation has been developed, enabling 3D magnetostatic field analysis. According to this method, the magnetic field strength **H** is conveniently partitioned to a rotational and an irrotational part as follows:

$$\mathbf{H} = \mathbf{K} - \nabla \Phi , \qquad (6.4)$$

where Φ is a scalar potential extended all over the solution domain, while **K** is a vector quantity (fictitious field distribution), that satisfies the following conditions (Kladas and Tegopoulos 1992):

1. **K** is limited in a simply connected subdomain comprising the conductor

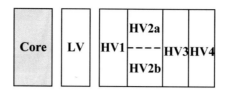

Fig. 6.20 Winding arrangement (yz-plane) for the production of dual primary voltage levels 20 kV and 15 kV

2. $\nabla \times \mathbf{K} = \mathbf{J}$ in the conductor and $\nabla \times \mathbf{K} = 0$ outside it

3. $\mathbf{K}$ is perpendicular on the subdomain boundary

The above formulation satisfies Ampere's law for an arbitrary contour in the subdomain.

6.4.2.2 Transformer Representation

The transformer under consideration is a 630 kVA, rated primary voltage 20 kV and 15 kV delta connected (dual primary voltage 20 and 15 kV), rated secondary voltage 400 V star connected, three-phase, wound core, oil-immersed, power transformer. The transformer magnetic circuit is of shell type and is assembled from two small and two large iron wound cores shown in Fig. 6.1. Figure 6.19 illustrates the perspective view of the transformer one phase part modeled.

The model of Fig. 6.19 comprises the low voltage (LV) and high voltage (HV) windings of one phase, as well as the small and large iron core that surrounds them. An air box, whose dimensions are equal to the transformer tank dimensions, surrounds the active part, therefore confining the field calculation to this domain. The xy-plane of the Cartesian coordinate system used is the transformer symmetry plane, and the z-axis crosses the cores symmetry plane. Due to the symmetries of the problem, the solution domain is reduced to one fourth of the device. These symmetries were taken into account by the imposition of Dirichlet boundary condition ($\Phi = 0$) along the xy-plane and Neumann boundary condition ($\frac{\partial \Phi}{\partial n} = 0$) along the yz-plane, xz-plane and the three outer faces of the air box.

The use of this one-phase model instead of the whole three-phase transformer model was implemented for the following reasons:

1. The smaller model size enables the construction of a denser tetrahedral finite element mesh without great computational cost (given that the exact representation of the transformer magnetic field requires great accuracy which is dependent on the mesh density and the total execution time of the finite element calculations).

6.4 Impedance Voltage Evaluation with Numerical Models

2. The representation of one phase of the active part does not affect the accuracy of the calculation of the equivalent circuit parameters.

The HV winding is divided into four subcoils. This division models the winding arrangement that produces the second primary voltage level, shown in Fig. 6.20. The second one of the four HV subcoils consists of two sections (HV2a, HV2b) with the same number of turns. When these sections are connected in parallel and then in series with the rest of the HV subcoils, the lower voltage level (15 kV) is obtained. For the production of the higher voltage rating (20 kV), the two sections (HV2a, HV2b) are connected in series and then in series with the rest of the HV subcoils. Hence, in the case of the first primary voltage level (20 kV) the nominal current is considered to flow through all the subcoils, while in the second one (15 kV), the current of subcoil HV2 (whose sections HV2a and HV2b are connected in parallel) is half of the current flowing through subcoils HV1, HV3 and HV4.

In the FEM model presented in the next sections, the magnetic nonlinearity as well as the magnetic anisotropy of the iron cores are ignored. This assumption is justified on the basis that flux densities during short-circuit are very low, therefore confining the transformer operation below the saturation region of the magnetization curve.

6.4.2.3 Simplified Modeling of Transformer Windings

The representation of the magnetic field sources, i.e., the winding currents in the case of the transformer magnetic field, is carried out with the use of a fictitious field distribution **K**, which must satisfy the conditions described in Sect. 6.4.2.1. For the calculation of **K**, a simply connected subdomain must be defined for each winding, comprising its conductors. Figure 6.21 shows the bottom view of the subdomain corresponding to the LV winding of Fig. 6.19. This subdomain is divided into four regions ($\Omega_1, \Omega_2, \Omega_3$ and Ω_4), in order to facilitate the calculation. The symbols shown in Fig. 6.21 are described in the following:

X_{W11MIN}, X_{W11MAX}	:	boundaries of the coil area along the x-axis inside the small core window
X_{W1MIN}, X_{W1MAX}	:	boundaries of the coil area along the x-axis inside the large core window
Y_{W1MIN}, Y_{W1MAX}	:	boundaries of the coil area along the y-axis
J_x, J_y	:	x, y components of winding current density
X_C	:	x-coordinate of the winding center

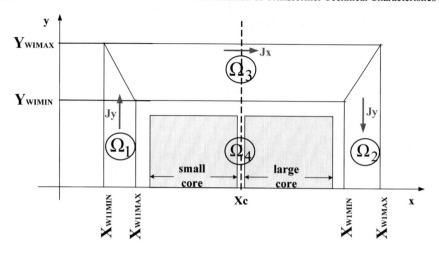

Fig. 6.21 Regions (xy-plane) of the subdomain used in the calculation of the fictitious field distribution Kz corresponding to the LV winding (orthogonal approximation of the winding corners)

The calculation of **K** is quite straightforward, given the winding dimensions along the x-, y- and z-axis:

1. Region Ω_1: In this region, $J_x = J_z = 0$. The current density J_y is given by:

$$J_y = NI \cdot \left[\frac{Z}{X_{W1MAX} - X_{W1MIN}} \right], \qquad (6.5)$$

where NI are the ampere turns of the LV winding. The distribution **K** must be perpendicular to the Ω_1 boundary (third condition described in Sect. 6.4.2.1). Therefore, **K** consists of component K_z only, while $K_x = K_y = 0$. The second condition described in Sect. 6.4.2.1 yields:

$$\nabla \times \mathbf{K} = \mathbf{J} \Rightarrow K_z = -\int_{\Omega_1} J_y \, dx \Rightarrow$$

$$K_z = NI \cdot Z \cdot \left[\frac{X_{W1MIN} - X}{X_{W1MIN} - X_{W1MAX}} \right]. \qquad (6.6)$$

2. Region Ω_2: In this region, $J_x = J_z = 0$ and

6.4 Impedance Voltage Evaluation with Numerical Models

$$J_y = -\text{NI} \cdot \left[\frac{Z}{X_{\text{W1MAX}} - X_{\text{W1MIN}}} \right], \quad (6.7)$$

while K_z is derived from:

$$K_z = -\int_{\Omega_2} J_y \, dx = \text{NI} \cdot Z \cdot \left[\frac{X_{\text{W1MAX}} - X}{X_{\text{W1MAX}} - X_{\text{W1MIN}}} \right]. \quad (6.8)$$

3. Region Ω_3: In this region, $J_y = J_z = 0$ and

$$J_x = \text{NI} \cdot \left[\frac{Z}{Y_{\text{W1MAX}} - Y_{\text{W1MIN}}} \right], \quad (6.9)$$

while K_z derives from:

$$K_z = \int_{\Omega_3} J_x \, dy = \text{NI} \cdot Z \cdot \left[\frac{Y_{\text{W1MAX}} - Y}{Y_{\text{W1MAX}} - Y_{\text{W1MIN}}} \right]. \quad (6.10)$$

4. Region Ω_4: The application of the continuity boundary condition for K_z between regions Ω_3 and Ω_4 yields:

$$K_z(Y = Y_{\text{W1MIN}}^-) = K_z(Y = Y_{\text{W1MIN}}^+) \Rightarrow$$

$$K_z(Y = Y_{\text{W1MIN}}^-) = \text{NI} \cdot Z. \quad (6.11)$$

The application of the continuity boundary condition between regions Ω_1 and Ω_4 or Ω_2 and Ω_4 results in the same equation for K_z in region Ω_4.

Consequently, the overall equation describing the fictitious field distribution corresponding to the LV winding is of the form:

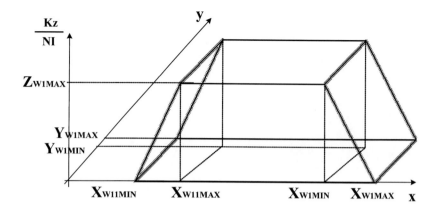

Fig. 6.22 3D graphical representation of the fictitious field distribution corresponding to the LV winding of Fig. 6.19 (orthogonal approximation of the winding corners)

$$K_z = \begin{cases} NI \cdot Z \cdot \left[\dfrac{X_{W11MIN} - X}{X_{W11MIN} - X_{W11MAX}} \right] & , \text{ region } \Omega_1 \\ NI \cdot Z \cdot \left[\dfrac{X_{W1MAX} - X}{X_{W1MAX} - X_{W1MIN}} \right] & , \text{ region } \Omega_2 \\ NI \cdot Z \cdot \left[\dfrac{Y_{W1MAX} - Y}{Y_{W1MAX} - Y_{W1MIN}} \right] & , \text{ region } \Omega_3 \\ NI \cdot Z & , \text{ region } \Omega_4 \end{cases} \quad (6.12)$$

Figure 6.22 gives the 3D graphical representation of the K_z component corresponding to LV winding. The symbol Z_{W1MAX} appearing in Fig. 6.22 denotes the boundary of the coil area along the z-axis. Figure 6.23 shows the K_z distribution along a plane parallel to the y-axis, crossing the center X_C of the winding.

Derivation of the distribution **K** for the HV winding is similar, resulting in an equation identical to (6.12), with the respective boundaries of the winding along the x-, y- and z-axis. The representation of current sources through distribution **K** has the advantage of being compatible with the discrete scheme of first order tetrahedral elements so that it does not suffer from cancellation errors, present when using the Biot-Savart law to determine source field distribution.

6.4 Impedance Voltage Evaluation with Numerical Models 309

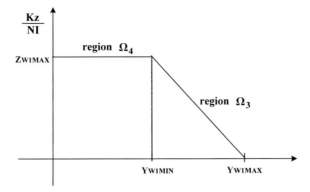

Fig. 6.23 Fictitious field distribution corresponding to the LV winding along the plane $X = X_C$ of Fig. 6.21

6.4.2.4 Detailed Modeling of Transformer Windings

The construction of the transformer model with detailed winding geometry is realized in two steps: first, an elliptical approximation of the winding corners is considered, and, afterwards, the winding cooling ducts are inserted into the model.

The simplicity of the calculation of the fictitious field distribution presented in Sect. 6.4.2.3 relies on the orthogonal approximation of the winding corners. The orthogonal approximation of the winding corners, which are in fact curved, is likely to result in significant overestimation of the current density and thus the derived magnetic field density.

For a more detailed representation of the winding geometry, their corners were considered to be part of ellipses with known center coordinates. Under this consideration, the fictitious field distribution corresponding to the LV winding is the one shown in Fig. 6.24.

The calculation of the K_z component shown in Fig. 6.24 is more complicated: the subdomain comprising the LV winding has to be divided into six regions, whose bottom view is shown in Fig. 6.25: regions $\Omega_1, \Omega_2, \Omega_3$ and Ω_4 (used in the calculation described in Sect. 6.4.2.3) and regions Ω_5 and Ω_6, which correspond to the winding corners and are bounded by inner ellipses (ε_2), (ε_2') and outer ellipses (ε_1), (ε_1').

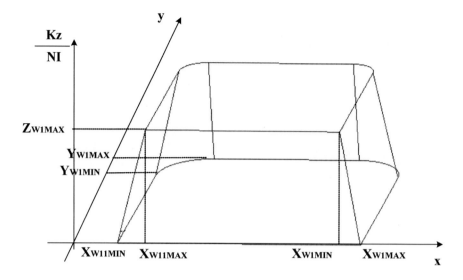

Fig. 6.24 3D graphical representation of the fictitious field distribution corresponding to the LV winding (elliptical approximation of the winding corners)

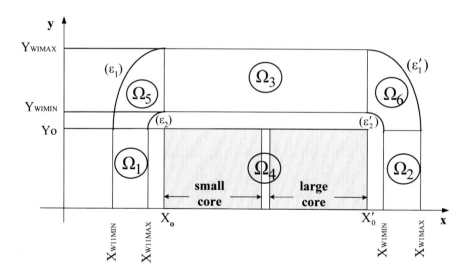

Fig. 6.25 Regions (xy-plane) of the subdomain used in the calculation of the fictitious field distribution K_z corresponding to the LV winding (elliptical approximation of the winding corners)

6.4 Impedance Voltage Evaluation with Numerical Models

In Fig. 6.25, point (X_o, Y_o) is the center of (ε_1) and (ε_2) while point (X'_0, Y_o) is the center of (ε'_1) and (ε'_2). The coordinates X_o, X'_0 and Y_o are derived from the transformer dimensions and can easily be calculated. The equations of the ellipses are:

$$(\varepsilon_1): \frac{(X-X_o)^2 + (Y-Y_o)^2}{(X_{W11MIN} - X_o)^2 + (Y_{W1MAX} - Y_o)^2} = 1, \qquad (6.13)$$

$$(\varepsilon_2): \frac{(X-X_o)^2 + (Y-Y_o)^2}{(X_{W11MAX} - X_o)^2 + (Y_{W1MIN} - Y_o)^2} = 1, \qquad (6.14)$$

$$(\varepsilon'_1): \frac{(X-X'_o)^2 + (Y-Y_o)^2}{(X_{W1MAX} - X'_o)^2 + (Y_{W1MAX} - Y_o)^2} = 1, \qquad (6.15)$$

$$(\varepsilon'_2): \frac{(X-X'_o)^2 + (Y-Y_o)^2}{(X_{W1MIN} - X'_o)^2 + (Y_{W1MIN} - Y_o)^2} = 1. \qquad (6.16)$$

The equation describing K_z is of the form (6.17). The ellipse symbols appearing in (6.17) refer to the left hand side of (6.13)–(6.16), an expression that was chosen for the sake of simplicity.

$$K_z = \begin{cases} NI \cdot Z \cdot \left[\dfrac{X_{W11MIN} - X}{X_{W11MIN} - X_{W11MAX}} \right] & , \text{ region } \Omega_1 \\[6pt] NI \cdot Z \cdot \left[\dfrac{X_{W1MAX} - X}{X_{W1MAX} - X_{W1MIN}} \right] & , \text{ region } \Omega_2 \\[6pt] NI \cdot Z \cdot \left[\dfrac{Y_{W1MAX} - Y}{Y_{W1MAX} - Y_{W1MIN}} \right] & , \text{ region } \Omega_3 \\[6pt] NI \cdot Z & , \text{ region } \Omega_4 \\[6pt] NI \cdot Z \cdot \left[\dfrac{(\varepsilon_1) - 1}{(\varepsilon_1) - (\varepsilon_2)} \right] & , \text{ region } \Omega_5 \\[6pt] NI \cdot Z \cdot \left[\dfrac{(\varepsilon'_1) - 1}{(\varepsilon'_1) - (\varepsilon'_2)} \right] & , \text{ region } \Omega_6 \end{cases} \qquad (6.17)$$

Note that (6.17) is derived under the assumption that the current flows through the whole area of the considered winding. This assumption does not take into account the existence of cooling ducts in the winding area outside the core windows, where the current density is in fact equal to zero, because of the oil flowing through them. This approximation is another factor contributing to magnetic field overestimation, as it increases the magnetic field source area. Therefore, the existence of cooling ducts must be considered in the analysis in order to obtain more reliable results.

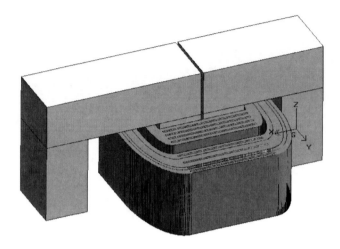

Fig. 6.26 Perspective view of the active part of the transformer, one phase part modeled (detailed winding geometry)

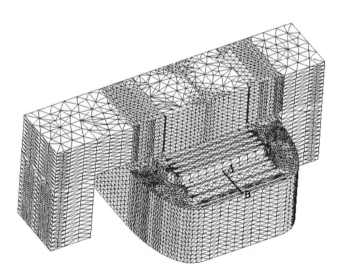

Fig. 6.27 Tetrahedral finite element mesh representing the transformer active part (detailed winding geometry)

6.4 Impedance Voltage Evaluation with Numerical Models

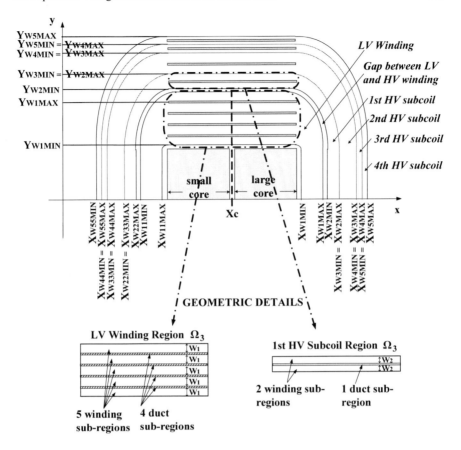

Fig. 6.28 Division of the windings used in the representation of cooling ducts (xy-plane).

Figure 6.26 shows the perspective view of the transformer model with detailed representation of the transformer windings, including the cooling ducts. The modeling of cooling ducts affects the calculation of the fictitious field distribution K_z as well as the construction of the finite element mesh of the transformer active part, shown in Fig. 6.27.

With respect to Fig. 6.25, the ducts are located in the region Ω_3 of the winding; thus, the distribution must be recalculated in this region only. The new bottom view of both LV and HV windings is shown in Fig. 6.28. The LV winding comprises four cooling ducts while each one of the HV subcoils comprises one cooling duct. All the ducts have the same thickness, equal to W_{DUCT}.

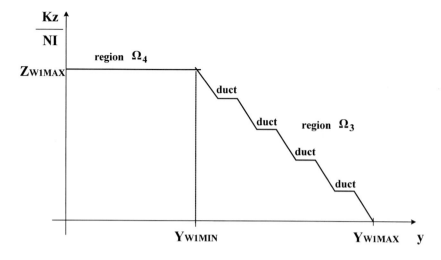

Fig. 6.29 Fictitious field distribution corresponding to the LV winding along the plane $X = X_C$ of Fig. 6.28 (similar to Fig. 6.23, but with consideration of cooling ducts)

The region Ω_3 of the LV winding is divided into nine sub-regions: five winding sub-regions and four duct sub-regions. The distribution K_z of the region Ω_3 corresponding to the LV winding has the form of Fig. 6.29, which shows the K_z distribution along a plane parallel to the y-axis, crossing the center X_C of the winding (similar to Fig. 6.23).

The following equation gives the distribution K_z for region Ω_3 of LV winding (Fig. 6.29). The first branch refers to the winding sub-regions, while the second branch refers to the duct sub-regions.

$$\frac{K_z}{NI \cdot Z} = \begin{cases} -\dfrac{1}{5 \cdot W_1}\left[Y + Y_{W1MIN} + j \cdot W_1 + (j-1) \cdot W_{DUCT}\right] + \dfrac{|j-5|}{5} &, j=1,\ldots,5 \\ \dfrac{|j-5|}{5} &, j=1,\ldots,4 \end{cases} \quad (6.18)$$

where W_1 is the width of each winding sub-region, given by:

$$W_1 = \frac{LV\ Width}{5}, \tag{6.19}$$

where LV Width is the width of the LV winding inside the core windows.

6.4 Impedance Voltage Evaluation with Numerical Models

Similarly, the region Ω_3 of the first HV subcoil is divided into three sub-regions: two winding sub-regions and one duct sub-region. The respective equation for K_z is:

$$\frac{K_z}{NI \cdot Z} = \begin{cases} -\frac{1}{3 \cdot W_2}\left[Y + Y_{W2MIN} + j \cdot W_2 + (j-1) \cdot W_{DUCT}\right] + \frac{|j-3|}{3} &, j = 1,...,3 \\ \frac{|j-3|}{3} &, j = 1 \end{cases} \quad (6.20)$$

where W_2 is the width of each winding sub-region, given by:

$$W_2 = \frac{\text{HVWidth1}}{3}, \quad (6.21)$$

where HVWidth1 is the width of the first HV subcoil inside the core windows. The first branch of (6.20) refers to winding sub-regions, while the second one corresponds to the duct sub-region. The form of the equations for K_z of the next three HV subcoils are identical to (6.20) with the respective boundaries along the x-, y- and z-axis.

6.4.2.5 Impedance Voltage Evaluation Using FEM

Impedance voltage evaluation is based on the magnetic field energy of the finite element model. In particular, having calculated the total magnetic field energy, W_m, of the finite element model, the total leakage inductance, L_{tot}, of the windings is calculated as follows:

$$L_{tot} = \frac{2 \cdot W_m}{(NI)^2}, \quad (6.22)$$

where NI are the ampere turns of the secondary winding (LV winding).

Having calculated the total leakage inductance from (6.22), the percentage leakage reactance (referred to the secondary winding), IX, is calculated from the equation:

$$IX = \frac{I_2 \cdot 2 \cdot \pi \cdot f \cdot N_2^2 \cdot L_{tot}}{V_2} \cdot 100, \quad (6.23)$$

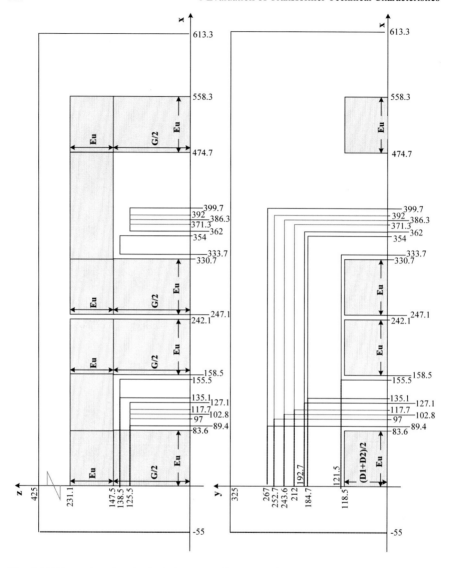

Fig. 6.30 Dimensions (mm) of 630 kVA transformer model

where I_2 is the current in the secondary winding, N_2 is the number of turns of the secondary winding, V_2 is the voltage of the secondary winding, and f is the frequency.

Finally, the impedance voltage, U_k, is calculated as follows:

$$U_k = \sqrt{(IX)^2 + (IR)^2} \,, \tag{6.24}$$

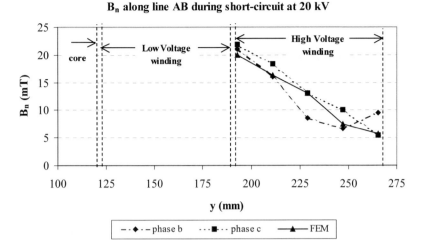

Fig. 6.31 Comparison of measured and computed field values along the line AB during short-circuit at 20 kV.

where *IR* is the percentage resistive component of the impedance voltage, which is calculated by dividing the transformer load losses by the transformer rated power.

6.4.3 Results and Discussion

6.4.3.1 Transformer Model Dimensions

As already mention in Sect. 6.4.2.2, the transformer under consideration is a 630 kVA, 20-15/0.4 kV, three-phase transformer. The secondary winding comprises 16 layers (per phase) of copper sheet, while the primary consists of 1385 turns (per phase) of insulated copper wire. Figure 6.30 shows the dimensions of this 630 kVA transformer model.

6.4.3.2 Local Field Values

The field values computed by the 3D FEM model (detailed winding geometry) have been compared to those measured by a Hall effect probe during short-circuit test. Figures 6.31 and 6.32 give the variation of the perpendicular flux density component B_n along the line AB, positioned as shown in Fig. 6.27, in the case of short-circuit, with the high voltage winding connections corresponding to 20 kV and 15 kV voltage supply, respectively. These figures illustrate the good correlation of the simulated results with the local leakage field measured by Hall effect probes.

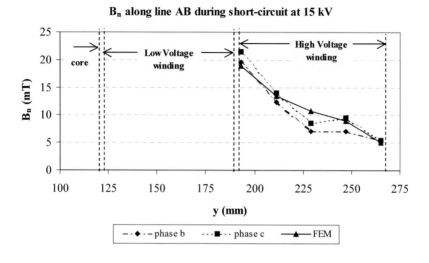

Fig. 6.32 Comparison of measured and computed field values along the line AB during short-circuit at 15 kV

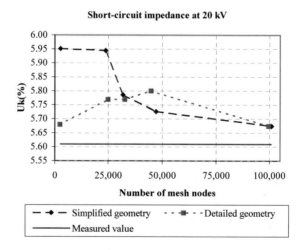

Fig. 6.33 Short-circuit impedance results (primary voltage 20 kV) for the 630 kVA transformer (simplified and detailed winding geometry)

6.4 Impedance Voltage Evaluation with Numerical Models

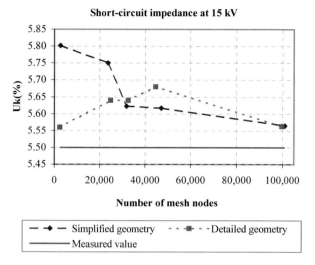

Fig. 6.34 Short-circuit impedance results (primary voltage 15 kV) for the 630 kVA transformer (simplified and detailed winding geometry)

6.4.3.3 Short-circuit Impedance

The finite element method results were used to calculate the transformer short-circuit impedance. Both models with simplified and detailed winding geometry were used.

The results were compared with the short-circuit impedance measured after the transformer construction. Figures 6.33 and 6.34 show the deviation of the calculated short-circuit impedance from the measured value for different mesh densities of the two models used in the case of the first (20 kV) and second (15 kV) primary voltage level, respectively.

The results of Figs. 6.33 and 6.34 are tabulated in Tables 6.23 and 6.24. The deviation appearing in these tables is defined by:

$$\text{deviation}(\%) = \frac{|U_k^{\text{calculated}} - U_k^{\text{measured}}|}{U_k^{\text{measured}}} \cdot 100\%, \qquad (6.25)$$

where $U_k^{\text{calculated}}$ is the short-circuit impedance calculated using the FEM model, while U_k^{measured} is the measured short-circuit impedance value.

Comparison of the curves shown in Figs. 6.33 and 6.34 leads to the following conclusions:

1. The error in the calculation of the short-circuit impedance using the detailed winding geometry model begins from a very small value for a sparse mesh (instead of the great error given by the simplified geometry model) and rises with increase in the number of mesh nodes. This may be attributed to the fact that detailed geometry representation using a small number of unknowns leads to some kind of compensating errors for the source field representation.

2. At an intermediate mesh density (30,000 nodes approximately) the error of the detailed model approaches that of the simplified model.

3. The two models converge to the same error at high mesh densities (beyond 90,000 nodes).

4. The variation of the error is similar for the two high voltage levels.

5. The minimum deviation appearing in Tables 6.23 and 6.24 is less than 1.5%, obtained using the detailed winding model, for both first (lowest) and last (highest) mesh density and when using the simplified winding model with the larger number of nodes.

Table 6.23 Short-circuit impedance results using simplified winding geometry for the 630 kVA transformer

Primary voltage level (kV)	Number of mesh nodes	$U_k^{calculated}$ (%)	$U_k^{measured}$ (%)	Deviation (%)
20	2613	5.97	5.61	6.42
	23696	5.95		6.06
	31818	5.77		2.85
	47044	5.73		2.14
	100999	5.67		1.07
15	2613	5.82	5.50	5.82
	23696	5.75		4.55
	31818	5.63		2.36
	47044	5.62		2.18
	100999	5.57		1.27

6.4 Impedance Voltage Evaluation with Numerical Models

Table 6.24 Short-circuit impedance results using detailed winding geometry for the 630 kVA transformer

Primary voltage level (kV)	Number of mesh nodes	$U_k^{calculated}$ (%)	$U_k^{measured}$ (%)	Deviation (%)
20	3260	5.69	5.61	1.43
	24862	5.77		2.85
	32555	5.77		2.85
	44660	5.90		3.39
	99567	5.67		1.07
15	3260	5.56	5.50	1.09
	24862	5.65		2.73
	32555	5.63		2.36
	44660	5.68		3.27
	99567	5.57		1.27

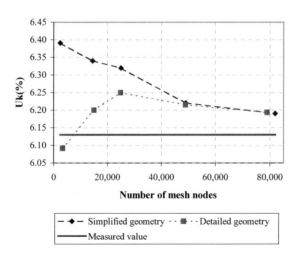

Fig. 6.35 Short-circuit impedance results (primary voltage 20 kV) for the 1000 kVA transformer (simplified and detailed winding geometry)

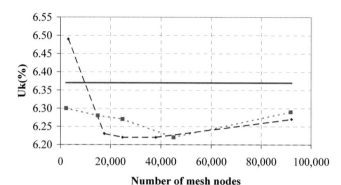

Fig. 6.36 Short-circuit impedance results (primary voltage 20 kV) for the 400 kVA transformer (simplified and detailed winding geometry)

In order to validate the above conclusions, the same analysis was conducted for two more cases of transformers of rated primary voltage 20 and 15 kV, rated secondary voltage 400 V and rated power 400 and 1000 kVA. Figures 6.35 and 6.36 give the respective error curves in the short-circuit impedance calculation, using both the simplified and the detailed model, at 20 kV. The shape of the error curves is similar to those of Figs. 6.33 and 6.34, as the greater accuracy is achieved in the lowest mesh density for the detailed model in both cases.

The detailed model of the winding geometry is therefore appropriate for very accurate calculation of the short-circuit impedance using a sparse mesh. This ability is quite important for the finite element method as it overcomes one of its main drawbacks, namely the computation time required to obtain reliable results.

6.4.3.4 Generalization of Results

The FEM methodology has proven to be cost effective and quite accurate in the prediction of the leakage field and the short-circuit impedance of the transformer examined in Sect. 6.4.2.2. However, further work was necessary to enable its implementation in several three-phase dual voltage wound core transformers. The design of dual voltage transformers is of special interest, as their windings are arranged in specific ways and their characteristics may vary widely with the change in connection, depending on winding arrangement (Ebert 1995). Therefore, accurate calculation techniques such as the finite element analysis for both connections

6.4 Impedance Voltage Evaluation with Numerical Models

must be conducted to ensure reliable performance and to maintain design margins during short-circuit and other transient phenomena (Digby and Sim 2002).

For the above purpose, a computer code was developed, performing the finite element calculations that provide the value of the short-circuit impedance (Tsili et al. 2008). A process of mesh parameterization was adopted, which modifies the coordinates of initial tetrahedral meshes of various densities in accordance with the geometric data of the examined transformers. In this way, the program user does not interfere with 3D model construction, a time-consuming procedure that demands specific computer aided design knowledge. This interface has overbalanced another major deficiency that has so far restrained the proliferation of the use of 3D FEM techniques in the transformer manufacturing industry.

The use of the detailed geometry model of the transformer, described in Sect. 6.4.2.4, enables the representation of transformers of different power ratings and voltage levels in the primary winding. Single or dual primary voltage transformers can be modeled. The windings connection (delta, star or zig-zag) does not affect the model characteristics. The initial division of the HV winding area into four sub-regions was selected, as it can also model other possible winding arrangements for dual primary voltage production, apart from the one described in Fig. 6.20.

An alternative way of obtaining dual primary voltage 20-15 kV is shown in Fig. 6.37. In this case, the two intermediate HV subcoils (HV2, HV3) are connected in parallel and then in series with the remaining two subcoils (HV1, HV4), in order to obtain the second HV level (15 kV). Therefore, to model this connection, one needs to consider that the current flowing through HV2 and HV3 is half of the current flowing through HV1 and HV4.

Figure 6.38 illustrates a HV winding arrangement that gives dual primary voltage 20-10 kV. In this case, the HV winding is divided into two subcoils only (HV1 and HV2), which consist of two sections (HV1a-HV1b and HV2a-HV2b), connected in parallel for the production of the lower HV level (10 kV). Consequently, at primary voltage level equal to 10 kV, subcoils HV1 and HV2 carry half of the nominal current. In order to model the connection described above, the four HV winding sub-regions of the FEM model of Fig. 6.26 are arranged as follows: the first two sub-regions represent subcoil HV1, while the two last sub-regions represent subcoil HV2.

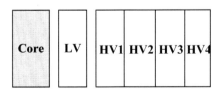

Fig. 6.37 Alternative winding arrangement (yz-plane) for the production of dual primary voltage levels 20 kV and 15 kV

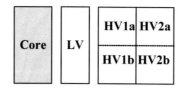

Fig. 6.38 Winding arrangement (yz-plane) for the production of dual primary voltage levels 20 kV and 10 kV

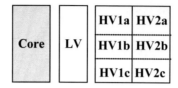

Fig. 6.39 Winding arrangement (yz-plane) for the production of dual primary voltage levels 20 kV and 6.6 kV

Table 6.25 Application of FEM to several transformer cases

	Rating (kVA)	Primary voltage level (kV)	$U_k^{calculated}$ (%)	$U_k^{measured}$ (%)	Deviation (%)
1	1000	20	6.26	6.27	0.16
		15	6.30	6.17	2.06
2	630	20	3.83	3.77	1.57
		6.6	3.81	3.75	1.57
3	400	20	6.37	6.22	2.35
		15	6.08	5.95	2.14
4	100	33	4.14	4.09	1.21
5	100	20	4.27	4.16	2.58
		15	4.19	4.17	0.48

In Fig. 6.39, an arrangement that produces dual primary voltage levels 20-6.6 kV is shown. This arrangement resembles that of Fig. 6.38, except for the fact that HV1 and HV2 are divided into three sections. Thus, at primary voltage level equal to 6.6 kV, one third of the nominal current flows through HV1 and HV2.

Table 6.25 summarizes the results of the calculated short-circuit impedance values in five cases along with values measured after construction of the transformers. The secondary voltage levels of the transformers are equal to 400 V, except for transformer 4, whose nominal secondary voltage is equal to 433 V. The computed results compare favorably with the measured values, as the difference between them is less than 2.5% in most cases. The maximum deviation is 2.64%, while the minimum deviation is less than 0.5%. The calculations were conducted with the detailed transformer model and a sparse mesh of approximately 3,500 nodes, requiring less than 2.5 minutes execution time on an average performance personal computer (Pentium, 133 MHz, 64 MB RAM).

References

Amoiralis EI, Tsili MA, Georgilakis PS (2008) The state of the art in engineering methods for transformer design and optimization: a survey. Journal of Optoelectronics and Advanced Materials 10(5):1149–1158

Andersen OW (1973) Transformer leakage flux program based on the finite element method. IEEE Transactions on Power Apparatus and Systems 92(2):682–689

Annakkage UD, McLaren PG, Dircks E, Jayasinghe RP, Parker AD (2000) A current transformer model based on the Jiles-Atherton theory of ferromagnetic hysteresis. IEEE Transactions on Power Delivery 15:57–61

Basak A, Bonyar AA (1992) Effects of transformer core assembly on building factors. Journal of Magnetism and Magnetic Materials 112:406–408

Basak A, Yu CH, Lloyd G (1994) Core loss computation of a 1000 kVA distribution transformer. Journal of Magnetism and Magnetic Materials 133:564–567

Bertotti G (1988) General properties of power losses in soft ferromagnetic materials. IEEE Transactions on Magnetics 24:621–630

Digby SH, Sim HJ (2002) Transformer design for dual-voltage applications. Proc IEEE Rural Electric Power Conference

Djurovic M, Carpenter CJ (1975) 3-dimensional computation of transformer leakage fields and associated losses. IEEE Transactions on Magnetics 11(5):1535–1537

Doulamis ND, Doulamis AD, Georgilakis PS, Kollias SD, Hatziargyriou ND (2002) A synergetic neural network-genetic scheme for optimal transformer construction. Integrated Computer-Aided Engineering 9:37–56

Dymkov A (1975) Transformer design. Mir Publishers, Moscow

Ebert JA (1995) Criteria for reliable dual voltage power transformers. IEEE Transactions on Power Delivery 10(2):845–852

Elleuch M, Poloujadoff M (1996) A contribution to the modeling of three phase transformers using reluctances. IEEE Transactions on Magnetics 32:3335–3343

Elleuch M, Poloujadoff M (2003) Analytical model of iron losses in power transformers. IEEE Transactions on Magnetics 39(2):973–980

Enokizono M, Soda N (1997) Finite element analysis of transformer model core with measured reluctivity tensor. IEEE Transactions on Magnetics 33(5):4110–4112

Enokizono M, Soda N (1998) Direct magnetic loss analysis by FEM considering vector magnetic properties. IEEE Transactions on Magnetics 34(5):3008–3011

Enokizono M, Soda N (1999) Core loss analysis of transformer by improved FEM. Journal of Magnetism and Magnetic Materials 196-197:910–912

Enokizono M, Yuki K, Kawano S (1995) An improved magnetic field analysis in oriented steel sheet by finite element method considering tensor reluctivity. IEEE Transactions on Magnetics 31(3):1797–1800

Escarela-Perez R, Kulkarni SV, Kodela NK, Olivares-Galvan JC (2007) Asymmetry during load-loss measurement of three-phase three-limb transformers. IEEE Transactions on Power Delivery 22(3):1566–1574

Fecich D, Balmer L (1985) Process factors of transformer cores made from 0.18 mm, 0.23 mm, and 0.28 mm thick grain oriented steel. IEEE Transactions on Magnetics 21(5):1915–1917

Fiorello F, Novikov M (1990) An improved approach to power loss in magnetic laminations under nonsinusoidal induction waveform. IEEE Transactions on Magnetics 26:2904–2910

Georgilakis PS (2000) Contribution of artificial intelligence techniques in the reduction of distribution transformer iron losses. PhD dissertation. National Technical University of Athens, Athens, Greece

Georgilakis PS, Hatziargyriou ND (1999) Machine learning applications in the transformer manufacturing industry. Proc Advanced Course on Artificial Intelligence 57–65

Georgilakis P, Hatziargyriou N (2002) On the application of artificial intelligence techniques to the quality improvement of industrial processes. Lecture Notes in Computer Science 2308:473–484

Georgilakis PS, Bakopoulos JA, Hatziargyriou ND (1997) A decision tree method for prediction of distribution transformer iron losses. Proc Universities Power Engineering Conference 1:257–260

Georgilakis PS, Hatziargyriou ND, Souflaris AT (1998a) Artificial intelligence approaches to distribution transformer core quality improvement. Proc International Conference on Electrical Machines 1:541–546

Georgilakis PS, Hatziargyriou ND, Doulamis ND, Doulamis AD, Kollias SD (1998b) Prediction of iron losses of wound core distribution transformers based on artificial neural networks. Neurocomputing 23:15–29

Georgilakis PS, Hatziargyriou ND, Paparigas D (1999a) AI helps reduce transformer iron losses. IEEE Computer Applications in Power 12(4):41–46

Georgilakis PS, Hatziargyriou ND, Doulamis ND, Doulamis AD, Bakopoulos JA (1999b) An efficient PC-based environment for the improvement of magnetic cores industrial process. In: Tzafestas SG (ed.) Advances in manufacturing: decision, control, and information technology. Springer, London

Georgilakis P, Hatziargyriou N, Paparigas D, Bakopoulos J, Elefsiniotis S (1999c) Automatic learning techniques for on-line control and optimization of transformer core manufacturing process. Proc IEEE Industry Applications Society Annual Meeting 1:311–322

Georgilakis P, Hatziargyriou N, Paparigas D, Bakopoulos J (1999d) On-line combined use of neural networks and genetic algorithms to the solution of transformer iron loss reduction problem. Proc IEEE PowerTech

Georgilakis PS, Hatziargyriou ND, Doulamis AD, Doulamis ND, Kollias SD (1999e) A neural network framework for predicting transformer core losses. Proc IEEE International Conference on Power Industry Computer Applications 1:301–308

Georgilakis PS, Doulamis ND, Doulamis AD, Hatziargyriou ND, Kollias SD (2001a) A novel iron loss reduction technique for distribution transformers based on a genetic algorithm-neural network approach. IEEE Transactions on Systems, Man, and Cybernetics, Part C 31(1):16–34

Georgilakis P, Hatziargyriou N, Paparigas D, Elefsiniotis S (2001b) Effective use of magnetic materials in transformer manufacturing. Journal of Materials Processing Technology 108:209–212

Girgis RS, teNyenhuis EG, Gramm K, Wrethag JE (1998) Experimental investigations on effect of core production attributes on transformer core loss performance. IEEE Transactions on Power Delivery 13(2):526–531

Godec Z (1977) Influence of slitting on core losses and magnetization curve of grain-oriented electrical steels. IEEE Transactions on Magnetics 13(4):1053–1057

Hatziargyriou ND, Prousalidis JM, Papadias BC (1993) A generalized transformer model based on the analysis of its magnetic circuit. IEE Proceedings, Part C 140(4):269–278

Hatziargyriou N, Georgilakis P, Spiliopoulos D, Bakopoulos J (1998a) Quality improvement of individual cores of distribution transformers using decision trees. International Journal of Engineering Intelligent Systems for Electrical Engineering and Communications 6(3):141–146

Hatziargyriou ND, Georgilakis PS, Paparigas DG, Bakopoulos JA (1998b) Prediction of distribution transformer no-load losses using the learning vector quantization neural network. Proc IEEE Mediterranean Electrotechnical Conference 2:1180–1184

References

Ilo A, Weiser B, Booth T, Pfutzner H (1996) Influence of geometric parameters on the magnetic properties of model transformer cores. Journal of Magnetism and Magnetic Materials 160:38–40

Im CH, Kim HK, Lee CH, Jung HK (2001) Analysis of the three-phase transformer considering the nonlinear and anisotropic properties using the transmission line modeling method and FEM. IEEE Transactions on Magnetics 37(5):3490–3493

Jiles DC, Atherton DL (1986) Theory of ferromagnetic hysteresis. Journal of Magnetism and Magnetic Materials 61:48–60

Kaimori H, Kameari A, Fujiwara K (2007) FEM computation of magnetic field and iron loss in laminated iron core using homogenization method. IEEE Transactions on Magnetics 43(4):1405–1408

Kalokiris GK, Kladas AG, Hatzilau IK, Cofinas S, Gyparis IK (2007) Advances in magnetic materials and their impact on electric machine design. Journal of Materials Processing Technology 181(1-3):148–152

Kanada T, Enokizono M, Kawamura K, Sievert JD (1996) Distributions on localized iron loss of three-phase amorphous transformer model core by using two-dimensional magnetic tensor. IEEE Transactions on Magnetics 32(5):4797–4799

Kefalas TD, Georgilakis PS, Kladas AG, Souflaris AT, Paparigas, DG (2008) Multiple grade lamination wound core: a novel technique for transformer iron loss minimization using simulated annealing with restarts and an anisotropy model. IEEE Transactions on Magnetics 44(6):1082–1085

Kladas A, Tegopoulos J (1992) A new scalar potential formulation for 3D magnetostatics necessitating no source field calculation. IEEE Transactions on Magnetics 28:1103–1106

Kladas AG, Papadopoulos MP, Tegopoulos JA (1994) Leakage flux and force calculation on power transformer windings under short-circuit: 2D and 3D models based on the theory of images and the finite element method compared to measurements. IEEE Transactions on Magnetics 30(5):3487–3490

Kobayashi H (1981) Modeling and analysis. Addison-Welsey, Reading, MA

Kulkarni SV, Khaparde SA (2004) Transformer engineering: design and practice. Marcel-Dekker, New York

Kulkarni SV, Olivares JC, Escarela-Perez R, Lakhiani VK, Turowski J (2004) Evaluation of eddy current losses in the cover plates of distribution transformers. IEE Proc Science, Measurement and Technology 151(5):313–318

Ling PCY, Moses AJ, McQuade F, Grimmond W, Fox D (1992) Investigation of magnetic degradation of wound cores due to adhesive bonding. Journal of Magnetism and Magnetic Materials 112:77–80

Logothetis N (1992) Managing for total quality. Prentice-Hall International, UK

Lu J, Yuan J, Chen L, Sheng J, Ma X (1998) Calculation of short-circuit reactance of transformers by a line integral based on surface magnetic charges. IEEE Transactions on Magnetics 34(5):3483–3486

Mechler GF, Girgis RS (1998) Calculation of spatial loss distribution in stacked power and distribution transformer cores. IEEE Transactions on Power Delivery 13(2):532–537

Mechler GF, Girgis RS (2000) Magnetic flux distributions in transformer core joints. IEEE Transactions on Power Delivery 15(1):198–203

Miti GK, Moses AJ, Derebasi N, Fox D (2003) A neural network-based tool for magnetic performance prediction of toroidal cores. Journal of Magnetism and Magnetic Materials 254-255:262–264

Mittle VN, Mittal A (1996) Design of electrical machines, 4th edn. Standard Publishers Distributors, Delhi

Mork BA (1999) Five-legged wound-core transformer model: derivation, parameters, implementation, and evaluation. IEEE Transactions on Power Delivery 14:1519–1526

Moses AJ (1984) Factors affecting localized flux and iron loss distribution in laminated cores. Journal of Magnetism and Magnetic Materials 41:409–414

Moses AJ (1992) Development of alternative magnetic core materials and incentives for their use. Journal of Magnetism and Magnetic Materials 112:150–155

Moses AJ (1998) Comparison of transformer loss prediction from computed and measured flux density distribution. IEEE Transactions on Magnetics 34(4):1168–1170

Moses AJ (2003) Prediction of core losses of three phase transformers from estimation of the components contributing to the building factor. Journal of Magnetism and Magnetic Materials 254-255:615–617

Nakata T, Takahashi N, Kawase Y, Nakano M (1984) Influence of lamination orientation and stacking on magnetic characteristics of grain-oriented silicon steel laminations. IEEE Transactions on Magnetics 20:1774–1776

Nussbaum C, Booth T, Ilo A, Pfutzner H (1996) A neural network for the prediction of performance parameters of transformer cores. Journal of Magnetism and Magnetic Materials 160:81–83

Nussbaum C, Pfutzner H, Booth Th, Baumgartinger N, Ilo A, Clabian M (2000) Neural networks for the prediction of magnetic transformer core characteristics. IEEE Transactions on Magnetics 36(1):313–329

Olivares JC, Cañedo J, Moreno P, Driesen J, Escarela R, Palanivasagam S (2002) Experimental study to reduce the distribution-transformers stray losses using electromagnetic shields. Electric Power Systems Research 63(1):1–7

Olivares JC, Yilu L, Canedo JM, Escarela-Perez R, Driesen J, Moreno P (2003a) Reducing losses in distribution transformers. IEEE Transactions on Power Delivery 18(3):821–826

Olivares JC, Kulkarni SV, Canedo J, Escarela R, Driesen J, Moreno P (2003b) Impact of the joint design parameters on transformer losses. International Journal of Power & Energy Systems 23(3):151–157

Olivares JC, Escarela-Perez R, Kulkarni SV, de Leon F, Melgoza-Vasquez E, Hernandez-Anaya O (2004a) Improved insert geometry for reducing tank-wall losses in pad-mounted transformers. IEEE Transactions on Power Delivery 19(3):1120–1126

Olivares JC, Escarela-Perez R, Kulkarni SV, de León F, Venegas-Vega MA (2004b) 2D finite-element determination of tank wall losses in pad-mounted transformers. Electric Power Systems Research 71(2):179–185

Olivares-Galván JC, Georgilakis PS, Ocon-Valdez R (2009) A review of transformer losses. Electric Power Components and Systems, accepted for publication

Proussalidis J, Hatziargyriou N, Kladas A (2001) Iron lamination efficient representation in power transformers. Journal of Materials Processing Technology 108:217–220

Raitsios P (2001) Leakage field of a transformer under conventional and superconducting condition. Journal of Materials Processing Technology 102:246–252

Rajakovic N, Semlyen A (1989a) Harmonic domain analysis of field variables related to eddy current and hysteresis losses in saturated laminations. IEEE Transactions on Power Delivery 4(2):1111–1116

Rajakovic N, Semlyen A (1989b) Investigation of the inrush phenomenon: a quasi-stationary approach in the harmonic domain. IEEE Transactions on Power Delivery 4(4):2114–2120

Rovolis P, Kladas A, Tegopoulos J (2007) Laminated iron core losses evaluation and measurements. Journal of Materials Processing Technology 181(1-3):182–185

Sato T, Sakaki Y (1990) Physical meaning of equivalent loss resistance of magnetic cores. IEEE Transactions on Magnetics 26:2894–2897

Semlyen A, Rajakovic N (1989) Harmonic domain modeling of laminated iron core. IEEE Transactions on Power Delivery 4(1):382–390

Soda N, Enokizono M (2000) Improvement of T-joint part constructions in three-phase transformer cores by using direct loss analysis with E&S model. IEEE Transactions on Magnetics 36(4):1285–1288

Taguchi G, Konishi S (1987) Taguchi methods: orthogonal arrays and linear graphs; tools for quality engineering. ASI, Dearborn, MI

References

Tatis KB, Kladas AG, Tegopoulos JA (2004) Harmonic iron loss determination in laminated iron cores by using a particular 3-D finite-element model. IEEE Transactions on Magnetics 40(2):860–863

teNyenhuis EG, Girgis RS (2006) Measured variability of performance parameters of power and distribution transformers. Proc IEEE Power Engineering Society Transmission and Distribution Conference and Exposition, 523–528

teNyenhuis EG, Mechler GF, Girgis RS (2000) Flux distribution and core loss calculation for single phase and five limb three phase transformer core designs. IEEE Transactions on Power Delivery 15(1):204–209

teNyenhuis EG, Girgis RS, Mechler GF (2001) Other factors contributing to the core loss performance of power and distribution transformers. IEEE Transactions on Power Delivery 16(4):648–653

Tomczuk B (1988) Analysis of 3D magnetic fields in high leakage reactance transformers. IEEE Transactions on Magnetics 30:94–97

Tsili MA (2005) Development of mixed finite element – boundary element numerical techniques for the design of power transformers. PhD dissertation. National Technical University of Athens, Athens, Greece

Tsili MA, Kladas AG, Georgilakis PS, Souflaris AT, Pitsilis CP, Bakopoulos JA, Paparigas DG (2004) Hybrid numerical techniques for power transformer modeling: a comparative analysis validated by measurements. IEEE Transactions on Magnetics 40(2):842–845

Tsili M, Kladas A, Georgilakis P, Souflaris A, Paparigas D (2005) Numerical techniques for design and modeling of distribution transformers. Journal of Materials Processing Technology 161(1-2):320–326

Tsili MA, Kladas AG, Georgilakis PS, Souflaris AT, Paparigas DG (2006) Advanced design methodology for single and dual voltage wound core power transformers based on a particular finite element model. Electric Power Systems Research 76:729–741

Tsili MA, Kladas AG, Georgilakis PS (2008) Computer aided analysis and design of power transformers. Computers in Industry 59(4):338–350

Turowski J, Turowski M, Kopec M (1990) Method of three-dimensional network solution of leakage field of three-phase transformers. IEEE Transactions on Magnetics 26(5):2911–2919

Valkovic Z (1982) Influence of transformer core design on power losses. IEEE Transactions on Magnetics 18:801–804

Valkovic Z (1984) Additional losses in three-phase transformer cores. Journal of Magnetism and Magnetic Materials 41:424–426

Valkovic Z, Rezic A (1992) Improvement of transformer core magnetic properties using the step-lap design. Journal of Magnetism and Magnetic Materials 112:413–415

Xiang C, Jinsha Y, Guoquiang Z, Yuanlu Z, Qifan H (1997) Analysis of leakage magnetic problems in shell-form power transformer. IEEE Transactions on Magnetics 33(2):2049–2051

Zakrzewski K, Kukaniszyn M (1992) Three-dimensional model of one- and three-phase transformer for leakage field calculation. IEEE Transactions on Magnetics 28(2):1344–1347

Zakrzewski K, Tomczuk B (1996) Magnetic field analysis and leakage inductance calculation in current transformer by means of 3D integral method. IEEE Transactions on Magnetics 32(3):1637–1640

7 Transformer Design Optimization

Abstract This chapter deals with modern design optimization of wound core type transformers. Four methods are presented that solve important transformer design problems. First, genetic algorithms are combined with artificial neural networks to optimally group $4 \cdot N$ available individual cores into N transformers so as to minimize the total no-load loss of N transformers. This method significantly reduces the no-load loss design margin as well as the cost of transformer main materials. Second, decision trees and artificial neural networks successfully solve the winding material selection problem, thus avoiding the need to optimize the transformer twice, once with copper and once with aluminum windings. Third, a mixed integer programming–finite element method (MIP-FEM) technique is developed for the solution of the transformer design optimization (TDO) problem. Finally, a recursive genetic algorithm–finite element method technique is developed to solve the TDO problem and is compared with MIP-FEM. The recursive genetic algorithm approach can be also very useful for the solution of other optimization problems in electric machines and power systems.

7.1 Introduction

This chapter solves three very important design problems of wound core type transformers:

1. *No-load loss reduction problem* or optimal individual core grouping problem. After the completion of core manufacturing and before assembly of the transformer active part, $2 \cdot N$ small individual cores and $2 \cdot N$ large individual cores are available that have to be optimally combined into N transformers so as to minimize the total no-load loss of N transformers. This problem is solved using genetic algorithms in combination with artificial neural networks (Sect. 7.2).

2. *Winding material selection problem.* The transformer windings material can be copper or aluminum. Since copper and aluminum are stock exchange commodities, their prices can change significantly through time. Thus, in some transformer designs, it is more economical to use copper instead of aluminum windings, and in others the opposite is true. The winding material selection problem is solved using decision trees and artificial neural networks (Sect. 7.3).

3. *Transformer design optimization problem.* The objective of transformer design optimization (TDO) is to design the transformer so as to optimize an objective

function subject to constraints imposed by international standards and transformer specifications. The most commonly used objective functions are: (a) the minimization of transformer manufacturing cost; and (b) the minimization of transformer total owning cost. This chapter solves the TDO problem using a mixed integer programming–finite element method (Sect. 7.4) as well as a recursive genetic algorithm–finite element method (Sect. 7.5).

7.2 No-Load Loss Reduction with Genetic Algorithms

7.2.1 Introduction

Construction of distribution transformers of high quality at minimum possible cost is crucial for any transformer manufacturing industry facing market competition. A critical measure of transformer quality is transformer no-load loss. The less the transformer no-load loss, the higher the transformer quality and efficiency become. The transformer designer can reduce no-load loss by using lower loss core materials or reducing core flux density or flux path length (Kennedy 1998).

Electric utilities use more generating capacity to produce additional electrical energy to compensate for transformer energy losses. The production of this additional electrical energy increases electrical energy cost as well as greenhouse gas emissions. Although distribution transformers inherently have high energy transfer efficiencies, the accumulated transformer energy losses in an electric utility distribution network are high since a large number of distribution transformers is installed. In addition, transformer no-load loss appears 24 hours per day, every day, for a continuously energized transformer. Thus, it is in general preferable to design a transformer for minimum no-load loss (Heathcote 2007).

Transformer actual (measured) no-load loss deviates from designed no-load loss due to variability in the production process (teNyenhuis and Girgis 2006). Reduction of transformer actual no-load loss is a very important task for any manufacturing industry, since (1) it helps the manufacturer not to pay no-load loss penalties, and (2) it reduces the material cost (since a smaller no-load loss design margin is used).

7.2.2 Conventional Core Grouping Process

The three-phase wound core distribution transformer is composed of two small and two large individual cores (Fig. 7.1a). Industrial experiments have shown that if the position of one core within the transformer changes, then the transformer no-load loss also changes (Georgilakis et al. 1999, 2001; Georgilakis 2000), e.g., the

7.2 No-Load Loss Reduction with Genetic Algorithms

transformer with core arrangement $S_1 - L_1 - L_2 - S_2$ has different transformer no-load loss in comparison with the transformer with core arrangement $S_2 - L_1 - L_2 - S_1$ (Fig. 7.1b). The small cores S_1 and S_2 theoretically have the same technical characteristics (e.g., individual core no-load loss), however, in practice their characteristics are different (due to the variability in production process), so the above two mentioned core arrangements have different transformer no-load loss due to the non homogeneous electromagnetic field of the individual cores (Georgilakis 2000, Georgilakis et al. 2001).

Figure 7.1a shows that one transformer requires two small individual cores and two large individual cores. Assuming that $2 \cdot N$ small individual cores and $2 \cdot N$ large individual cores are available, then N transformers can be assembled. Each small (resp. large) individual core can be put to any of the two outer (resp. middle) positions and to any of N transformers. From all possible combinations of grouping N transformers, only one combination, providing the optimum (minimum) total no-load loss, should be selected. In brief, the transformer no-load loss reduction (TNLLR) problem seeks the optimum arrangement of the four individual cores in each one of N transformers so as to minimize the total no-load loss of N transformers.

The current industrial practice to solve the TNLLR problem is to pre-measure and assign a grade (quality category) to each individual core and then combine higher and lower graded individual cores to achieve an "average" value for the entire transformer (Georgilakis et al. 2001). This is referred to as the *conventional core grouping process* (CGP).

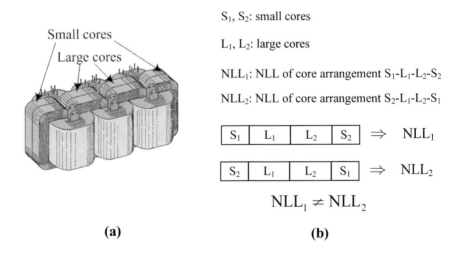

Fig. 7.1 Three-phase wound core distribution transformer: (a) assembled active part, (b) impact of core position on transformer no-load loss (NLL)

7.2.3 Genetic Algorithm Solution to the TNLLR Problem

7.2.3.1 Formulation of the TNLLR Problem

N three-phase transformers are constructed from $2 \cdot N$ small individual cores and $2 \cdot N$ large individual cores. Let us denote as V_s (V_l) the set of all $2 \cdot N$ small (large) cores. A transformer is represented by a vector $\mathbf{t}_i$, the elements of which correspond to the four individual cores that assemble the transformer:

$$\mathbf{t}_i = \begin{bmatrix} s_i^l & l_i^l & l_i^r & s_i^r \end{bmatrix}^T. \tag{7.1}$$

Variables $s_i^l, s_i^r \in V_s$ represent the left and right small core of transformer $\mathbf{t}_i$, while $l_i^l, l_i^r \in V_l$ the left and right large core, respectively. Since only one core (small or large) can be assigned to one transformer and one position (left or right), the following restrictions hold:

$$s_i^l \neq s_i^r, \quad l_i^l \neq l_i^r, \tag{7.2}$$

$$s_k^{\{l,r\}} \neq s_i^{\{l,r\}}, \quad l_k^{\{l,r\}} \neq l_i^{\{l,r\}}, \quad \text{with} \quad k \neq i, \tag{7.3}$$

where $s_i^{\{l,r\}}$ ($l_i^{\{l,r\}}$) indicates the small (large) core in the left or right position for transformer $\mathbf{t}_i$.

Let us denote as $\mathbf{c}$ a vector containing one possible combination of N three-phase transformers $\mathbf{t}_i$, $i = 1, 2, ..., N$, that can be constructed by $2 \cdot N$ small individual cores and $2 \cdot N$ large individual cores:

$$\mathbf{c} = \begin{bmatrix} \mathbf{t}_1^T & \mathbf{t}_2^T & \cdots & \mathbf{t}_N^T \end{bmatrix}^T, \tag{7.4}$$

where T indicates the transpose of a vector.

Vector $\mathbf{c}$ is of $4N \times 1$ dimensions since each transformer $\mathbf{t}_i$ is represented by a 4×1 vector as (7.1) indicates. A specific arrangement (combination) of all small and large cores for constructing the N three-phase transformers corresponds to a given value of vector $\mathbf{c}$. Therefore, any reordering of the elements of vector $\mathbf{c}$ results in different arrangement of individual cores, i.e., different three-phase transformers. Figure 7.2 presents an example of vector $\mathbf{c}$ in the case that six small and six large cores are available. The serial numbers 1 to 6 correspond to small cores, while the numbers from 7 to 12 to large cores. A randomly selected arrangement of these cores is also presented in Fig. 7.2 for constructing three different transformers. For example, the first transformer consists of the small cores 5 and 1 and

7.2 No-Load Loss Reduction with Genetic Algorithms

large cores 10 and 12. This is represented by the vector $\begin{bmatrix} 5 & 10 & 12 & 1 \end{bmatrix}^T$ in accordance with (7.1). Then, vector **c** is constructed by concatenating the vectors of the three transformers. The core arrangement for the other two transformers is generated accordingly and depicted in Fig. 7.2.

It is clear that the estimation of N transformers with optimal quality (minimum total no-load loss) is equivalent to the estimation of vector **c**, which minimizes the following:

$$\mathbf{c}_{opt} = \arg\min_{\mathbf{c}} \left\{ \sum_{i=1}^{N} NLL_{\mathbf{t}_i} \right\}, \tag{7.5}$$

where $NLL_{\mathbf{t}_i}$ is the actual no-load loss of transformer $\mathbf{t}_i$ and $\mathbf{c}_{opt}$ is a vector that contains the optimal arrangement of all available small and large cores so that the actual total no-load loss over all N transformers is minimized.

The actual no-load loss of each transformer $\mathbf{t}_i$ is computed as follows:

$$NLL_{\mathbf{t}_i} = w_{\mathbf{t}_i} \cdot SNLL_{\mathbf{t}_i}, \quad \forall \, \mathbf{t}_i, \quad i = 1, 2, \ldots, N, \tag{7.6}$$

where $w_{\mathbf{t}_i}$ is the actual (measured) weight of the four individual cores of transformer $\mathbf{t}_i$ and $SNLL_{\mathbf{t}_i}$ is the specific no-load loss (W/kg) of transformer $\mathbf{t}_i$ estimated by the multi-layer perceptron (MLP) architecture of Sect. 6.3.4.1.

The actual no-load loss of each transformer $\mathbf{t}_i$ must be smaller than a maximum no-load loss, NLL_{max}:

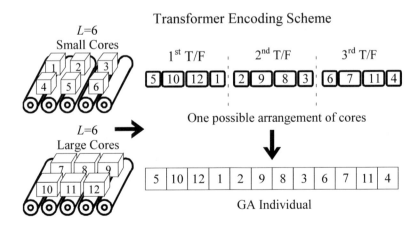

Fig. 7.2 Example of adopted encoding scheme in the case of six small and six large cores

$$NLL_{t_i} < NLL_{max} \quad , \quad \forall \, \mathbf{t}_i \, , \, i = 1, 2, ..., N \, . \tag{7.7}$$

In brief, the transformer no-load loss reduction problem is mathematically formulated as follows: minimize the objective function (7.5) subject to the constraints (7.2), (7.3), and (7.7).

As observed from (7.5), estimation of the optimal core arrangement results in a combinatorial optimization problem. For a typical number of small/large cores, direct minimization of (7.5) is practically infeasible since the computational complexity for an exhaustive search is very high. For example, assuming that 100 small and 100 large individual cores are available, about 5.35×10^{22} combinations of core arrangements should be considered (Doulamis et al. 2002). For this reason, the TNLLR problem is solved using the genetic algorithm method (Goldberg 1989; Michalewicz 1994).

7.2.3.2 Genetic Algorithm Solution

In the genetic algorithm (GA) approach, possible solutions of the optimization problem are represented by chromosomes whose "genetic material" corresponds to a specific arrangement of individual cores. This means that the vector **c** of (7.4) is represented by a chromosome, while the serial numbers of individual cores are considered as the genetic material of the chromosome. An integer number scheme is adopted for encoding the chromosome elements (genes) as is illustrated in Fig. 7.2.

Initially, M different chromosomes, say $\mathbf{c}_1, \mathbf{c}_2, ..., \mathbf{c}_M$, are created to form a population. In our case, M possible solutions of the grouping method used in the conventional core grouping process are selected for the initial population. This is performed so that the genetic material of the initial chromosomes is of good quality and thus fast convergence of the GA is achieved. The performance of each chromosome representing a particular core arrangement is evaluated by the sum of the predicted actual no-load loss of all transformers corresponding to this chromosome. The neural network model is used as no-load loss predictor. For each chromosome, a fitness function is used to map its performance to a fitness value, following a rank-based normalization scheme. In particular, all chromosomes $\mathbf{c}_i$, $i = 1, 2, ..., M$ are ranked in ascending order according to their performance, i.e., the sum of the predicted transformer no-load loss. Let $rank(\mathbf{c}_i) \in \{1, 2, ..., M\}$ be the rank of chromosome $\mathbf{c}_i$ ($rank = 1$ corresponds to the best chromosome and $rank = M$ to the worst). Defining an arbitrary fitness value F_0 for the best chromosome, the fitness $F(\mathbf{c}_i)$ of the ith chromosome is given by the function:

$$F(\mathbf{c}_i) = F_0 - [rank(\mathbf{c}_i) - 1] \cdot \mu, \quad i = 1, 2, ..., N, \tag{7.8}$$

7.2 No-Load Loss Reduction with Genetic Algorithms

where μ is a decrement rate and is computed in such a way that the fitness function $F(\mathbf{c}_i)$ always takes positive values, that is $\mu < F_0 /(M-1)$. The major advantage of the rank-based normalization is that it prevents the generation of super chromosomes, avoiding premature convergence to local minima, since fitness values are uniformly distributed (Goldberg 1989; Fonseca and Fleming 1993).

The parent selection mechanism then begins by selecting appropriate chromosomes (parents) from the current population. The roulette wheel (Michalewicz 1994) is used as the parent selection procedure. This is accomplished by assigning to each chromosome a selection probability equal to the ratio of the fitness value of the respective chromosome to the sum of fitness values of all chromosomes, i.e.:

$$p_p(\mathbf{c}_i) = \frac{F(\mathbf{c}_i)}{\sum_{i=1}^{M} F(\mathbf{c}_i)}, \qquad (7.9)$$

where $p_p(\mathbf{c}_i)$ is the probability of the chromosome $\mathbf{c}_i$ being selected as parent. The equation (7.9) means that chromosomes of high quality have a higher chance of survival in the next generation. Using this scheme, M chromosomes are selected as candidate parents to generate the next population. Obviously, some chromosomes would be selected more than once, which is in accordance with the Schema Theorem (Michalewicz 1994); the best chromosomes have more copies, the "average" stay even, while the worst die off. Consequently, each chromosome has a growth rate proportional to its fitness value.

In the following step of the algorithm, couples of chromosomes (two parents) are randomly selected from the set of candidates, obtained from the parent selection mechanism. Then, their genetic material is mated to generate new chromosomes (offspring). The number of couples selected depends on a crossover rate. A crossover mechanism is also used to define how the genes should be exchanged to produce the next generation. Several crossover mechanisms have been reported in the literature. In the TNLLR problem, a modification of the uniform crossover operator (Goldberg 1989; Michalewicz 1994) has been adopted. As is explained in Sect. 7.2.3.3, this modification does not spoil the GA convergence. In this case, each parent gene, i.e., an individual core, is considered as a potential crossover point. In particular, a gene is exchanged (undergoes crossover), if a random variable, uniformly distributed in the interval $[0 \ 1]$, is smaller than a pre-determined threshold. Otherwise, the gene remains unchanged. It is possible however for an individual core to appear more than once in the genetic material of the generated chromosome. This means that one individual core is placed in more than one position of the same transformer or in more than one transformer, which corresponds to an unacceptable core arrangement that violates the constraint (7.2) or (7.3), respectively. For this reason, the following modification of the uniform crossover operator is adopted. After the exchange of one gene between the two parents, it is highly possible that the gene appears twice in the chromosome. In this case the gene coinciding with the new gene is replaced with the gene before the exchange. Figure 7.3 illustrates an example of the adopted crossover mechanism in the case that six small and six large cores are assembled to generate three transformers. In this example, the two parents exchange their genes only between the crossover points 2, 3 and 4 for simplicity. As observed, the genes $\{10, 12, 1\}$ of the 1st parent are exchanged with the genes $\{8, 12, 3\}$ of the 2nd parent. By applying this exchange of genes, in the 1st chromosome the genes 8 and 3 appear twice, while genes 10 and 1 disappear. An equivalent problem occurs in the 2nd chromosome. For this reason, in the 1st chromosome the genes $\{10, 1\}$ are one-by-one exchanged with genes $\{8, 3\}$ as Fig. 7.3 depicts. The same happens for the 2nd chromosome.

The next step is to apply mutation to the newly created population, introducing random gene variations that are useful for restoring lost genetic material, or for producing new material that corresponds to new search areas (Michalewicz 1994). Uniform mutation is the most common mutation operator and is selected for the TNLLR problem. In particular, for each gene a uniform number is generated in the interval $[0\ 1]$ and if this number is smaller than the mutation rate the respective gene is swapped for other randomly selected gene of the same category, i.e., small or large core. Otherwise, the gene remains unchanged. In the TNLLR problem, the mutation rate is selected to be 5%. Swapping genes of the same category is necessary to create a valid core arrangement.

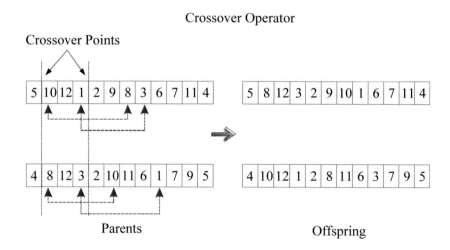

Fig. 7.3 Example of the modification of the crossover operator

7.2 No-Load Loss Reduction with Genetic Algorithms

Table 7.1 Summary of the main steps in the GA method to solve the TNLLR problem

Step 1:	Based on customer requirements and several techno-economical criteria, design the transformers of a specific production batch. From the transformer design, the environment type (i.e., supplier, thickness and grade of magnetic material) is defined.
Step 2:	Based on transformer design, construct the small and large individual cores and measure all necessary parameters (i.e., the actual no-load loss and weight) so that the 19 attributes of Table 6.12 for a specific core arrangement can be calculated.
Step 3:	Use the GA to minimize the total no-load loss for all transformers of the production batch (7.5). At each GA cycle, the neural network architecture of Sect. 6.3.4.1 is used to estimate transformer specific no-load loss. The attributes of Table 6.12 are used as input to the neural network,
Step 4:	Assemble the transformers using the optimal core arrangement $\mathbf{c}_{opt}$ that has been computed by the GA.
Step 5:	Measure the actual no-load loss for all constructed transformers of the production batch. Then, compare them with the predicted no-load losses, which are provided by the neural network structure.
Step 6:	In the case of large differences between the measured and predicted no-load loss, retrain the artificial neural network (Sect. 6.3.4.4). Then, store the new estimated weights in the neural network database to be used for the following production batches. Otherwise, retain the same neural network weights.

At each iteration, a new population is created by inserting the new chromosomes, generated by the crossover mechanism, and deleting their respective parents, so that each population always consists of M chromosomes. Several GA cycles including fitness evaluation, parent selection, crossover and mutation are repeated, until the population converges to an optimal solution. The GA terminates when the best chromosome fitness remains constant for a large number of generations, indicating that further optimization is unlikely.

Table 7.1 summarizes the main steps of the GA method for solution of the TNLLR problem.

7.2.3.3 Genetic Algorithm Convergence

The modifications of the crossover and mutation operators of Sect. 7.2.3.2 do not affect the convergence property of the GA. To show this, an analysis is presented in the following, by modeling the GA as a Markov Chain. In particular, each state of the Markov state corresponds to a possible solution of the GA, i.e., a specific vector $\mathbf{c}$. Let us denote as D a set that contains all possible Markov states. Then, for two arbitrary states, say, $i, j \in D$, we denote as p_{ij} the transition probability from state i to state j. Gathering transition probabilities for all states in D, the transition matrix of the chain is formed as $\mathbf{P} = (p_{ij})$. Since in the GA, transition from one state to another is obtained by applying the crossover and mutation operator, matrix $\mathbf{P}$ can be decomposed as follows (Rudolph 1994):

$$\mathbf{P} = \mathbf{C} \cdot \mathbf{M}, \tag{7.10}$$

where matrix $\mathbf{C}$ indicates the effect of the crossover operator and matrix $\mathbf{M}$ the effect of the mutation operator.

Let us denote as c_{ij} the elements of matrix $\mathbf{C} = (c_{ij})$. c_{ij} expresses the transition probability from state $i \in D$ to state $j \in D$, if only the effect of the crossover operator is taken into consideration. Since the crossover operator probabilistically maps any state of D to any other state of D, matrix $\mathbf{C}$ is a stochastic matrix. More specifically, a matrix is said to be stochastic if its elements c_{ij} satisfy the following property:

$$\sum_j c_{ij} = 1 \Rightarrow \text{Matrix } \mathbf{C} = (c_{ij}) \text{ is stochastic}. \tag{7.11}$$

The previous equation means that from a valid solution (i.e., a state of D), the crossover operator produces another valid solution (i.e., another state of D). This exactly happens with the proposed modification of the crossover operator, since only valid solutions are permitted.

On the other hand, matrix $\mathbf{M}$ is positive. This is due to the fact that the mutation operator is applied independently to each gene of a chromosome. Furthermore, each gene can potentially undergo mutation. Consequently, the elements m_{ij} of matrix $\mathbf{M}$, which express the transition probabilities from state $i \in D$ to state $j \in D$ taking into account only the effect of the mutation operator, are strictly positive:

$$m_{ij} > 0 \Rightarrow \text{Matrix } \mathbf{M} = (m_{ij}) \text{ is positive}. \tag{7.12}$$

It has been proven (Rudolph 1994) that if matrix $\mathbf{C}$ is stochastic and matrix $\mathbf{M}$ is positive, the transition matrix $\mathbf{P} = \mathbf{C} \cdot \mathbf{M}$ [see (7.10)] of the Markov chain is primitive (i.e., there exists $k > 0 : \mathbf{P}^k$ is positive). In this case, it has been shown (Rudolph 1994) that the GA converges to the optimum solution if the best solution is maintained over time. This means that, starting from any arbitrary state (valid solution), the algorithm visits any other state (valid solution) within a finite number of transitions.

7.2 No-Load Loss Reduction with Genetic Algorithms 341

Table 7.2 Optimization results for grouping 100 small and 100 large cores of the same transformer design with 100 kVA rated power

Method	Total no-load loss (W)	Reduction of total no-load loss (% of CGP)	Execution time (min)
CGP	11221	0.00	0.23
GA	10846	3.34	2.45

7.2.4 Results

7.2.4.1 No-Load Loss Prediction

The MLP based no-load loss prediction technique has been extensively tested in the transformer industry, the results have been presented in Sect. 6.3.4.1, and the conclusion is that the MLP method improves the no-load loss prediction accuracy by more than 65% in comparison with the no-load loss curve method (current practice).

7.2.4.2 No-Load Loss Reduction

Table 7.2 presents results from the application of the GA technique of Sect. 7.2.3.2 in order to group 100 small and 100 large cores of the same production batch of 50 transformers, 100 kVA, 50 Hz. Table 7.2 shows that the GA method finds the global optimum (minimum) no-load loss as it combines the cores in such a way that the 50 transformers have total no-load loss equal to 10846 W, i.e., the GA method reduces the total no-load loss by 3.34% in comparison with the conventional core grouping process (CGP). As can be seen from Table 7.2, the execution time of both methods (CGP and GA) is very low, so both methods are applicable in an actual industrial environment, however, the GA method is better, since it finds the global optimum no-load loss. It should be noted that the superiority of the GA method has been verified on numerous transformer production batches of several power and voltage ratings (Georgilakis 2000; Georgilakis et al. 2001).

7.2.4.3 Exploitation of Results

It was found in Sect. 7.2.4.2 that the best technique for solution of the TNLLR problem is the GA method. That is why several core production batches of various power and voltage ratings have been grouped using the GA method and the average absolute relative error (*AARE*) of these batches is 1.63%, as Table 7.3 shows. This is compared with 5.22% *AARE* for the CGP, as Table 7.3 shows. It should be noted that the low *AARE* is due to the accuracy of the MLP used to estimate transformer no-load loss during the GA based grouping process. On the other hand, in case of the CGP, the *AARE* derives from the accuracy of the no-load loss curve (e.g., Fig. 2.8) used to estimate transformer no-load loss. The *AARE* is computed using (6.2). The transformer designer can use a no-load loss design margin, e.g., 15% higher than the respective *AARE* value, so no-load loss design margins of 1.87% and 6.00% have been used when the core grouping process is based on GA and CGP, respectively, as can be seen from Table 7.3.

The significant reduction of no-load loss design margin due to the effectiveness of the GA method in solving the TNLLR problem yields significant reduction in the cost of transformer main materials. As an example, the design of the same transformer (same specification) with 160 kVA rated power and 315 W no-load loss is implemented twice. The first design, denoted Design 1, uses 6.00% no-load loss design margin (the core grouping is based on CGP), so the designed no-load loss is $315 \times (1-0.06)$, i.e., 296 W. The second design, denoted Design 2, uses 1.87% no-load loss design margin (the core grouping is based on GA), so the designed no-load loss is $315 \times (1-0.0187)$, i.e., 309 W. With the help of appropriate software that is based on a parallel mixed integer programming–finite element method (Amoiralis et al. 2008), the above two transformer designs are optimized, i.e., their main materials cost is minimized and the results are shown in Table 7.4. Using the cost data of Table 7.4, Fig. 7.4 is created to compare the cost of materials of the two different designs of Table 7.4. Figure 7.4 shows that a $100 - 92.3$, i.e., 7.7% cost saving on the four main materials is obtained due to the reduced no-load loss design margin, made possible by the use of a GA based grouping method. It can also be seen from Fig. 7.4 that the cost saving of magnetic material is 12%, while the cost saving of winding material (copper) is 3.3%.

Table 7.3 Accuracy of TNLLR methods and no-load loss design margin

Method	*AARE* (%)	No-load loss design margin (%)
CGP	5.22	6.00
GA	1.63	1.87

Table 7.4 Comparison of the cost of main materials of two different designs for the same 160 kVA transformer specification

Description	Design 1 (CGP)	Design 2 (GA)
Rated power (kVA)	160	160
Specified no-load loss (W)	315	315
No-load loss design margin (%)	6.00	1.87
Designed no-load loss (W)	296	309
Magnetic material cost ($)	1825.18	1606.14
Winding material cost ($)	1626.32	1573.16
Insulating material cost ($)	194.01	171.98
Oil cost ($)	298.45	289.21
Main materials cost ($)	3943.96	3640.49

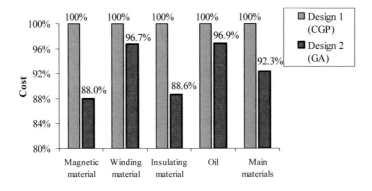

Fig. 7.4 Comparison of the cost of main materials of the two different designs of Table 7.4 for the same 160 kVA transformer specification

7.3 Winding Material Selection with Decision Trees and Artificial Neural Networks

7.3.1 Introduction

The variation in the cost of the materials used in transformer manufacturing has a direct impact on the design of the technically and economically optimum transformer. The material of transformer windings can be copper (Cu) or aluminum (Al). Since copper and aluminum are stock exchange commodities, their price can significantly change through time (Tilton 2002; Sullivan 2008). Thus, in some transformer designs, it is more economical to use copper instead of aluminum windings, and in others the opposite is true (Amoiralis et al. 2007; Georgilakis and Amoiralis 2007; Georgilakis et al. 2007a). In addition, both materials have different technical characteristics. The choice of copper or aluminum has not only been of interest to transformer designers, but the advantages and disadvantages of selecting copper versus aluminum have been investigated for induction motors (Craggs 1976; Poloujadoff et al. 1995; Finley and Hodowanec 2001) and transmission lines (Wheeler 1980; Mohtar et al. 2004).

In order to check which winding material results in a more economical solution, there is a need to optimize the transformer twice (once with copper and once with aluminum windings) and afterwards to select the most economical design. Solution of the winding material selection problem can be implemented using artificial intelligence, since artificial intelligence has been proven very efficient in solving problems in the transformer industry. In this section, decision trees and adaptive trained neural networks (ATNNs) are combined with the aim of automatically selecting the appropriate winding material to design an optimum distribution transformer (Georgilakis and Amoiralis 2007).

7.3.2 Creation of Knowledge Base

One of the most crucial steps in artificial intelligence methodologies is undoubtedly the creation of a knowledge base, which is composed of the learning, validation (in the case of the ATNN), and test set. In order to generate these sets, six transformer power ratings (250, 400, 630, 800, 1000 and 1600 kVA) are considered. For each transformer, nine categories of losses are taken into account, namely AA', AB', AC', BA', BB', BC', CA', CB', CC' (CENELEC 1992). For example, a 250 kVA transformer with AC' category losses has 3250 W of load losses and 425 W of no-load losses. Seven different unit costs ($/kg) are considered for Cu and Al winding. Based on the above, $6 \cdot 9 \cdot 7 = 378$ transformer design optimizations with Cu winding (Cu designs) and 378 transformer design optimizations with Al winding (Al designs) are realized. For each of them, either the Cu design or the Al design is the final optimum design, i.e., the one having the least manufacturing cost. In total, $6 \cdot 9 \cdot 7^2 = 2646$ final optimum designs (FODs) are collected and stored into the database or other knowledge base. The knowledge base is composed of sets of FODs, and each of them is composed of a collection of input/output pairs. The input pairs or attributes are the parameters affecting the selection of transformer winding material. Thirteen attributes are selected based on extensive research and transformer design experience, as shown in Table 7.5. The output pairs comprise the type of winding (Cu or Al) that corresponds to each FOD.

Table 7.5 Attributes affecting the selection of transformer winding material

Symbol	Definition
I_1	Copper unit cost ($/kg)
I_2	Aluminum unit cost ($/kg)
I_3	I_1/I_2
I_4	Magnetic material unit cost ($/kg)
I_5	I_4/I_1
I_6	I_4/I_2
I_7	Guaranteed no-load loss (W)
I_8	Guaranteed load loss (W)
I_9	I_7/I_8
I_{10}	Rated power (kVA)
I_{11}	Guaranteed impedance voltage (%)
I_{12}	I_7/I_{10}
I_{13}	I_8/I_{10}

7.3 Winding Material Selection with Decision Trees and Artificial Neural Networks

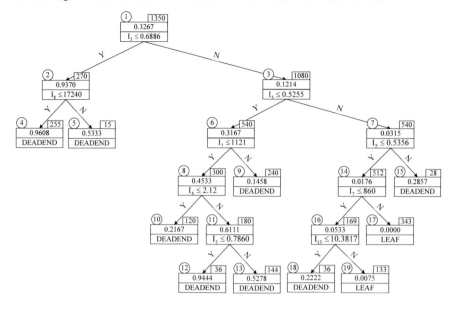

Fig. 7.5 Decision tree for selection of winding material in distribution transformers

Table 7.6 Rules of the decision tree of Fig. 7.5

Node	Rule description
4	If $I_3 \leq 0.6886$ and $I_8 \leq 17240 \Rightarrow$ Cu
5	If $I_3 \leq 0.6886$ and $I_8 > 17240 \Rightarrow$ Cu
9	If $I_3 > 0.6886$ and $I_5 \leq 0.5255$ and $I_7 > 1121 \Rightarrow$ Al
10	If $I_3 > 0.6886$ and $I_5 \leq 0.5255$ and $I_7 \leq 1121$ and $I_4 \leq 2.12 \Rightarrow$ Al
12	If $0.6886 < I_3 \leq 0.7860$ and $I_5 \leq 0.5255$ and $I_7 \leq 1121$ and $I_4 > 2.12 \Rightarrow$ Cu
13	If $I_3 > 0.7860$ and $I_5 \leq 0.5255$ and $I_7 \leq 1121$ and $I_4 > 2.12 \Rightarrow$ Cu
15	If $I_3 > 0.6886$ and $I_5 > 0.5356 \Rightarrow$ Al
17	If $I_3 > 0.6886$ and $0.5255 < I_5 \leq 0.5356$ and $I_7 > 860 \Rightarrow$ Al
18	If $I_3 > 0.6886$ and $0.5255 < I_5 \leq 0.5356$ and $I_7 \leq 860$ and $I_{13} \leq 10.3817 \Rightarrow$ Al
19	If $I_3 > 0.6886$ and $0.5255 < I_5 \leq 0.5356$ and $I_7 \leq 860$ and $I_{13} > 10.3817 \Rightarrow$ Al

Table 7.7 Classification success rate on the test set for the decision tree of Fig. 7.5

Node	Cu index	Transformer designs	Correctly classified transformer designs	Classification success rate (%)
4	0.9608	250	244	97.60
5	0.5333	8	5	62.50
9	0.1458	226	196	86.73
10	0.2167	110	93	84.55
12	0.9444	37	36	97.30
13	0.5278	125	72	57.60
15	0.2857	26	21	80.77
17	0.0000	347	347	100.00
18	0.2222	30	27	90.00
19	0.0075	137	137	100.00
	Total	1296	1178	90.90

7.3.3 Decision Trees

The learning set is composed of 1350 sets of FODs and the test set has 1296 independent sets of FODs. Figure 7.5 illustrates the decision tree for the selection of transformer winding material, which is automatically constructed using the learning set of 1350 FODs with the 13 attributes of Table 7.5. The notation used for the decision tree nodes is explained in Fig. 4.6. Each terminal node of the decision tree of Fig. 7.5 produces one decision rule, on the basis of its Cu index, i.e., the ratio of Cu designs over the FODs of that node. For example, from terminal node 17 of the decision tree of Fig. 7.5 the following decision rule is derived: if $I_3 > 0.6886$ and $I_5 > 0.5255$ and $I_5 \leq 0.5356$ and $I_7 > 860$ then select aluminum winding, since the Cu index of node 17 is 0.0. Table 7.6 presents the ten decision rules of the ten terminal nodes of the decision tree of Fig. 7.5.

It is also important to note that, among the 13 attributes, the decision tree method automatically selects the six most important ones (attributes I_3, I_4, I_5, I_7, I_8, and I_{13}) that appear at the various test nodes of the decision tree of Fig. 7.5. The selection of the above six attributes is reasonable and expected, since they are all related to the selection of the winding material (Cu or Al) in distribution transformers. Thus, taking for granted the values of the six above-mentioned attributes, the decision tree of Fig. 7.5 estimates the appropriate winding material (Cu or Al) from which the distribution transformer has to be designed.

7.3 Winding Material Selection with Decision Trees and Artificial Neural Networks 347

The decision tree of Fig. 7.5 achieves a total classification success rate of 90.9% on the independent test set, i.e., the decision tree correctly classifies 1178 of the 1296 FODs of the test set (Table 7.7). This high classification success rate makes the decision tree method suitable for the selection of winding material in distribution transformers.

7.3.3.1 Example 7.1

Using the appropriate input data for the transformer design example of Sect. 2.5 as well as the decision tree of Fig. 7.5, select the winding material of this transformer. Consider that the aluminum unit cost is 6.3 $/kg.

Solution

Six attributes (I_3, I_4, I_5, I_7, I_8, and I_{13}) appear at the various test nodes of the decision tree of Fig. 7.5, so we need the values of these six attributes in order to use the Fig. 7.5 to select the transformer winding material.

The transformer windings are made of copper (Sect. 2.5). It can be seen from Table 2.9 that the copper unit cost is $I_1 = 12.01\,\$/kg$ and the magnetic material unit cost is $I_4 = 6.01\,\$/kg$. The value of attribute I_3 (defined in Table 7.5) is computed as follows:

$$I_3 = \frac{I_1}{I_2} = \frac{12.01\,\$/kg}{6.3\,\$/kg} \Rightarrow I_3 = 1.91.$$

The value of attribute I_5, which is defined in Table 7.5, is:

$$I_5 = \frac{I_4}{I_1} = \frac{6.01\,\$/kg}{12.01\,\$/kg} \Rightarrow I_5 = 0.50.$$

It can be seen from Table 2.8 that the guaranteed no-load loss is $I_7 = 1100\,W$ and the guaranteed load loss is $I_8 = 8900\,W$. The transformer rated power is $I_{10} = 630\,kVA$ (Table 2.6). The value of attribute I_{13} (defined in Table 7.5) is:

$$I_{13} = \frac{I_8}{I_{10}} = \frac{8900\,W}{630\,kVA} \Rightarrow I_{13} = 14.13\,\frac{W}{kVA}.$$

The values of the six attributes are shown in Table 7.8.
We start from node 1 of the decision tree of Fig. 7.5. The transformer has $I_3 = 1.91$, so the node 1 test $I_3 \leq 0.6886$ is not satisfied, so we go to node 3.

Table 7.8 Values of attributes for the transformer design example of Section 2.5

Symbol	Value	Unit
I_3	1.91	-
I_4	6.01	$/kg
I_5	0.50	-
I_7	1100	W
I_8	8900	W
I_{13}	14.13	W/kVA

The transformer has $I_5 = 0.50$, so the node 3 test $I_5 \leq 0.5255$ is satisfied, so we go to node 6.

The transformer has $I_7 = 1100$, so the node 6 test $I_7 \leq 1121$ is satisfied, so we go to node 8.

The transformer has $I_4 = 6.01$, so the node 8 test $I_4 \leq 2.12$ is not satisfied, so we go to node 11.

The transformer has $I_3 = 1.91$, so the node 11 test $I_3 \leq 0.7860$ is not satisfied, so we go to node 13.

Node 13 is a terminal node with Cu index of 0.5278, so 52.78% of transformers of this node are made of copper and the rest 47.22% are made of aluminum. Consequently, using the decision tree of Fig. 7.5, we found that the transformer design example of Sect. 2.5 has to be designed with copper windings.

7.3.3.2 Example 7.2

Remove the attributes I_3, I_5, I_6, I_9, I_{12}, and I_{13} from the attribute list of Table 7.5 and compute the classification success rate of the new decision tree for winding material selection.

Solution

We remove the attributes I_3, I_5, I_6, I_9, I_{12}, and I_{13} from the attribute list of Table 7.5. Next, we update the knowledge base and split it into two independent sets: the learning set and the test set. We build the new decision tree for winding material selection based on the learning set using an appropriate computer program. Finally, we find that the classification success rate of the new decision tree on the test set is only 84.26% (Amoiralis et al. 2007). The conclusion is that, although attributes I_3, I_5, I_6, I_9, I_{12}, and I_{13} depend on other attributes (as can be seen from Table 7.5), it is very important to consider them, since the classification success rate on the test set would be 90.90%, as Table 7.7 shows. Otherwise, the decision tree would not select the attributes I_3, I_5, I_6, I_9, I_{12}, and I_{13}, which implies a lower classification success rate (84.26%).

7.3.4 Adaptive Trained Neural Networks

Artificial neural networks, due to their highly nonlinear capabilities and universal approximation properties, are proposed to select the appropriate winding material that results in optimum distribution transformer design. This means that the considered problem is a problem of classification into two classes: Cu or Al. At the training stage, the proper artificial neural network architecture (e.g., number and type of neurons and layers) is selected. In addition, an adaptive training mechanism allows the artificial neural network to learn from its mistakes and correct its output by adjusting the parameters (weights) of its neurons. The adaptive training process enhances the performance of the proposed method as additional training data become available. It is important to note that normalization of data is a crucial stage for training the adaptive trained neural network (ATNN). In doing so, it not only facilitates the training process but also helps in shaping the activation function. It should be done so that the higher values do not suppress the influence of lower values, and the symmetry of the activation function is retained.

Moreover, the knowledge base is divided into training, validation, and test sets. The selection of the size of these sets is important for the ATNN behavior. Since the goal is to find the neural network having the best performance on new data, the simplest approach to the comparison of different neural networks is to evaluate the error function using data that is independent of that used for training. Various neural networks are trained by minimization of an appropriate error function defined with respect to a training data set. This function defines the classification failure rate of the winding material selection problem. The performance of the neural networks is then compared by evaluating the error function using an independent validation set, and the network having the smallest error with respect to the validation set is selected. Since this procedure can itself lead to some over-fitting to the validation set, the performance of the selected network should be confirmed by measuring its performance on a third independent set of data called a test set (Fig. 7.6). Consequently, a learning set is used that is always split into a training set and a validation set. After training, each of the different neural network architectures is tested on the basis of an independent test set. Finally, the neural network architecture with the minimum classification failure rate (maximum classification success rate) on the test set is selected, which is the optimum ATNN for the winding material selection problem. When new FODs occur, the neural network is retrained following the above-mentioned training mechanism.

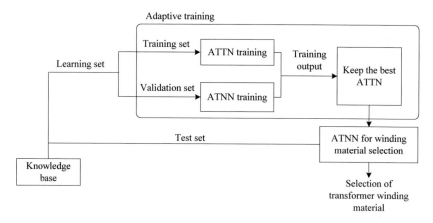

Fig. 7.6 Adaptive training mechanism for transformer winding material selection

7.3.4.1 Optimum Training and Transfer Function

The best training and transfer function for the neural network is selected as follows: taking into consideration an initial investigation that showed that the 13-13-1 architecture (13 input neurons, 13 neurons in the hidden layer, and one single neuron in the output layer) achieved the highest classification success rate on the test set, the training and transfer functions of MATLAB Neural Network Toolbox are combined in order to find the combination that gives the best result, i.e., the highest classification success rate on the test set. Figures 7.7 and 7.8 show the training and transfer function results, respectively, using 1350 FODs from which 675 FODs composed the learning test and the remaining 675 FODs the test set. As shown in Figs. 7.7 and 7.8, the best three training functions are traincgb, traingdx, and trainbfg, and the best three transfer functions are tansig, satlins, and logsig, respectively, since they achieve the highest classification success rates. Next, new experiments are conducted among all the possible combinations of these three best training and transfer functions. As shown in Fig. 7.9, the highest classification success rate is achieved by using the traincgb as training function and the satlins as transfer function. Traincgb is a network training function that updates weight and bias values according to the conjugate gradient backpropagation with Powell–Beale restarts, and satlins is a symmetric saturating linear transfer function. Figure 7.9 shows that this combination achieved 95.71% classification success rate on the test set, which is not only the best classification performance but also this classification success rate is considered very high for the transformer winding material selection problem. It should be noted that the classification success rates of Fig. 7.9 resulted from averaging ten different executions of the algorithm.

7.3 Winding Material Selection with Decision Trees and Artificial Neural Networks

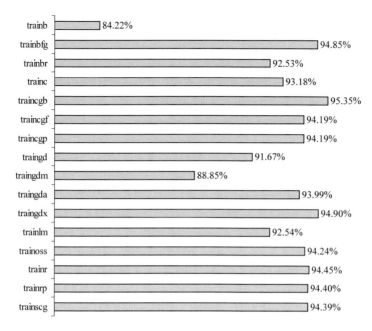

Fig. 7.7 Classification success rate on the test set using 16 different training functions of Table 4.10

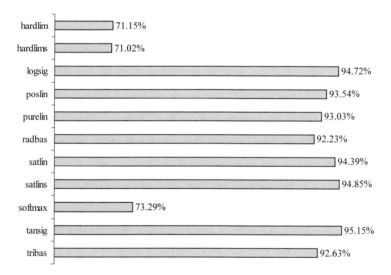

Fig. 7.8 Classification success rate on the test set using 11 different transfer functions of Table 4.11

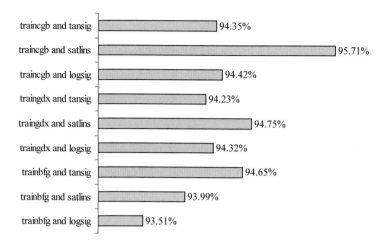

Fig. 7.9 Classification success rate on the test set of the combination of the best three training and transfer functions

7.3.4.2 Optimum ATNN

The primary goal is to find the optimum architecture of the ATNN, which has the highest classification success rate on test set, solving successfully the problem of winding material selection. In order to achieve this goal, extensive research on ATNN has been conducted. To be more precise, experiments were carried out by studying ATNN behavior for two different scenarios as input neurons: (a) all 13 attributes of Table 7.5; and (b) only the six attributes (I_3, I_4, I_5, I_7, I_8, and I_{13}) that stem from decision tree methodology (Sect. 7.3.3). Both cases have one single neuron in the output layer that represents the winding material (Cu or Al) that corresponds to each FOD. Regarding the hidden layer(s), numerous possible topologies have been investigated. To be more precise, one and two hidden layers are explored by trying a wide range of candidate number of neurons. More specifically, in the cases of 13 attributes and six decision tree attributes with one hidden layer, 18 different numbers of neurons (2, 4, 6, 8, 10, 13, 15, 17, 19, 22, 24, 26, 28, 30, 32, 34, 36, and 39 neurons) have been examined. In the case of two hidden layers and 13 attributes, 81 different combinations for the number of neurons of the two hidden layers (namely all possible combinations of the following numbers of neurons: 10, 13, 16, 19, 22, 26, 30, 34, and 39 neurons) have been investigated. In the case of two hidden layers and six decision tree attributes, 64 different combinations (namely all possible combinations of the following numbers of neurons: 3, 6, 9, 12, 15, 18, 21, and 24 neurons) have been examined. It should be noted

7.3 Winding Material Selection with Decision Trees and Artificial Neural Networks 353

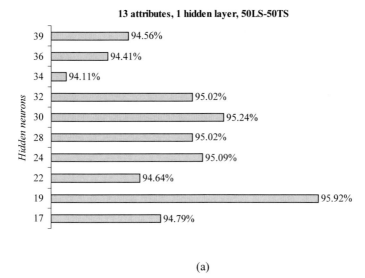

(a)

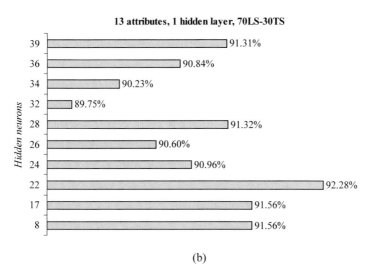

(b)

Fig. 7.10 The best ten classification success rates on the test set using 13 attributes, one hidden layer, while the knowledge base is split as (**a**) 50LS-50TS (i.e., 50% learning set and 50% test set), and (**b**) 70LS-30TS (i.e., 70% learning set and 30% test set)

that the classification success rate in each case on the learning and test set resulted in the average of five different executions of the algorithm.

The impact on the classification success rate of the split of the knowledge base into learning set (training set and validation set) and test set has been investigated through study of the following two different cases: (a) 50LS-50TS case, i.e., the learning set is composed of 50% of the FODs and the remaining 50% compose the test set (different from the FODs of the learning set); and (b) 70LS-30TS case, i.e., the learning set is composed of 70% of the FODs and the remaining 30% compose the test set. In addition, it is important to mention that a test case having 30% of the FODs as learning set and 70% of the FODs as test set has also been checked, however, the behavior of the ATNN was unstable and the classification success rate on the test set was approximately 80%, which is quite low in comparison with the other test cases. This observation is reasonable due to the fact that the ATNN does not have enough data in the learning set, implementing poor training of the ATNN.

Figure 7.10 presents the best ten classification success rate results for 13 attributes, one hidden layer, and two different splits of the knowledge base (50LS-50TS and 70LS-30TS). As shown in Fig. 7.10, the highest classification success rate on test set (95.92%) is achieved using a fully connected three-layer feed-forward neural network with the following topology: 13-19-1 (i.e., 13 neurons in the input layer, 19 neurons in the hidden layer, and one single neuron in the output layer), for 50LS-50TS split of the knowledge base. Figure 7.10 shows that the 50LS-50TS split of the knowledge base provides better classification results than the 70LS-30TS split.

Figure 7.11 shows the best ten classification success rate results for 13 attributes, two hidden layers, and the two different splits of the knowledge base. In Figures 7.11 and 7.13, the number of hidden neurons is denoted as x_y, where x is the number of neurons of the first hidden layer and y is the number of neurons of the second hidden layer. As shown in Fig. 7.11, the best ATNN topology is 13-13-26-1 (i.e., 13 neurons in the input layer, 13 neurons in the first hidden layer, 26 neurons in the second hidden layer, and one single neuron in the output layer) with 95.73% classification success rate on test set, for 50LS-50TS split of the knowledge base.

Figures 7.12 and 7.13 show the best ten classification success rate results for the six decision tree attributes and the two different splits of the knowledge base, in the case of one and two hidden layers, respectively. In the case of one hidden layer, the best ATNN topology is 6-8-1 for 50LS-50TS split of the knowledge base and provides 94.28% classification success rate (CSR), as Fig. 7.12 shows. In the case of two hidden layers, the best ATNN topologies are 6-3-15-1 and 6-6-9-1 for 50LS-50TS split of the knowledge base with 94.58% CSR, as Fig. 7.13 shows.

Table 7.9 contains a summary of the results. Among the eight optimum ATNNs of Table 7.9, the global optimum ATNN for the winding material selection problem is the first one, since it has the highest CSR on the test set (95.92%). The global optimum ATNN is shown in Fig. 7.14 and its characteristics are given in Table 7.10.

7.3 Winding Material Selection with Decision Trees and Artificial Neural Networks

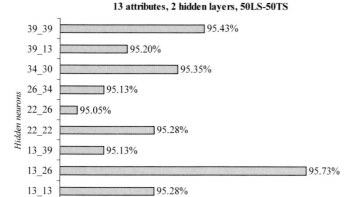

(a)

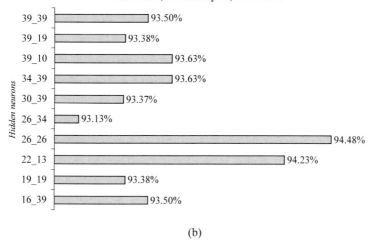

(b)

Fig. 7.11 The best ten classification success rates on test set using 13 attributes, two hidden layers, while the knowledge base is split as (**a**) 50LS-50TS, and (**b**) 70LS-30TS

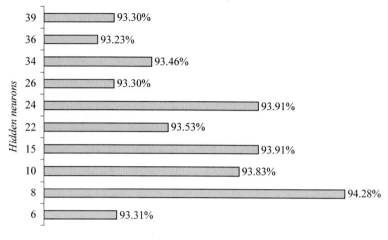

(a)

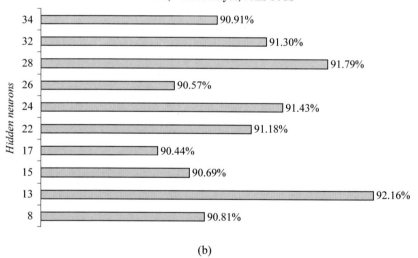

(b)

Fig. 7.12 The best ten classification success rates on the test set using six decision tree attributes, one hidden layer, while the knowledge base is split as (**a**) 50LS-50TS, and (**b**) 70LS-30TS

7.3 Winding Material Selection with Decision Trees and Artificial Neural Networks

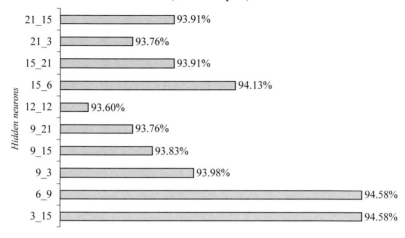

(a)

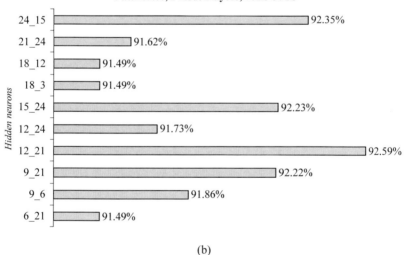

(b)

Fig. 7.13 The best ten classification success rates on the test set using six decision tree attributes, two hidden layers, while the knowledge base is split as (**a**) 50LS-50TS, and (**b**) 70LS-30TS

Table 7.9 ATNN classification success rate (CSR) results for the winding material selection problem

Attributes	Hidden layers	Knowledge base split	Optimum ATNN Topology	CSR on test set (%)
13	1	50LS-50TS	13-19-1	95.92
13	1	70LS-30TS	13-22-1	92.28
13	2	50LS-50TS	13-13-26-1	95.73
13	2	70LS-30TS	13-26-26-1	94.48
6	1	50LS-50TS	6-8-1	94.28
6	1	70LS-30TS	6-13-1	92.16
6	2	50LS-50TS	6-3-15-1	94.58
6	2	70LS-30TS	6-12-21-1	92.59

Table 7.10 Global optimum ATNN for winding material selection problem

Parameter	Value
Training function	traincgb
Transfer function	satlins
Number of input neurons	13
Number of hidden layers	1
Number of hidden neurons	19
Number of output neurons	1
Knowledge base split	50LS-50TS
Classification success rate on test set (%)	95.92

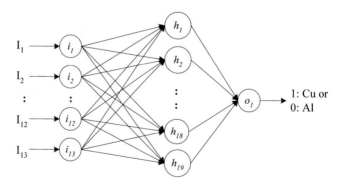

Fig. 7.14 Global optimum ATNN for winding material selection problem

Table 7.11 Comparison of methods for solving the winding material selection problem

Method	Classification success rate on test set (%)
Decision tree	90.90
Adaptive trained neural network	95.92

7.3.5 Synthesis

Of the two methods examined, the best technique for solving the winding material selection problem is the ATNN, since it provides the highest classification success rate on the test set, as Table 7.11 shows.

7.4 Transformer Design Optimization with Branch-and-Bound

7.4.1 Introduction

The objective of TDO is to design the transformer so as to optimize an objective function subject to constraints imposed by international standards and the transformer specification. The most commonly used objective functions are: (a) the minimization of transformer manufacturing cost; and (b) the minimization of transformer total owning cost. The transformer design requires knowledge of electromagnetism, magnetic circuit analysis, electric circuit analysis, loss mechanisms, and heat transfer (Mittle and Mittal 1996).

The TDO problem, because of its importance and complexity, has attracted the interest of many researchers. There are two different methodologies for solution of the TDO problem: (a) the multiple design method; and (b) the mathematical programming method.

The multiple design method (Andersen 1991; Georgilakis et al. 2007b) is a heuristic technique that assigns many alternative values to the design variables so as to generate a large number of alternative designs, and finally to select the design that satisfies all the problem constraints with the minimum manufacturing cost; however, this technique is not able to find the global optimum. Chapter 2 presented a conventional transformer design methodology based on a multiple design technique.

The geometric programming method is the most representative mathematical programming method for solution of the TDO problem (Del Vecchio et al. 2002; Jabr 2005), however, it has two drawbacks: (a) it requires development of a mathematical model for each specific transformer type and configuration in advance; and (b) because of the large number of coefficients in polynomial approximations, the geometric programming method lacks flexibility and cannot easily be combined with more general transformer performance verification or cost estimation algorithms.

This section presents another mathematical programming method, more specifically a mixed integer programming–finite element method (MIP-FEM) technique (Amoiralis 2008; Amoiralis et al. 2008, 2009) for solution of the TDO problem.

7.4.2 MIP-FEM Methodology

7.4.2.1 Method Overview

The MIP-FEM technique solves the TDO problem using the following ten steps (Fig. 7.15):

1. The objective function is selected. There are three different options for the objective function:

    a. Minimization of transformer active part cost using (2.2).
    b. Minimization of transformer manufacturing cost using (2.4).
    c. Minimization of transformer total owning cost using (2.5).

2. Values are given to transformer design input data, e.g.:

    a. Values for description variables (rated power, voltages, frequency, and vector group). An example is shown in Table 2.6.
    b. Values for special variables. An example is shown in Table 2.7.
    c. Values for default variables. An example is shown in Table 2.8.
    d. Values for cost variables. An example is shown in Table 2.9.
    e. Values for various variables. An example is shown in Table 2.10. It should be noted that some of various variables shown in Table 2.10 could also be computed during transformer design optimization, e.g., the height of the corrugated panel.

3. The number of design variables, n, is selected. There are two options for the number of design variables:

    a. Selection of the following four design variables ($n = 4$):

        i. Number of turns of low voltage winding, x_1. This is an integer type variable.
        ii. Width of core leg (mm), x_2. This is usually a variable of integer type.

7.4 Transformer Design Optimization with Branch-and-Bound

      iii.   Height of core window (mm), x_3. This is practically a variable of integer type.

      iv.   Magnetic induction (Gauss), x_4. This is usually an integer type variable.

  b. Selection of the following six design variables ($n = 6$):

      i.   Number of turns of low voltage winding, x_1.

      ii.   Width of core leg (mm), x_2.

      iii.   Height of core window (mm), x_3.

      iv.   Magnetic induction (Gauss), x_4.

      v.   Current density of low voltage winding, x_5 (A/mm^2). This is a variable of real type.

      vi.   Current density of high voltage winding, x_6 (A/mm^2). This is a variable of real type.

4. Values for the lower bound lb_j, the upper bound ub_j, and the initial value x_j^0 are given for each one of the n design variables x_j. It should be noted that $lb_j \leq x_j^0 \leq ub_j$, $\forall\ j = 1, ..., n$.

5. The TDO problem is formulated. The dimensioning of the transformer and the value of the objective function are computed using the approximate formulas of Chap. 2, e.g.:

  a. The calculations of core weight and no-load loss are based on Sect. 2.9.

  b. The computation of load loss is based on Sect. 2.11.

  c. The computation of the inductive part of the impedance voltage is based on Sect. 2.10.

  d. The calculation of transformer manufacturing cost and the computation of transformer total owning cost are based on Sect. 2.23.

6. The TDO problem is a constrained mixed-integer nonlinear programming problem. The TDO problem is solved using the branch-and-bound (BB) method (Sect. 5.4), according to which all integer restrictions of the TDO problem are relaxed and the resulting nonlinear programming problem is solved using the sequential quadratic programming (SQP) method (Sect. 5.3). In brief, the TDO problem is solved using the BB-SQP method, which is an efficient mixed-integer programming (MIP) method.

7. If no solution has been found for the TDO problem, then go to step 2, else go to step 8.

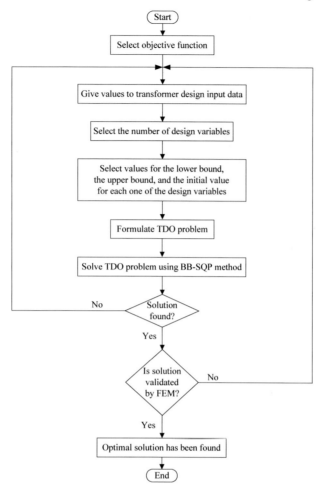

Fig. 7.15 Flowchart of parallel MIP-FEM methodology

8. Check if the solution found by BB-SQP is validated by FEM. This is done as follows:

   a. Calculate again the no-load loss using a permeability tensor finite element model (Kefalas 2008; Kefalas et al. 2008), which is more accurate than the approximate no-load loss curve method of Sect. 2.9.
   b. Calculate again the inductive part of the impedance voltage using an efficient finite element model with detailed representation of winding geometry and cooling ducts (Sect. 6.4.2.4), which is more accurate than the approximate method of Sect. 2.10 (Tsili 2005; Tsili et al. 2006). Next, compute the impedance voltage (Sect. 6.4.2.5).

7.4 Transformer Design Optimization with Branch-and-Bound

c. If the no-load loss, computed during step 8a satisfies the no-load loss specification and if the impedance voltage, computed during step 8b satisfies the impedance voltage specification, then the solution found by BB-SQP is validated by FEM.

9. If the solution found by BB-SQP is validated by FEM, then go to step 10, else go to step 2.

10. The solution found by BB-SQP and validated by FEM is the optimal solution to the TDO problem.

7.4.2.2 Determination of Current Density

The impact of cross-section area of conductors on the optimal transformer design is crucial (Georgilakis et al. 2007b). The cross-section area (mm^2) of the low voltage (LV) conductor is derived from the current density (A/mm^2) of the LV conductor. Similarly, the cross-section area of the high voltage (HV) conductor is derived from the current density of the HV conductor.

Three approaches have been investigated with the aim of determining the values of LV and HV current density that produce the optimum transformer design:

1. *Method 1: constant value for LV and HV current densities.* Using this approach, the transformer designer has to directly define one value for the LV winding current density and one value for the HV winding current density. In this case, four design variables are considered, as defined in Sect. 7.4.2.1. The main drawback of this approach is that the transformer designer requires sufficient experience to correctly set the LV and HV current densities and direct the MIP-FEM method to the optimal solution.

2. *Method 2: set of discrete values for LV and HV current densities.* Using this approach, c_{LV} values for the LV winding current density and c_{HV} values for the HV winding current density are defined. It means that four design variables are considered, as defined in Sect. 7.4.2.1. In this case, the MIP-FEM method is executed $c_{LV} \cdot c_{HV}$ times, so $c_{LV} \cdot c_{HV}$ optimum designs are computed, and the best of them is considered to be the final optimum design. Although this approach is time-consuming, it assures a global optimum design.

3. *Method 3: addition of LV and HV current densities to design variables.* Using the third approach, the LV winding current density and the HV winding current density are added to the design vector, which means that the number of design variables is increased from four to six, as defined in Sect. 7.4.2.1. In particular, the correct definition of the current density value is under the rules (supervision) of the MIP-FEM optimization method.

7.4.3 Results and Discussion

7.4.3.1 Impact of Current Density on Optimum Design

The MIP-FEM method is used to investigate the impact of current density on optimum design. As objective function, the minimization of transformer manufacturing cost, has been considered.

As an example, a three-phase transformer with rated power 400 kVA, rated frequency 50 Hz, rated primary voltage 20 kV, rated secondary voltage 400 V, guaranteed no-load loss 750 W, and guaranteed load loss 4600 W, has been designed. The loss tolerances are according to IEC 60076-1 (Table 1.6), i.e., the maximum total loss is 5885 W, the maximum no-load loss is 862.5 W, and the maximum load loss is 5290 W. Using the heuristic method of Chap. 2, the manufacturing cost of the optimum design was found to be $ 10987.

The three different methods of Sect. 7.4.2.2 for determining the current density have been investigated. More specifically:

1. *Method 1*: it was considered that the LV winding current density is 3 A/mm^2 and the HV winding current density is also 3 A/mm^2. The MIP-FEM method finds that the manufacturing cost of the optimum design is $ 10889. The main parameters of this design are shown in the second column of Table 7.12.

2. *Method 2*: it was considered that the LV winding current density (A/mm^2) takes the four discrete values of the set {3, 3.2, 3.4, 3.6}, while the HV winding current density (A/mm^2) takes the five discrete values of the set {3, 3.15, 3.30, 3.45, 3.6}. In this case, the MIP-FEM method is executed $4 \cdot 5 = 20$ times, it detects 20 optimum transformer designs and selects the most cost effective among them, the main parameters of which are shown in the third column of Table 7.12. It is concluded that the second method of determining the current density in combination with MIP-FEM provides a solution with manufacturing cost of $ 10363 (Table 7.12), i.e., a solution that is 5.68% cheaper than the optimum transformer found using the heuristic method of Chap. 2.

3. *Method 3*: the LV winding current density and the HV winding current density are added to the design vector. In this case, the MIP-FEM method found an optimum design with manufacturing cost of $ 10500. The main parameters of the optimum design can be seen in the fourth column of Table 7.12.

Table 7.12 shows that Method 2 is the best for selecting the LV and the HV current density, since it helps find the optimum transformer with the minimum manufacturing cost, i.e., $ 10363.

7.4 Transformer Design Optimization with Branch-and-Bound

Table 7.12 Impact of current density on optimum design

Parameter	Current density method		
	Method 1	Method 2	Method 3
Number of low voltage turns	18	18	19
Width of core leg (mm)	239	219	230
Height of core window (mm)	248	237	261
Magnetic induction (Gauss)	18000	18000	18000
LV current density (A/mm^2)	3	3.4	3.4
HV current density (A/mm^2)	3	3.6	3.3
Designed no-load loss (W)	859	841	818
Designed load loss (W)	4288	4945	4890
Manufacturing cost ($)	10889	10363	10500
Cost saving versus conventional method (%)	0.89	5.68	4.43

7.4.3.2 Impact of Objective Function on Optimum Design

In Sect. 7.4.3.1, the 400 kVA transformer design example was optimized having as an objective the minimization of manufacturing cost, and the parameters of the optimum design are shown in the second column of Table 7.13, where this design is denoted as Design 1. The same transformer design example has also been optimized having as an objective the minimization of total owning cost, and the parameters of the optimum design are shown in the third column of Table 7.13, where this design is denoted as Design 2. The fourth column of Table 7.13 shows the difference between Design 2 and Design 1, where the difference is computed in relation to manufacturing cost, bid price, cost of losses, and total owning cost. As can be seen from Table 7.13:

1. The manufacturing cost of Design 2 is 8.46% higher than that of Design 1. This is due to the fact that the objective of Design 1 is the minimization of manufacturing cost.

2. The bid price of Design 2 is 8.46% higher than that of Design 1.

3. The cost of losses of Design 2 is 9.54% lower than that of Design 1.

4. The total owning cost of Design 2 is 1.39% lower than that of Design 1. This is due to the fact that the objective of Design 2 is the minimization of total owning cost.

Table 7.13 Impact of objective function on optimum design

	Objective function		
	Minimum manufacturing cost	Minimum total owning cost	
Parameter	Design 1	Design 2	Difference (%)
Manufacturing cost ($)	10363	11240	8.46
Sales margin (%)	35	35	
Bid price ($)	15943	17292	8.46
Designed no-load loss (W)	841	719	
Designed load loss (W)	4945	4613	
A ($/W)	8.31	8.31	
B ($/W)	2.49	2.49	
Cost of losses ($)	19302	17461	-9.54
Total owning cost ($)	35245	34754	-1.39

Table 7.14 Comparison of MIP-FEM with the conventional TDO method of Chap. 2

Rated power (kVA)	Number of designs	Average cost reduction of MIP-FEM method in comparison with the conventional TDO method (%)
1600	14	1.09
1000	24	0.29
800	20	0.47
630	48	2.28
400	28	0.73
250	16	1.83
160	24	3.69
100	14	1.47
	Total: 188	Average: 1.60%

It is concluded from the above that although Design 2 has higher manufacturing cost, it has lower total owning cost, attributed to the lower cost of losses as well as the fact that the objective of Design 2 is the minimization of total owning cost.

7.4 Transformer Design Optimization with Branch-and-Bound

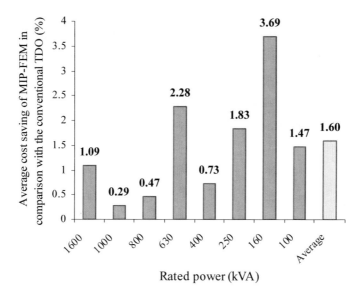

Fig. 7.16 Average cost reduction of MIP-FEM method in comparison with the conventional TDO method of Chap. 2

7.4.3.3 Generalization of Results

The MIP-FEM method has been tested on a wide spectrum of real transformers of different voltage ratings and loss categories. In particular, 188 transformer designs were optimized using the MIP-FEM as well as the conventional TDO method of Chap. 2. The minimization of transformer manufacturing cost has been considered as objective function. It should be noted that constant values for LV and HV current densities (i.e., Method 1 of Sect. 7.4.2.2) has been used for MIP-FEM because the conventional TDO method of Chap. 2 could not support the other two methods for determining current density. Figure 7.16 and Table 7.14 show the results. For example, Table 7.14 shows that among the 28 transformer designs with 400 kVA rated power, the average reduction in the manufacturing cost of the MIP-FEM method in comparison with the conventional TDO method is 0.73%. Moreover, Table 7.14 shows that among the 188 transformer designs, the average cost reduction of the MIP-FEM method in comparison with the conventional TDO method is 1.60%.

7.5 Transformer Design Optimization with Genetic Algorithms

7.5.1 Introduction

Chapter 2 presented a conventional transformer design methodology based on a multiple design technique. Section 7.4 solved the TDO problem using a MIP-FEM methodology. This section solves the TDO problem using a recursive genetic algorithm–finite element method (GA-FEM) technique (Georgilakis 2008, 2009).

7.5.2 Recursive GA-FEM Methodology

7.5.2.1 Introduction and Configuration of a Recursive Genetic Algorithm

This section introduces an improved GA for solution of the TDO problem. This section presents the contributions, features, and optimal parameter settings of the improved GA.

Since the GA is a stochastic optimization method, in general, it converges to different solution each time the GA is executed. That is why it is proposed to implement a novel recursive GA approach, i.e., to run the GA N times and to introduce an *external elitism strategy* that copies the best solution found at the end of each GA run to the initial population of the next GA run. This innovative external elitism strategy assures that after the completion of each GA run, a solution is provided that is better or at least the same as the solution of the previous GA run. As will be shown in Sect. 7.5.3.1, after 7 to 10 GA runs, the global optimum is reached for the TDO problem.

An *internal elitism strategy* is also adopted, i.e., the best solution of every generation is copied to the next generation so that the possibility of its destruction through a genetic operator is eliminated.

The initial population of candidate solutions is created randomly. However, in the initial population of the initial GA run, the worst solution (i.e., the one with the maximum manufacturing cost) is substituted by a solution computed using the MIP-FEM method of Sect. 7.4. Incorporation of the MIP-FEM solution into the initial population of the initial GA run in combination with the external and internal elitism strategies assure that the recursive GA-FEM method converges to a better or at least the same solution as the MIP-FEM method.

In order to improve the GA search by assuring good exploration at the beginning of evolution, and more and more exploitation capability while optimization goes on, different crossover and mutation rates were tested. After sufficient experimentation it was found that the best results were obtained with the following crossover and mutation probabilities:

7.5 Transformer Design Optimization with Genetic Algorithms

Table 7.15 Determination of the number of bits for the GA chromosome

Design variable	Symbol	Unit	Possible values	Type	Bits
Number of LV turns	x_1	-	$8 \leq x_1 \leq 1000$	Integer	10
Magnetic material type	x_2	-	$1 \leq x_2 \leq 12$	Integer	4
Magnetic induction	x_3	Gauss	$10000 \leq x_3 \leq 18500$	Integer	15
Width of core leg	x_4	Mm	$80 \leq x_4 \leq 500$	Integer	9
Core window height	x_5	mm	$80 \leq x_5 \leq 500$	Integer	9
LV current density	x_6	A/mm²	$1.5 \leq x_6 \leq 5.5$	Real	7
HV current density	x_7	A/mm²	$1.5 \leq x_7 \leq 5.5$	Real	7
			Number of bits of GA chromosome		61

$$P_{ck} = 0.35 + 0.45 \cdot \left[\frac{k-1}{N_g - 1}\right], \quad (7.13)$$

$$P_{mk} = 0.055 - 0.045 \cdot \left[\frac{k-1}{N_g - 1}\right], \quad (7.14)$$

where P_{ck} is the crossover probability at generation k, P_{mk} is the mutation probability at generation k, and N_g is the number of generations.

The first column of Table 7.15 presents the seven design variables that have been used for solution of the TDO problem by the recursive GA-FEM. The fifth column of Table 7.15 shows that the first five design variables are of integer type, while the remaining two design variables are of real type. The fourth column of Table 7.15 shows the range of possible values that each design variable can take. This range of possible values has been determined from a large database of actual transformer designs with the following main characteristics: three-phase, oil-immersed, wound core distribution transformers from 25 kVA up to 2000 kVA, with voltages up to 36 kV. Binary coding is used for chromosome representation. The last column of Table 7.15 presents the number of bits used for each design variable. As can be seen from the last row of Table 7.15, the GA chromosome has 61 bits.

After trial and error, it was found that a population size of 40 chromosomes and 30 generations provide very good results for TDO.

Among the four different selection schemes tested, i.e., roulette wheel, tournament, deterministic sampling, and stochastic remainder sampling (Goldberg 1989), the tournament selection scheme produced the best results and convergence for TDO.

7.5.2.2 Finite Element Models

For the TDO problem of wound core type transformers, two finite element (FE) models are used, the first to compute the transformer no-load loss and the second to evaluate the transformer impedance voltage. In particular, a permeability tensor FE model is adopted for computation of the no-load loss, since this model accurately represents the core material and the geometry of wound cores (Kefalas 2008; Kefalas et al. 2008). Moreover, an efficient FE model with detailed representation of winding geometry and cooling ducts (Sect. 6.4.2.4) is adopted for impedance voltage evaluation (Tsili 2005; Tsili et al. 2006). Both FE models are based on a particular magnetic scalar potential formulation (Kladas and Tegopoulos 1992), which is advantageous in terms of computational speed in comparison with FEM based on magnetic vector potential, as there is only one unknown at each node of the FE mesh. Accuracy and computation speed are the main advantages of the above two FE models, making them ideal for solution of the TDO problem.

7.5.2.3 Method Overview

The flowchart of the proposed optimization model for solution of the TDO problem, shown in Fig. 7.17, is composed of two submodels:

1. *MIP-FEM submodel*. Initially, the MIP-FEM deterministic optimization method of Sect. 7.4 is used to solve the TDO problem. Let S_0 be the solution provided by that method.

2. *Recursive GA-FEM submodel* (N GA-FEM runs). After execution of the MIP-FEM submodel, N runs of the proposed recursive GA-FEM submodel are executed. Each run of the GA-FEM submodel requires two internal runs:

    – *GA run*. The recursive GA based optimization model, described in Sect. 7.5.2.1, is executed to solve the TDO problem. The solution S_0 provided by the MIP-FEM submodel is included in the initial population of the initial GA run. In all other GA runs, the best solution S_i provided by the previous GA-FEM run is included at the initial population of the next GA run. This approach assures that the solution S_i is better or at least the same as the solution S_{i-1} (see Sect. 7.5.2.1).

    – *FEM run*. The two FE models of Sect. 7.5.2.2 are used for computation of the transformer no-load loss and impedance (unlike the analytical formulas used in the GA run) in order to provide more accurate results and better convergence to the optimal solution.

7.5 Transformer Design Optimization with Genetic Algorithms

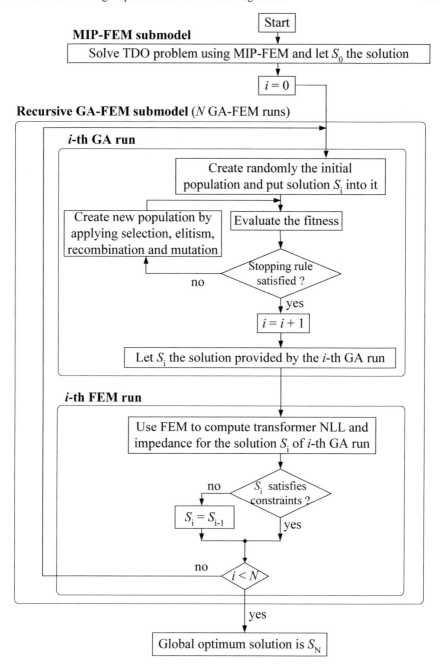

Fig. 7.17 Flowchart of the recursive GA-FEM method for solution of the TDO problem

7.5.3 Results and Discussion

7.5.3.1 Application to 1600 kVA Transformer Design

The recursive GA-FEM method has been used for solution of the TDO problem for a real 1600 kVA transformer design with the following main specifications: rated frequency 50 Hz, rated HV 20 kV, rated LV 0.4 kV, prescribed no-load loss 1700 W, prescribed load loss 20000 W, prescribed impedance voltage 6%. The no-load loss, load loss, and impedance voltage tolerances are according to IEC 60076-1 international standard, i.e., the maximum no-load loss is 1955 W, the maximum load loss is 23000 W, the maximum total loss is 23870 W, the minimum impedance voltage is 5.4%, and the maximum impedance voltage is 6.6%. Minimization of manufacturing cost has been considered as objective function. Table 7.16 compares the results of the recursive GA-FEM method with the heuristic (Chap. 2) and the MIP-FEM method (Sect. 7.4). As can be seen from Table 7.16, the three techniques converged to three different solutions. In particular, the recursive GA-FEM method, after seven GA-FEM runs implemented in 3.42 minutes, provides the best result, since it converges to the global minimum manufacturing cost (MC) of $23271.

Table 7.16 Comparison of GA-FEM with heuristic (Chap. 2) and MIP-FEM (Sect. 7.4) design methods for a 1600 kVA transformer design

Parameter	Heuristic	MIP-FEM	GA-FEM
Number of LV turns	10	10	11
Magnetic material type	1 (i.e., HiB)	2 (i.e., M4)	1 (HiB)
Magnetic induction (Gauss)	16012	16991	18000
Width of core leg (mm)	290	322	325
Core window height (mm)	338	322	354
LV current density (A/mm^2)	4.3	4.6	4.3
HV current density (A/mm^2)	4.0	3.8	4.6
No-load loss (W)	1581	1952	1791
Load loss (W)	19035	18767	21151
Total loss (W)	20616	20719	22942
Impedance voltage (%)	5.89	6.41	6.20
Manufacturing cost ($)	24814	24446	23271
Number of algorithm runs	1	1	7
Total execution time (minutes)	0.45	0.79	3.42

7.5 Transformer Design Optimization with Genetic Algorithms

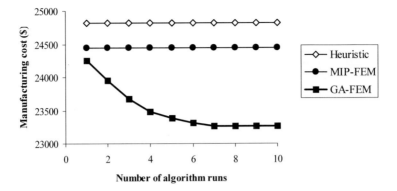

Fig. 7.18 Comparative results for a 1600 kVA transformer design

Table 7.17 Comparison of average manufacturing cost saving of GA-FEM versus heuristic (Chap. 2) and MIP-FEM (Sect. 7.4) design method

Rated power (kVA)	Number of designs	Cost saving of GA-FEM versus heuristic (%)	Cost saving of GA-FEM versus MIP-FEM (%)
100	25	5.3	1.8
160	25	4.9	2.6
250	25	8.0	4.4
400	25	6.5	3.0
630	25	6.0	2.8
800	25	5.9	2.0
1000	25	3.7	3.2
1600	25	6.3	4.9
	Average	5.8	3.1

Figure 7.18 compares the minimum manufacturing cost computed by the above three techniques for solution of the 1600 kVA TDO problem. Since the heuristic and the MIP-FEM are both deterministic optimization techniques, they always converge to the same minimum MC, i.e., $24814 for the heuristic and $24446 for the MIP-FEM. On the other hand, the recursive GA-FEM, because of its special design presented in Sect. 7.5.2.1, manages to progressively reduce the MC, as the number of GA-FEM algorithm runs is increased. In particular, after seven GA-FEM runs, the global minimum MC is achieved, which is 4.8% cheaper than the MC computed by the MIP-FEM method and 6.2% cheaper than the MC computed by the heuristic method. As can be seen from Fig. 7.18, after the seventh GA-FEM

run, the MC is not decreased further, which means that seven GA-FEM runs are sufficient to obtain the global optimum solution to the TDO problem.

7.5.3.2 Generalization of Results

The recursive GA-FEM method has been tested on 200 real transformer designs, of eight power ratings and various loss categories and voltage ratings. As can be seen from Table 7.17, the recursive GA-FEM method finds the global optimum solution, which is, on average, (a) 5.8% cheaper than the solution provided by the heuristic technique, and (b) 3.1% cheaper than the solution found using the MIP-FEM method.

References

Amoiralis EI (2008) Energy savings in electric power systems by development of advanced uniform models for the evaluation of transformer manufacturing and operating cost. PhD dissertation. Technical University of Crete, Chania, Greece

Amoiralis EI, Georgilakis PS, Kefalas TD, Tsili MA, Kladas AG (2007) Artificial intelligence combined with hybrid FEM-BE techniques for global transformer optimization. IEEE Transactions on Magnetics 43(4):1633–1636

Amoiralis EI, Tsili MA, Georgilakis PS, Kladas AG, Souflaris AT (2008) A parallel mixed integer programming-finite element method technique for global design optimization of power transformers. IEEE Transactions on Magnetics 44(6):1022–1025

Amoiralis EI, Georgilakis PS, Tsili MA, Kladas AG (2009) Global transformer optimization method using evolutionary design and numerical field computation. IEEE Transactions on Magnetics 45(3):1720–1723

Andersen OW (1991) Optimized design of electric power equipment. IEEE Computer Applications in Power 4(1):11–15

CENELEC (1992) Three phase oil-immersed distribution transformers 50 Hz, from 50 to 2500 kVA with highest voltage for equipment not exceeding 36 kV. CENELEC Harmonization Document 428.1 S1, CENELEC, Brussels, Belgium

Craggs JL (1976) Fabricated aluminum cage construction in large induction motors. IEEE Transactions on Industry Applications 12(3):261–267

Del Vecchio RM, Poulin B, Feghali PT, Shah DM, Ahuja R (2002) Transformer design principles with applications to core-form power transformers. CRC Press, Boca Raton, FL

Doulamis ND, Doulamis AD, Georgilakis PS, Kollias SD, Hatziargyriou ND (2002) A synergetic neural network-genetic scheme for optimal transformer construction. Integrated Computer-Aided Engineering 9:37–56

Finley WR and Hodowanec MM (2001) Selection of copper versus aluminum rotors for induction motors. IEEE Transactions on Industry Applications 37(6):1563–1573

Fonseca CM and Fleming PJ (1993) Genetic algorithms for multiobjective optimization: formulation, discussion and generalization. Proc International Conference on Genetic Algorithms, 416–423

Georgilakis PS (2000) Contribution of artificial intelligence techniques in the reduction of distribution transformer iron losses. PhD dissertation. National Technical University of Athens, Athens, Greece

References

Georgilakis PS (2008) A recursive genetic algorithm–finite element method technique for the solution of transformer manufacturing cost minimization problem. Proc IEEE Conference on Electromagnetic Field Computation

Georgilakis PS (2009) A recursive genetic algorithm–finite element method technique for the solution of transformer manufacturing cost minimization problem. IET Electric Power Applications, accepted for publication

Georgilakis PS and Amoiralis EI (2007) Spotlight on transformer design. IEEE Power and Energy 5(1):40–50

Georgilakis PS, Hatziargyriou ND, Paparigas D (1999) AI helps reduce transformer iron losses. IEEE Computer Applications in Power 12(4):41–46

Georgilakis PS, Doulamis ND, Doulamis AD, Hatziargyriou ND, Kollias SD (2001) A novel iron loss reduction technique for distribution transformers based on a genetic algorithm-neural network approach. IEEE Transactions on Systems, Man, and Cybernetics, Part C 31(1):16–34

Georgilakis PS, Gioulekas AT, Souflaris AT (2007a) A decision tree method for the selection of winding material in power transformers. Journal of Materials Processing Technology 181(1-3):281–285

Georgilakis PS, Tsili MA, Souflaris AT (2007b) A heuristic solution to the transformer manufacturing cost optimization problem. Journal of Materials Processing Technology 181(1-3):260–266

Goldberg DE (1989) Genetic algorithms in search, optimization, and machine learning. Addison-Wesley, Reading, MA

Heathcote MJ (2007) The J&P transformer book, 13th edn. Elsevier, Oxford

Jabr RA (2005) Application of geometric programming to transformer design. IEEE Transactions on Magnetics 41(11):4261–4269

Kefalas TD (2008) Loss analysis of power transformers with advanced materials. PhD dissertation. National Technical University of Athens, Athens, Greece

Kefalas TD, Georgilakis PS, Kladas AG, Souflaris AT, Paparigas, DG (2008) Multiple grade lamination wound core: a novel technique for transformer iron loss minimization using simulated annealing with restarts and an anisotropy model. IEEE Transactions on Magnetics 44(6):1082–1085

Kennedy BW (1998) Energy efficient transformers. McGraw-Hill, New York

Kladas A, Tegopoulos J (1992) A new scalar potential formulation for 3D magnetostatics necessitating no prior source field calculation. IEEE Transactions on Magnetics 28(2):1103-1106

Michalewicz Z (1994) Genetic algorithms + data structures = evolution programs. Springer, Berlin

Mittle VN, Mittal A (1996) Design of electrical machines, 4th edn. Standard Publishers Distributors, Nai Sarak, Delhi

Mohtar SN, Jamal N, Sulaiman M (2004) Analysis of all aluminum conductor (AAC) and all aluminum alloy conductor (AAAC). Proc IEEE TENCON Conference, 3:409–412

Poloujadoff M, Mipo JC, Nurdin M (1995) Some economical comparisons between aluminium and copper squirrel cages. IEEE Transactions on Energy Conversion 10(3):415–418

Rudolph G (1994) Convergence analysis of canonical genetic algorithms. IEEE Transactions on Neural Networks 5(1):96–101

Sullivan CR (2008) Aluminum windings and other strategies for high-frequency magnetics design in an era of high copper and energy costs. IEEE Transactions on Power Electronics 23(4):2044–2051

teNyenhuis EG and Girgis RS (2006) Measured variability of performance parameters of power and distribution transformers. Proc IEEE Power Engineering Society Transmission and Distribution Conference and Exposition, 523–528

Tilton JE (2002) Long-term trends in copper prices. Mining Engineering 54(7):25–32

Tsili MA (2005) Development of mixed finite element – boundary element numerical techniques for the design of power transformers. PhD dissertation. National Technical University of Athens, Athens, Greece

Tsili MA, Kladas AG, Georgilakis PS, Souflaris AT, Paparigas DG (2006) Advanced design methodology for single and dual voltage wound core power transformers based on a particular finite element model. Electric Power Systems Research 76:729–741

Wheeler HA (1980) Transmission-line conductors of various cross sections. IEEE Transactions on Microwave Theory and Techniques 28(2):73–83

8 Transformer Selection

Abstract This chapter deals with transformer selection by electric utilities and industrial transformer users. It reviews the classical total owning cost formula and it also introduces the external environmental cost due to transformer losses. Using the methodologies of this chapter, transformer users will save money by purchasing the most cost-effective and energy-efficient transformers.

8.1 Introduction

Distribution transformers have a significant impact on the losses of a utility's transmission and distribution system. Based on a study conducted in the USA, distribution transformers contributed: (a) about 40% of the losses for non-generating public utilities; and (b) over 16% of the losses for investor-owned utilities (Kennedy 1998). European Copper Institute studies indicated that improving the energy efficiency of the existing European stock of transformers by 40% would result in about 22 TWh annual energy savings equivalent to an annual reduction in greenhouse gas emissions of about 9 million tonnes of CO_2 equivalent (Targosz et al. 2005).

Energy-efficient transformers have reduced total losses, i.e., reduced load and no-load losses. Energy-efficient transformers reduce energy consumption and consequently reduce the generation of electrical energy and greenhouse gas emissions. In deregulated electricity markets, as the price of electrical energy varies every hour, so does the cost of transformer losses. The seasonal load variations also increase the benefits associated with efficient transformers, particularly if the season of maximum load is coincident with the maximum energy prices.

As system investment and energy costs continue to increase, electric utilities are more and more interested in installing energy-efficient transformers across their distribution networks. Transformer manufacturers have developed new manufacturing techniques and new types of core materials to provide cost-effective and energy-efficient transformers to transformer users.

Energy-efficient transformers cost more but use less energy than low efficiency transformers. The decision as to whether to purchase a low-cost, inefficient transformer or a more expensive, energy-efficient transformer, is primarily an economic one. The common practice used by electric utilities for determining the cost-effectiveness of distribution transformers is based on the total owning cost (*TOC*) method, where *TOC* is equal to the sum of transformer-purchasing price plus the cost of transformer losses throughout the transformer lifetime.

Section 8.2 presents the *TOC* method for industrial and commercial users (Merritt and Chaitkin 2003; Georgilakis 2007). Section 8.3 presents the *TOC* method for electric utilities (Nickel and Braunstein 1981a, 1981b; ANSI/IEEE 1992; Kennedy 1998). Section 8.4 introduces the external environmental cost into the *TOC* method (Amoiralis et al. 2007; Amoiralis 2008; Amoiralis et al. 2008; Georgilakis and Amoiralis 2009).

8.2 Total Owning Cost for Industrial and Commercial Users

8.2.1 Cost Evaluation Method

If S is the transformer actual load (kVA) and S_n is the transformer rated power (kVA), then the per-unit load, L, of the transformer is:

$$L = \frac{S}{S_n}. \tag{8.1}$$

If LL is the transformer load loss (W) at rated power S_n, then the transformer load loss LL_L (W) at per-unit load L is calculated from the formula:

$$LL_L = LL \cdot L^2. \tag{8.2}$$

If NLL is the transformer no-load loss (W), then the transformer total loss TL_L (W) at load L is:

$$TL_L = NLL + LL_L. \tag{8.3}$$

Combining (8.2) and (8.3), we find that the transformer total loss TL_L at load L is given by the formula:

$$TL_L = NLL + LL \cdot L^2. \tag{8.4}$$

The transformer operates *HPY* hours per year. If *EP* is the electricity price ($/kWh) that the industrial/commercial user pays for electricity, then the annual cost ($/year) of transformer total loss C_{TL} is:

$$C_{TL} = TL_L \cdot EP \cdot HPY \cdot 10^{-3}. \tag{8.5}$$

Substituting (8.4) to (8.5), we obtain:

8.2 Total Owning Cost for Industrial and Commercial Users

$$C_{TL} = (NLL + LL \cdot L^2) \cdot EP \cdot HPY \cdot 10^{-3}, \tag{8.6}$$

and finally:

$$C_{TL} = C_{NLL} + C_{LL}, \tag{8.7}$$

where C_{NLL} is the annual cost ($/year) of transformer no-load loss and C_{LL} is the annual cost ($/year) of transformer load loss, which are calculated from the following equations:

$$C_{NLL} = NLL \cdot EP \cdot HPY \cdot 10^{-3}, \tag{8.8}$$

$$C_{LL} = LL \cdot L^2 \cdot EP \cdot HPY \cdot 10^{-3}. \tag{8.9}$$

The industrial/commercial user pays the cost of transformer total loss C_{TL} ($/year) for each one of the N years of the transformer lifetime. If d is the discount rate, then the present value PV_{TL} of the cost ($) of transformer total loss throughout the transformer lifetime is:

$$PV_{TL} = \frac{C_{TL}}{(1+d)} + \frac{C_{TL}}{(1+d)^2} + \ldots + \frac{C_{TL}}{(1+d)^N} \Rightarrow$$

$$PV_{TL} = C_{TL} \cdot PV_m, \tag{8.10}$$

where PV_m is the present value multiplier, which is calculated as follows:

$$PV_m = \sum_{i=1}^{N} \frac{1}{(1+d)^i} = \frac{1 - 1/(1+d)^N}{1 - 1/(1+d)} \Rightarrow$$

$$PV_m = \frac{(1+d)^N - 1}{d \cdot (1+d)^{N-1}}. \tag{8.11}$$

The present value PV_{NLL} of the cost ($) of transformer no-load loss throughout the transformer lifetime is calculated as follows:

$$PV_{NLL} = C_{NLL} \cdot PV_m. \tag{8.12}$$

The present value PV_{LL} of the cost ($) of transformer load loss throughout the transformer lifetime is:

$$PV_{LL} = C_{LL} \cdot PV_m. \tag{8.13}$$

The following equation holds:

$$PV_{TL} = PV_{NLL} + PV_{LL}. \tag{8.14}$$

If the transformer is offered to the industrial/commercial user at a bid price BP (\$), then the total owning cost TOC (\$) of the transformer is equal to the sum of its bid price BP and the present value PV_{TL} of the cost (\$) of transformer total loss throughout the transformer lifetime:

$$TOC = BP + PV_{TL}. \tag{8.15}$$

Substituting (8.14) into (8.15), we obtain:

$$TOC = BP + PV_{NLL} + PV_{LL}. \tag{8.16}$$

Substituting (8.12) and (8.13) into (8.16), we have:

$$TOC = BP + C_{NLL} \cdot PV_m + C_{LL} \cdot PV_m. \tag{8.17}$$

Substituting (8.8) and (8.9) into (8.17), we obtain:

$$TOC = BP + (NLL + LL \cdot L^2) \cdot PV_m \cdot EP \cdot HPY \cdot 10^{-3}. \tag{8.18}$$

An equivalent and simpler expression for TOC is the following:

$$TOC = BP + A \cdot NLL + B \cdot LL, \tag{8.19}$$

where BP is the transformer bid price (\$) or purchasing price, NLL is the transformer no-load loss (W), LL is the transformer load loss (W), A is the no-load loss factor (\$/W) and B is the load loss factor (\$/W).

The loss factors A and B of (8.19) are calculated as follows:

$$A = PV_m \cdot EP \cdot HPY \cdot 10^{-3}, \tag{8.20}$$

$$B = A \cdot L^2. \tag{8.21}$$

The purchasing decision has to be based on the minimization of TOC. This means that if we have to evaluate m alternative transformer offers $O_i = \{BP_i, NLL_i, LL_i\}, i = 1, ..., m$, where BP_i, NLL_i, and LL_i is the bid price, the no-load loss, and the load loss, respectively, of the ith offer O_i, then for each one of the offers we calculate its total owning cost $TOC_i, i = 1, ..., m$, using (8.19) and

8.2 Total Owning Cost for Industrial and Commercial Users

the optimum transformer (to be purchased) is the one with the minimum total owning cost and not the transformer with the minimum purchasing price.

The loss factors A and B have to be part of the transformer specification, i.e., the loss factors A and B have to be supplied to transformer manufacturers in order to prepare their bids.

The transformer efficiency n is defined as follows:

$$n = \frac{S \cdot \cos\theta_L}{S \cdot \cos\theta_L + TL_L}, \tag{8.22}$$

where $\cos\theta_L$ is the power factor, S is the transformer actual load and TL_L is the transformer total loss at load L.

The transformer total loss (W) is calculated from (8.4). The transformer annual energy loss (kWh/year) is calculated as follows:

$$EL = TL_L \cdot HPY \cdot 10^{-3}. \tag{8.23}$$

The annual energy savings ES ($/year) by using transformer j instead of transformer i are given by the formula:

$$ES = (EL_i - EL_j) \cdot EP, \tag{8.24}$$

where EL_i is the annual energy loss of transformer i and EP is the electricity price.

The simple payback SP (years) by using transformer j instead of transformer i is calculated as follows:

$$SP = \frac{BP_j - BP_i}{ES}, \tag{8.25}$$

where BP_i is the bid price for transformer i.

Table 8.1 Data for two competing transformer offers

Parameter	Offer 1	Offer 2
Rated power (kVA)	250	250
No-load loss (W)	650	425
Load loss (W)	4200	2750
Loss category	BA'	CC'
Bid price ($)	11750	14800

8.2.2 Example 8.1

Table 8.1 shows two competing offers for three-phase, oil-immersed, distribution transformer of 250 kVA, 50 Hz. The electricity price is 0.12 $/kWh. The discount rate is 7%. The transformer operates 8760 hours per year. The transformer lifetime is 30 years. Compute the total owning cost of the two offers if the per-unit load is varied from 0.0 to 1.0 with step 0.1.

Solution

The loss category, shown in Table 8.1, is according to CENELEC HD 428.1 S1/1992 (see Tables 1.2 and 1.3).

The present value multiplier is computed using (8.11):

$$PV_m = \frac{(1+d)^N - 1}{d \cdot (1+d)^{N-1}} = \frac{(1+0.07)^{30} - 1}{0.07 \cdot (1+0.07)^{30-1}} \Rightarrow PV_m = 13.2777.$$

We will present the computation of the total owning cost for the case of Offer 1 and for per-unit load $L = 0.4$.

Equation 8.4 is used to compute transformer total loss TL_L at load L:

$$TL_L = NLL + LL \cdot L^2 \Rightarrow TL_L = 650 + 4200 \cdot 0.4^2 \Rightarrow TL_L = 1322 \text{ W}.$$

Equation 8.23 is used to compute the annual energy losses EL:

$$EL = TL_L \cdot HPY \cdot 10^{-3} = 1322 \cdot 8760 \cdot 10^{-3} \Rightarrow EL = 11581 \text{ kWh/year}.$$

The annual cost of losses is calculated by combining (8.5) and (8.23) as follows:

$$C_{TL} = TL_L \cdot EP \cdot HPY \cdot 10^{-3} \Rightarrow C_{TL} = EP \cdot EL = 0.12 \cdot 11581 \Rightarrow$$

$$C_{TL} = 1389.7 \text{ \$/year}.$$

Equation 8.10 is used to compute the present value PV_{TL} of the cost of transformer total losses throughout the transformer lifetime:

$$PV_{TL} = C_{TL} \cdot PV_m \Rightarrow PV_{TL} = 1389.7 \cdot 13.2777 \Rightarrow PV_{TL} = \$ \ 18452.$$

Equation 8.15 is used to compute the total owning cost of Offer 1 for per-unit load $L = 0.4$:

$$TOC = BP + PV_{TL} \Rightarrow TOC = 11750 + 18452 \Rightarrow TOC = \$ \ 30202.$$

8.2 Total Owning Cost for Industrial and Commercial Users

Table 8.2 Analysis of Offer 1 of Table 8.1

Per-unit load	Total losses (W)	Annual energy losses (kWh/year)	Annual cost of losses ($/year)	Present value of cost of losses ($)	Total owning cost ($)
1.0	4850	42486	5098	67694	79444
0.9	4052	35496	4259	56556	68306
0.8	3338	29241	3509	46590	58340
0.7	2708	23722	2847	37797	49547
0.6	2162	18939	2273	30176	41926
0.5	1700	14892	1787	23728	35478
0.4	1322	11581	1390	18452	30202
0.3	1028	9005	1081	14348	26098
0.2	818	7166	860	11417	23167
0.1	692	6062	727	9659	21409
0.0	650	5694	683	9072	20822

Table 8.3 Analysis of Offer 2 of Table 8.1

Per-unit load	Total losses (W)	Annual energy losses (kWh/year)	Annual cost of losses ($/year)	Present value of cost of losses ($)	Total owning cost ($)
1.0	3175	27813	3338	44315	59115
0.9	2653	23236	2788	37022	51822
0.8	2185	19141	2297	30497	45297
0.7	1773	15527	1863	24740	39540
0.6	1415	12395	1487	19750	34550
0.5	1113	9746	1169	15528	30328
0.4	865	7577	909	12073	26873
0.3	673	5891	707	9386	24186
0.2	535	4687	562	7467	22267
0.1	453	3964	476	6316	21116
0.0	425	3723	447	5932	20732

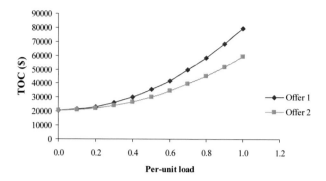

Fig. 8.1 Total owning cost for the two offers of Table 8.1 as a function of per-unit load

The total owning cost for the remaining values of per-unit load for Offer 1 is computed similarly and the results are shown in Table 8.2. The total owning cost of Offer 2 for the 11 different values of per-unit load is shown in Table 8.3. Figure 8.1 plots the total owning cost for Offer 1 and Offer 2 as a function of per-unit load. Figure 8.1 shows that for all the different values of per-unit load, the total owning cost of Offer 2 is lower than the total owning cost of Offer 1, which implies that the Offer 2 has to be selected by the industrial/commercial user, since it is the most energy-efficient transformer offer.

Although Offer 2 has higher purchasing price than Offer 1, as Table 8.1 shows, Offer 2 is the best choice since it has lower total owning cost than Offer 1, i.e., the cost to purchase and operate the transformer of Offer 2 is lower.

Equation 8.20 is used to compute the loss factor A:

$$A = PV_m \cdot EP \cdot HPY \cdot 10^{-3} \Rightarrow A = 13.2777 \cdot 0.12 \cdot 8760 \cdot 10^{-3} \Rightarrow$$

$$A = 13.9575 \; \$/W.$$

Equation 8.21 is used to compute the loss factor B:

$$B = A \cdot L^2 \Rightarrow B = 13.9575 \cdot 0.4^2 \Rightarrow B = 2.2332 \; \$/W.$$

Equation 8.19 can also be used to compute the total owning cost of Offer 1 for per-unit load $L = 0.4$:

$$TOC = BP + A \cdot NLL + B \cdot LL \Rightarrow$$

$$TOC = 11750 + 13.9575 \cdot 650 + 2.2332 \cdot 4200 \Rightarrow TOC = \$\,30202.$$

8.2 Total Owning Cost for Industrial and Commercial Users

8.2.3 Example 8.2

An industrial user has defined in the transformer specification that $A = 10$ \$/W, $B = 2.5$ \$/W, and the transformer selection will be based on the minimum *TOC*. The industrial user receives the two offers of Table 8.1. Compute the offer that the industrial user will select.

Solution
Equation 8.19 is used to compute the total owning cost of Offer 1:

$$TOC_1 = BP_1 + A \cdot NLL_1 + B \cdot LL_1 \Rightarrow$$

$$TOC_1 = 11750 + 10 \cdot 650 + 2.5 \cdot 4200 \Rightarrow TOC_1 = \$\,28750.$$

The total owning cost of Offer 2 is:

$$TOC_2 = BP_2 + A \cdot NLL_2 + B \cdot LL_2 \Rightarrow$$

$$TOC_2 = 14800 + 10 \cdot 425 + 2.5 \cdot 2750 \Rightarrow TOC_2 = \$\,25925.$$

Since $TOC_2 < TOC_1$, the industrial user will select Offer 2, although its bid price is higher.

8.2.4 Example 8.3

Table 8.4 shows nine competing offers for three-phase, oil-immersed, distribution transformer of 1000 kVA, 50 Hz, with loss categories according to CENELEC HD 428.1 S1/1992. The electricity price is 0.12 \$/kWh. The discount rate is 7%. The transformer operates 8760 hours per year. The transformer lifetime is 30 years. The power factor is 0.9 and the per-unit load is 0.5.

1. Rank the offers according to the bid price.

2. Rank the offers according to the total owning cost.

3. Compute the savings due to the selection of the offer with the lowest total owning cost instead of the offer with the lowest bid price.

4. If the electricity price is varied from 0.08 to 0.16 \$/kWh with step 0.01 \$/kWh, compute the savings due to the selection of the offer with the lowest total owning cost instead of the offer with the lowest bid price.

Table 8.4 Transformer offers

Offer code	Loss category	No-load loss (W)	Load loss (W)	Bid price ($)
O1	AA'	1700	10500	28630
O2	AB'	1400	10500	29600
O3	AC'	1100	10500	32820
O4	BA'	1700	13000	27150
O5	BB'	1400	13000	27420
O6	BC'	1100	13000	30650
O7	CA'	1700	9500	30500
O8	CB'	1400	9500	31550
O9	CC'	1100	9500	35300

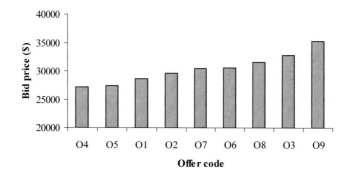

Fig. 8.2 Ranking of offers of Table 8.4 according to the bid price

Solution

1. Figure 8.2 ranks the offers of Table 8.4 from the lowest to the highest bid price.

If the purchasing criterion is simply the lowest bid price, then Offer O4 is the best choice, as can be seen from Fig. 8.2.

2. The present value multiplier is computed using (8.11):

$$PV_m = \frac{(1+d)^N - 1}{d \cdot (1+d)^{N-1}} = \frac{(1+0.07)^{30} - 1}{0.07 \cdot (1+0.07)^{30-1}} \Rightarrow PV_m = 13.2777.$$

We will present the computation of the total owning cost for Offer O4.

8.2 Total Owning Cost for Industrial and Commercial Users

Equation 8.4 is used to compute the transformer total loss TL_L at per-unit load $L = 0.5$:

$$TL_L = NLL + LL \cdot L^2 \Rightarrow TL_L = 1700 + 13000 \cdot 0.5^2 \Rightarrow TL_L = 4950 \text{ W}.$$

Equation 8.22 is combined with (8.1) to calculate the efficiency of the transformer of Offer O4:

$$n = \frac{S \cdot \cos\theta_L}{S \cdot \cos\theta_L + TL_L} \Rightarrow n = \frac{L \cdot S_n \cdot \cos\theta_L}{L \cdot S_n \cdot \cos\theta_L + TL_L} \Rightarrow$$

$$n = \frac{0.5 \cdot 1000000 \cdot 0.9}{0.5 \cdot 1000000 \cdot 0.9 + 4950} \Rightarrow n = 0.9891 \Rightarrow n = 98.91\%.$$

Equation 8.23 is used to compute the annual energy losses EL:

$$EL = TL_L \cdot HPY \cdot 10^{-3} = 4950 \cdot 8760 \cdot 10^{-3} \Rightarrow EL = 43362 \text{ kWh/year}.$$

The annual cost of losses is calculated by combining (8.5) and (8.23) as follows:

$$C_{TL} = TL_L \cdot EP \cdot HPY \cdot 10^{-3} \Rightarrow C_{TL} = EP \cdot EL = 0.12 \cdot 43362 \Rightarrow$$

$$C_{TL} = 5203.4 \text{ \$/year}.$$

Equation 8.10 is used to compute the present value PV_{TL} of the cost of transformer total losses throughout the transformer lifetime:

$$PV_{TL} = C_{TL} \cdot PV_m \Rightarrow PV_{TL} = 5203.44 \cdot 13.2777 \Rightarrow PV_{TL} = \$ \ 69090.$$

Equation 8.15 is used to compute the total owning cost of Offer O4:

$$TOC = BP + PV_{TL} \Rightarrow TOC = 27150 + 69090 \Rightarrow TOC = \$ \ 96240.$$

The total owning cost for the remaining offers of Table 8.4 is computed similarly and the results are shown in Table 8.5. Based on the TOC values of Table 8.5, Fig. 8.3 ranks the offers from the lowest to the highest total owning cost. If the purchasing criterion is the lowest total owning cost, then Offer O9 is the best choice, as can be seen from Fig. 8.3. It is concluded that, although Offer O9 has the highest bid price (Fig. 8.2), Offer O9 is the best choice since it has the lowest total owning cost, i.e., the lowest cost to purchase and operate the transformer.

Table 8.5 Total owning cost for the offers of Table 8.4

Offer	BP ($)	TL_L (W)	n (%)	EL (kWh/yr)	C_{TL} ($/yr)	PV_{TL} ($)	TOC ($)
O1	28630	4325	99.05	37887	4546	60366	88996
O2	29600	4025	99.11	35259	4231	56179	85779
O3	32820	3725	99.18	32631	3916	51992	84812
O4	27150	4950	98.91	43362	5203	69090	96240
O5	27420	4650	98.98	40734	4888	64902	92322
O6	30650	4350	99.04	38106	4573	60715	91365
O7	30500	4075	99.10	35697	4284	56877	87377
O8	31550	3775	99.17	33069	3968	52690	84240
O9	35300	3475	99.23	30441	3653	48502	83802

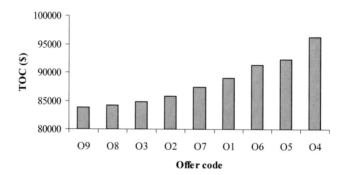

Fig. 8.3 Ranking of offers of Table 8.4 according to the total owning cost

3. Offer O9 has the lowest total owning cost (Fig. 8.3) and Offer O4 has the lowest bid price (Fig. 8.2). If Offer O9 is selected instead of Offer O4, then there are several savings.

The annual energy savings by using transformer with code 9 of Offer O9 instead of transformer with code 4 of Offer O4 are computed using (8.24):

$$ES = (EL_4 - EL_9) \cdot EP \Rightarrow ES = (43362 - 30441) \cdot 0.12 \Rightarrow$$

$$ES = 1550.52 \ \$/\text{year}.$$

8.2 Total Owning Cost for Industrial and Commercial Users

Table 8.6 Savings due to selection of the offer corresponding to the minimum *TOC* instead of Offer O4 corresponding to the minimum bid price

	Offer with minimum *TOC*			Offer O4		Selection of offer with minimum *TOC* instead of offer O4		
EP ($/kWh)	Offer	EL (kWh/yr)	TOC ($)	EL (kWh/yr)	TOC ($)	ES ($/yr)	SP (years)	S_TOC ($)
0.12	O9	30441	83802	43362	96240	1551	5.26	12437

Table 8.7 Savings due to selection of the offer corresponding to the minimum *TOC* instead of Offer O4 corresponding to the minimum bid price

	Offer with minimum *TOC*			Offer O4		Selection of offer with minimum *TOC* instead of offer O4		
EP ($/kWh)	Offer	EL (kWh/yr)	TOC ($)	EL (kWh/yr)	TOC ($)	ES ($/yr)	SP (years)	S_TOC ($)
0.08	O8	33069	66676	43362	73210	823	5.34	6533
0.09	O8	33069	71067	43362	78967	926	4.75	7900
0.10	O8	33069	75458	43362	84725	1029	4.27	9267
0.11	O9	30441	79760	43362	90482	1421	5.73	10722
0.12	O9	30441	83802	43362	96240	1551	5.26	12437
0.13	O9	30441	87844	43362	101997	1680	4.85	14153
0.14	O9	30441	91886	43362	107755	1809	4.51	15869
0.15	O9	30441	95928	43362	113512	1938	4.21	17584
0.16	O9	30441	99970	43362	119269	2067	3.94	19300

The simple payback (years) by using transformer with code 9 instead of transformer with code 4 is calculated using (8.25):

$$SP = \frac{BP_9 - BP_4}{ES} \Rightarrow SP = \frac{35300 - 27150}{1550.52} \Rightarrow SP = 5.26 \text{ years}.$$

The saving in total owning cost is:

$$S\_TOC = TOC_4 - TOC_9 = 96240 - 83802 \Rightarrow S\_TOC = \$\ 12437.$$

The results are shown in Table 8.6.

4. The case for electricity price of 0.12 $/kWh has already been studied and the results presented in Table 8.6. The same calculations are done again for all the considered electricity prices and the final results are presented in Table 8.7. The results of Table 8.7 are plotted in Figs. 8.4, 8.5, and 8.6.

It can be seen from Table 8.7 that if the electricity price is 0.08, or 0.09, or 0.10 $/kWh, then the most cost-effective offer is Offer O8, since it has the lowest total owning cost, while for all the other considered electricity prices, Offer O9 is the most cost-effective offer.

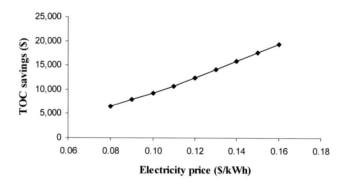

Fig. 8.4 Total owning cost savings due to selection of the offer corresponding to the minimum *TOC* instead of the offer corresponding to the minimum bid price. These savings are plotted as a function of electricity price

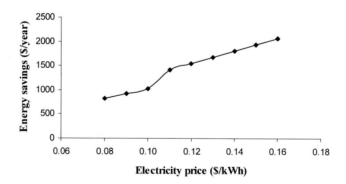

Fig. 8.5 Annual energy savings due to selection of the offer corresponding to the minimum *TOC* instead of the offer corresponding to the minimum bid price

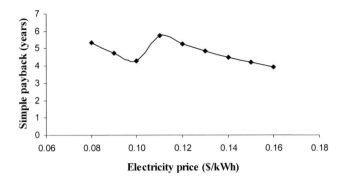

Fig. 8.6 Simple payback due to selection of the offer corresponding to the minimum *TOC* instead of the offer corresponding to the minimum bid price

8.3 Total Owning Cost for Electric Utilities

8.3.1 Cost Evaluation Method

Among the various transformer offers, the most cost-effective transformer is the one that minimizes the total owning cost, *TOC*, which is computed as follows:

$$TOC = BP + A \cdot NLL + B \cdot LL, \tag{8.26}$$

where *BP* is the transformer bid price ($) or purchasing price, *NLL* is the transformer no-load loss (W), *LL* is the transformer load loss (W), *A* is the no-load loss factor ($/W) and *B* is the load loss factor ($/W).

The total owning cost formula (8.26) can also be written as follows:

$$TOC = BP + CL, \tag{8.27}$$

where *CL* is the cost of transformer losses throughout the transformer lifetime. Equation 8.27 shows that the total owning cost includes the cost to purchase the transformer (*BP*) and the cost of losses (*CL*) throughout the transformer lifetime. The cost of losses is in fact the cost to operate the transformer, which is why sometimes in the literature the cost of losses *CL* is also called the *operating cost* throughout the transformer lifetime.

The cost of transformer losses *CL* is computed as follows:

$$CL = CNLL + CLL, \tag{8.28}$$

where:

$$CNLL = A \cdot NLL, \tag{8.29}$$

$$CLL = B \cdot LL, \tag{8.30}$$

where *CNLL* is the cost of transformer no-load loss throughout the transformer lifetime and *CLL* is the cost of transformer load loss throughout the transformer lifetime. It can be seen that by combining (8.27)–(8.30), then (8.26) is derived.

The *A* and *B* loss factors of (8.26) are computed according to ANSI/IEEE Standard C57.120-1991 (ANSI/IEEE 1992) as follows:

$$A = \frac{LIC + LECN}{ET \cdot FCR \cdot IF}, \tag{8.31}$$

$$B = \frac{LIC \cdot PRF^2 \cdot PUL^2 + LECL \cdot TLF^2}{ET \cdot FCR \cdot IF}, \tag{8.32}$$

where *LIC* is the levelized annual generation and transmission system investment cost ($/kW-yr), *LECN* is the levelized annual energy and operating cost of transformer no-load loss ($/kW-yr), *ET* is the efficiency of transmission, *FCR* is the fixed charge rate that represents the "cost of ownership", *IF* is the increase factor (it represents the total money that the user must pay to acquire the transformer, including the purchase price, overhead, fee, and tax), *PRF* is the peak responsibility factor that derives from the transformer load at the time of the power system peak load divided by the transformer peak load, *PUL* is the peak per-unit transformer load that derives from the average of the annual peaks throughout the transformer lifetime, *LECL* is the levelized annual energy and operating cost of load loss ($/kW-yr), and *TLF* is the transformer loading factor.

The basic concept for the derivation of the *A* and *B* loss factors is that each of these factors is the sum of the demand portion and the energy portion (ANSI/IEEE 1992), i.e.:

$$A = A^D + A^E, \tag{8.33}$$

$$B = B^D + B^E, \tag{8.34}$$

where A^D and B^D is the demand portion of the loss factors *A* and *B*, respectively, and A^E and B^E is the energy portion of the loss factors *A* and *B*, respectively. The demand portion ($/W) is the cost ($) of installing additional generation, transmission, and primary distribution capacity to supply 1 W of peak load to the distribution transformer (Kennedy 1998). The energy portion ($/W) is the present value ($) of the energy that will be used by 1 W of transformer loss throughout the transformer lifetime (ANSI/IEEE 1992).

8.3 Total Owning Cost for Electric Utilities

In line with (8.31), the A^D and A^E portions of the loss factor A are computed as follows:

$$A^D = \frac{LIC}{ET \cdot FCR \cdot IF}, \tag{8.35}$$

$$A^E = \frac{LECN}{ET \cdot FCR \cdot IF}. \tag{8.36}$$

In line with (8.32), the B^D and B^E portions of the loss factor B are computed as follows:

$$B^D = \frac{LIC \cdot PRF^2 \cdot PUL^2}{ET \cdot FCR \cdot IF}, \tag{8.37}$$

$$B^E = \frac{LECL \cdot TLF^2}{ET \cdot FCR \cdot IF}. \tag{8.38}$$

The levelized costs $LECN$ and $LECL$ are computed as follows:

$$LECN = CRF \cdot HPY \cdot AF \cdot \sum_{j=1}^{BL} CYEC \cdot \frac{(1+EIR)^j}{(1+d)^j}, \tag{8.39}$$

$$LECL = \frac{LECN}{AF}, \tag{8.40}$$

where CRF is the capital recovery factor computed using (8.43), HPY indicates the hours of transformer operation per year (typically 8760 hours), AF represents the transformer availability factor (i.e., the proportion of time that the transformer is predicted to be energized, which may be less than unity due to failures), BL is the number of years of transformer lifetime, EIR (%) is the annual escalation rate of the energy cost (cost of electricity), d (%) refers to the discount rate (interest rate), and $CYEC$ refers to the current year energy cost ($/kWh). It should be noted that throughout this chapter, the current year (or year zero) is defined as the year before the first year of transformer operation.

Since:

$$\sum_{j=1}^{BL} \frac{(1+EIR)^j}{(1+d)^j} = \left(\frac{1+EIR}{d-EIR}\right) \cdot \left[1 - \left(\frac{1+EIR}{1+d}\right)^{BL}\right], \tag{8.41}$$

Equation 8.39 can be further simplified as follows:

$$LECN = CRF \cdot HPY \cdot AF \cdot CYEC \cdot \left(\frac{1+EIR}{d-EIR}\right) \cdot \left[1 - \left(\frac{1+EIR}{1+d}\right)^{BL}\right]. \quad (8.42)$$

The capital recovery factor, *CRF*, is computed as follows:

$$CRF = \frac{d \cdot (1+d)^{BL}}{(1+d)^{BL} - 1}. \quad (8.43)$$

The peak per-unit load, *PUL*, is derived from the following equation (Nickel and Braunstein 1981a):

$$PUL = IP \cdot \sqrt{\left[\frac{(1+g)^{2 \cdot BL} - (1+d)^{BL}}{(1+g)^2 - (1+d)}\right] \cdot \left[\frac{d}{(1+d)^{BL} - 1}\right]}, \quad (8.44)$$

where *IP* (%) is the initial (year zero) transformer annual peak load as a percentage of transformer rated power and *g* (%) is the levelized annual compound peak load growth rate. *IP* and *g* are computed based on transformer load curve.

The transformer loading factor, *TLF*, is calculated as follows (ANSI/IEEE 1992):

$$TLF = PUL \cdot \sqrt{LSF}, \quad (8.45)$$

where *LSF* refers to the loss factor derived from the load factor *LDF*, i.e., the mean transformer loading throughout its lifetime, represented as an equivalent percentage of its nominal power, according to the following equation (Nickel and Braunstein 1981b):

$$LSF = 0.15 \cdot LDF + 0.85 \cdot LDF^2. \quad (8.46)$$

It is concluded from the above that for the computation of *A* and *B* loss factors, the following 14 parameters are involved: *AF*, *HPY*, *BL*, *CYEC*, *EIR*, *FCR*, *d*, *IP*, *g*, *PRF*, *LDF*, *IF*, *ET*, and *LIC*. More details for these 14 parameters can be found in the references (Nickel and Braunstein 1981a, 1981b; ANSI/IEEE 1992; Kennedy 1998).

8.3.2 Example 8.4

One electric utility has the following data: $AF = 0.97$, $HPY = 8760 \text{ h/yr}$, $BL = 30 \text{ yr}$, $CYEC = 0.084 \text{ \$/kWh}$, $EIR = 0.027$, $FCR = 0.10$, $d = 0.07$,

8.3 Total Owning Cost for Electric Utilities

$IP = 0.48$, $g = 0.025$, $PRF = 0.443$, $LDF = 0.678$, $IF = 1$, $ET = 0.95$, and $LIC = 270 \, \$/kW\text{-yr}$. Compute the A and B loss factors.

Solution

Equation 8.43 is used to compute the capital recovery factor:

$$CRF = \frac{d \cdot (1+d)^{BL}}{(1+d)^{BL} - 1} = \frac{0.07 \cdot (1+0.07)^{30}}{(1+0.07)^{30} - 1} \Rightarrow CRF = 0.0805864.$$

Equation 8.42 is used to compute the levelized annual energy and operating cost of transformer no-load loss:

$$LECN = CRF \cdot HPY \cdot AF \cdot CYEC \cdot \left(\frac{1+EIR}{d-EIR}\right) \cdot \left[1 - \left(\frac{1+EIR}{1+d}\right)^{BL}\right] \Rightarrow$$

$$LECN = 0.0805864 \cdot 8760 \cdot 0.97 \cdot 0.084 \cdot \left(\frac{1+0.027}{0.07-0.027}\right) \cdot \left[1 - \left(\frac{1+0.027}{1+0.07}\right)^{30}\right] \Rightarrow$$

$$LECN = 972.4394 \, \$/kW - yr.$$

Equation 8.40 is used to compute the levelized annual energy and operating cost of transformer load loss:

$$LECL = \frac{LECN}{AF} \Rightarrow LECL = \frac{972.4394}{0.97} \Rightarrow LECL = 1002.5149 \, \$/kW - yr.$$

Equation 8.44 is used to calculate the peak per-unit transformer load:

$$PUL = IP \cdot \sqrt{\left[\frac{(1+g)^{2 \cdot BL} - (1+d)^{BL}}{(1+g)^2 - (1+d)}\right] \cdot \left[\frac{d}{(1+d)^{BL} - 1}\right]} \Rightarrow$$

$$PUL = 0.48 \cdot \sqrt{\left[\frac{(1+0.025)^{2 \cdot 30} - (1+0.07)^{30}}{(1+0.025)^2 - (1+0.07)}\right] \cdot \left[\frac{0.07}{(1+0.07)^{30} - 1}\right]} \Rightarrow$$

$$PUL = 0.63594.$$

Equation 8.46 is used to calculate loss factor:

$$LSF = 0.15 \cdot LDF + 0.85 \cdot LDF^2 \Rightarrow$$

$$LSF = 0.15 \cdot 0.678 + 0.85 \cdot 0.678^2 \Rightarrow LSF = 0.49243.$$

Equation 8.45 is used to calculate transformer loading factor:

$$TLF = PUL \cdot \sqrt{LSF} \Rightarrow TLF = 0.63594 \cdot \sqrt{0.49243} \Rightarrow TLF = 0.44626.$$

Equation 8.31 is used to compute the A loss factor (no-load loss factor):

$$A = \frac{LIC + LECN}{ET \cdot FCR \cdot IF} \Rightarrow A = \frac{270 + 972.4394}{0.95 \cdot 0.10 \cdot 1} \Rightarrow A = 13078.31 \text{ \$/kW} \Rightarrow$$

$$A = 13.08 \text{ \$/W}.$$

Equation 8.32 is used to compute the B loss factor (load loss factor):

$$B = \frac{LIC \cdot PRF^2 \cdot PUL^2 + LECL \cdot TLF^2}{ET \cdot FCR \cdot IF} \Rightarrow$$

$$B = \frac{270 \cdot 0.443^2 \cdot 0.63594^2 + 1002.5149 \cdot 0.44626^2}{0.95 \cdot 0.10 \cdot 1} \Rightarrow$$

$$B = 2327.1 \text{ \$/kW} \Rightarrow B = 2.33 \text{ \$/W}.$$

8.3.3 Example 8.5

The electric utility of Example 8.4 has defined in the transformer specification that $A = 13.08$ \$/W, $B = 2.33$ \$/W, and the transformer selection will be based on the minimum *TOC*. The electric utility receives the three offers of Table 8.8 for a 1000 kVA transformer.

1. Construct the total loss curves for the three offers of Table 8.8, if the per-unit load ranges from 0.1 to 1.0 with step 0.1.

2. Construct the efficiency curves for the three offers of Table 8.8, if the power factor is 1.0 and the per-unit load ranges from 0.1 to 1.0 with step 0.1.

3. Determine the offer that the electric utility will select.

8.3 Total Owning Cost for Electric Utilities

Table 8.8 Transformer offers

Offer code	No-load loss (W)	Load loss (W)	Bid price ($)
D1	1700	13000	27150
D2	1400	10500	29600
D3	1100	9500	35300

Table 8.9 Transformer total loss and efficiency for the three offers of Table 8.8

Per-unit load	Transformer total loss (W)			Transformer efficiency (%)		
	Offer D1	Offer D2	Offer D3	Offer D1	Offer D2	Offer D3
0.1	1830	1505	1195	98.20	98.52	98.82
0.2	2220	1820	1480	98.90	99.10	99.27
0.3	2870	2345	1955	99.05	99.22	99.35
0.4	3780	3080	2620	99.06	99.24	99.35
0.5	4950	4025	3475	99.02	99.20	99.31
0.6	6380	5180	4520	98.95	99.14	99.25
0.7	8070	6545	5755	98.86	99.07	99.18
0.8	10020	8120	7180	98.76	99.00	99.11
0.9	12230	9905	8795	98.66	98.91	99.03
1.0	14700	11900	10600	98.55	98.82	98.95

Solution

1. We will show the calculation of the total loss and efficiency of Offer D1 for per-unit load $L = 0.4$.

   Equation 8.4 is used to compute the transformer total loss, TL_L, at per-unit load $L = 0.4$ for Offer D1:

   $$TL_L = NLL + LL \cdot L^2 \Rightarrow TL_L = 1700 + 13000 \cdot 0.4^2 \Rightarrow TL_L = 3780 \text{ W}.$$

   Equation 8.22 is combined with (8.1) to calculate the efficiency of the transformer of Offer D1 at per-unit load $L = 0.4$ and for power factor equal to 1.0:

   $$n = \frac{S \cdot \cos\theta_L}{S \cdot \cos\theta_L + TL_L} \Rightarrow n = \frac{L \cdot S_n \cdot \cos\theta_L}{L \cdot S_n \cdot \cos\theta_L + TL_L} \Rightarrow$$

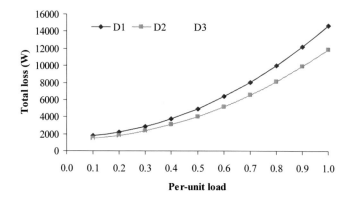

Fig. 8.7 Total loss curves for the three offers of Table 8.8

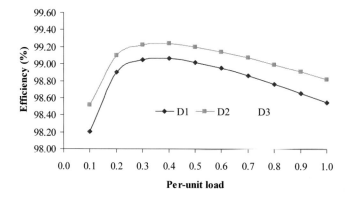

Fig. 8.8 Transformer efficiency curves for the three offers of Table 8.8. The most energy-efficient transformer is the one of Offer D3

$$n = \frac{0.4 \cdot 1000000 \cdot 1.0}{0.4 \cdot 1000000 \cdot 1.0 + 3780} \Rightarrow n = 0.9906 \Rightarrow n = 99.06\% \ .$$

Similarly we compute the total loss and the efficiency for the remaining values of per-unit load and for the three offers of Table 8.8, and the results are shown in Table 8.9. Based on Table 8.9, Fig. 8.7 presents the total loss curves for the three offers of Table 8.8.

8.3 Total Owning Cost for Electric Utilities

Table 8.10 *TOC* for the three offers of Table 8.8

Parameter	Offer D1	Offer D2	Offer D3
Bid price, *BP* ($)	27150	29600	35300
Cost of no-load loss, *CNLL* ($)	22236	18312	14388
Cost of load loss, *CLL* ($)	30290	24465	22135
Cost of losses, *CL* ($)	52526	42777	36523
Total owning cost, *TOC* ($)	79676	72377	71823
BP / *TOC* (%)	34.1	40.9	49.1
CL / *TOC* (%)	65.9	59.1	50.9

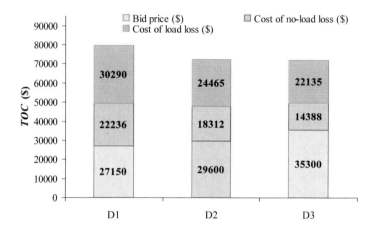

Fig. 8.9 Graphical representation of *TOC* results of Table 8.10. Each *TOC* bar is split into the three components of *TOC*, i.e., (1) the bid price, (2) the cost of no-load loss, and (3) the cost of load loss throughout the transformer lifetime

2. Based on Table 8.9, Fig. 8.8 presents the efficiency curves for the three offers of Table 8.8.

3. Equation 8.29 is used to compute the cost of transformer no-load loss throughout the transformer lifetime for Offer D1:

$$CNLL_1 = A \cdot NLL_1 \Rightarrow CNLL_1 = 13.08 \cdot 1700 \Rightarrow CNLL_1 = \$ \ 22236.$$

Equation 8.30 is used to compute the cost of transformer load loss throughout the transformer lifetime for Offer D1:

$$CLL_1 = B \cdot LL_1 \Rightarrow CLL_1 = 2.33 \cdot 13000 \Rightarrow CLL_1 = \$\ 30290.$$

Equation 8.28 is used to compute the cost of transformer losses throughout the transformer lifetime for Offer D1:

$$CL_1 = CNLL_1 + CLL_1 \Rightarrow CL_1 = 22236 + 30290 \Rightarrow CL_1 = \$\ 52526.$$

The total owning cost of Offer D1 is computed using (8.27):

$$TOC_1 = BP_1 + CL_1 \Rightarrow TOC_1 = 27150 + 52526 \Rightarrow TOC_1 = \$\ 79676$$

Since:

$$\frac{CL_1}{TOC_1} = \frac{52526}{79676} = 0.659 = 65.9\%,$$

it means that the cost of losses throughout the transformer lifetime is 65.9% of the TOC for Offer D1 and the remaining 34.1% of the TOC is the bid price for Offer D1.

The above calculations are repeated for Offers D2 and D3, and the results are shown in Table 8.10 and in graphical form in Fig. 8.9. The electric utility will select Offer D3 since it has the lowest TOC among the three offers, as can be seen from Table 8.10.

8.4 Proposed TOC Incorporating Environmental Cost

8.4.1 Introduction

Nowadays the reduction of greenhouse gas emissions is becoming a topical issue due to the growing concern for global warming and climate change. The need to undertake effective measures to protect the environment could be partially solved by improvements in energy efficiency of electrical equipment. The European Union (EU) has developed the EU Emission Trading System for cutting greenhouse gas (GHG) emissions cost-effectively (European Commission 2005), i.e., similar to the cost of energy, GHG emissions are also assigned a price by the energy markets (Bode 2006; Delarue et al. 2007). The price of GHG emissions varies as a function of supply and demand. In GHG emissions markets, those companies that do not use all their GHG emission credits can sell them to those companies that surpass them. Thus, companies who buy GHG emission credits should add this

8.4 Proposed TOC Incorporating Environmental Cost

environmental cost to the cost of transformer ownership. However, a methodology to incorporate this environmental cost into the transformer total owning cost (*TOC*) has not yet been developed.

This section proposes an innovative distribution transformer cost evaluation methodology (DTCEM) by introducing the environmental cost into the conventional total owning cost formula of Sect. 8.3.1. This environmental cost is from the cost to buy GHG emission credits because of the GHG emissions associated with supplying transformer losses throughout the transformer lifetime.

The proposed model is very important not only for energy management but also for electrical engineering for the following main reasons:

1. It has been proven that optimal transformer capacity planning based on *TOC* minimization offers significant cost reduction compared with the conventional practice of planning the transformers with initial capacity to cover the power loading for the peak operation in the target year (Saied et al. 1982; Schneider and Hoad 1992; Chen et al. 2002; Hong and Wu 2004). Consequently, electrical engineers in the planning departments of electric utilities can minimize the *TOC* of the proposed DTCEM for optimal transformer capacity planning.

2. Engineers in the purchasing departments not only of electric utilities but also of industrial users routinely use the *TOC* method to select distribution transformers (Kovacs 1980; Nickel and Braunstein 1981a, 1981b; Mamane 1984; Rasmusson 1984; Kennedy 1998; Nochumson 2002; Merritt and Chaitkin 2003; Targosz et al. 2005; Georgilakis 2007). These engineers can use the proposed DTCEM to select distribution transformers, since by applying the proposed DTCEM they can determine the relative economic benefit of a high-purchasing-cost, low-loss, low-GHG transformer versus one or more transformers with lower purchasing cost and higher losses and higher GHG emissions.

3. Electric utilities can benefit from the proposed DTCEM in order to increase power system efficiency, reduce energy costs and reduce GHG emissions by selecting and installing the most energy-efficient transformers.

4. The electricity regulatory framework has to consider the true cost of losses in the network so as to promote investments for energy-efficient transformers on the basis of the minimal *TOC* (Grenard and Strbac 2003; SEEDT 2008). The proposed DTCEM will be a valuable tool for engineers in regulatory authorities since it includes not only the cost of losses but also the environmental cost of losses.

5. Electrical engineers in the design departments of transformer manufacturers use *TOC* as an objective function when optimizing transformer design (Scofield 1982; Bins et al. 1986b; Baranowski and Hopkinson 1992). The usefulness of the *TOC* objective function is also very important when new transformer mate-

rials are introduced (Bins et al. 1986a; Lupi 1987; Ise and Murakami 1999; McShane 2001; Baldwin et al. 2003). Transformer manufacturers can use the proposed DTCEM to optimize transformer design and provide the most economical transformer to bid and manufacture.

8.4.2 Cost Evaluation Method

8.4.2.1 TOC Incorporating Environmental Cost

The objective of Sect. 8.4 is to redefine the *TOC* method to properly incorporate all the aspects of transformer life cycle, evaluating not only the transformer losses but also the environmental cost that is associated with various types of greenhouse gas emissions resulting from the combustion of fossil fuels so as to compensate for transformer losses. It is proposed to introduce an appropriate environmental cost parameter *EC* into the *TOC* formula of (8.26), resulting in the following proposed *TOCE* formula:

$$TOCE = TOC + EC, \qquad (8.47)$$

where *TOCE* is the total owning cost incorporating environmental cost, *TOC* is the total owning cost without environmental cost, and *EC* is the environmental cost of losses throughout the transformer lifetime that results from the combustion of fossil fuels so as to compensate for transformer energy losses. The *TOC* is computed using (8.26). The environmental cost *EC* is calculated as follows:

$$EC = ECNLL + ECLL, \qquad (8.48)$$

where:

$$ECNLL = A_e \cdot \Delta PNLL, \qquad (8.49)$$

$$ECLL = B_e \cdot \Delta PLL, \qquad (8.50)$$

where *ECNLL* is the environmental cost of no-load losses throughout the transformer lifetime, *ECLL* is the environmental cost of load losses throughout the transformer lifetime, A_e is the no-load loss environmental factor ($/W), $\Delta PNLL$ is the no-load loss difference (W) between an evaluated transformer and a reference transformer, B_e is the load loss environmental factor ($/W), and ΔPLL is the rated load loss difference (W) between an evaluated transformer and a reference transformer. The importance of the reference transformer is highlighted in Sect. 8.4.2.2.2.

8.4 Proposed TOC Incorporating Environmental Cost

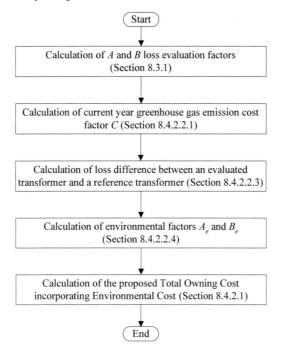

Fig. 8.10 Flowchart of the proposed DTCEM

By combining (8.26) with (8.47)–(8.50), the proposed *TOCE* formula is obtained:

$$TOCE = BP + A \cdot NLL + B \cdot LL + A_e \cdot \Delta PNLL + B_e \cdot \Delta PLL . \tag{8.51}$$

In the context of environmental protection, European Union countries have set GHG emission limits and electric utilities that violate these limits have to pay GHG emission penalties or to buy GHG emission credits from other utilities (European Commission 2005; Bode 2006; Delarue et al. 2007). This means that each electric utility has to assess this cost and to take care so as not to pay GHG emission penalties. This can be done by assessing the GHG emissions of its installed electrical equipment and specifying accordingly its new equipment. More specifically, in case of distribution transformers, the electric utility has to compute the reference transformer (see Sect. 8.4.2.2.2) for each power rating. When evaluating a transformer, it is important for the electric utility to compute the no-load loss difference between an evaluated transformer and the reference transformer, i.e., the term $\Delta PNLL$ using (8.54). The electric utility has to pay GHG emission penalties due to transformer no-load loss only if $\Delta PNLL > 0$. Similarly, the electric utility has to pay GHG emission penalties due to transformer load loss only if $\Delta PLL > 0$. That is why the terms $\Delta PNLL$ and ΔPLL are included in (8.51).

The A and B factors are computed according to (8.31) and (8.32), respectively. The A_e and B_e factors are computed according to (8.56) and (8.57), respectively. The values of $\Delta PNLL$ and ΔPLL are computed using (8.54) and (8.55), respectively.

Section 8.4 proposes that among all transformer offers, the most cost-effective and energy-efficient transformer is the one that minimizes the $TOCE$ (8.51). The flowchart of the proposed DTCEM is shown in Fig. 8.10.

8.4.2.2 Calculation of Environmental Factors A_e and B_e

In order to calculate the environmental factors A_e and B_e, the following steps should be followed: (1) calculation of the current year greenhouse gas (GHG) emission cost factor C; (2) computation of loss difference between an evaluated transformer and a reference transformer; and (3) calculation of environmental factors A_e and B_e.

8.4.2.2.1 Calculation of Current Year GHG Emission Cost Factor C

The current year GHG emission cost factor C ($/MWh) is computed as follows:

$$C = C_{cy} \cdot \sum_{i=1}^{N} f_i \cdot e_i, \qquad (8.52)$$

where C_{cy} is the current year GHG emission cost value in $/t_{CO_2}$, where t_{CO_2} denotes the tonnes of equivalent CO_2 emissions, e_i is the emission factor (t_{CO_2}/MWh) for fuel type i, f_i is the fraction (%) of end-use electricity coming from fuel i, and N is the number of fuels in the electricity mix.

In particular, three greenhouse gases: (1) carbon dioxide (CO_2), (2) methane (CH_4), and (3) nitrous oxide (N_2O) are considered (RETScreen 2002). According to the type of fuel (i.e., coal, diesel, natural gas, wind, nuclear, propane, solar, biomass, geothermal, etc.), GHG emissions are converted into equivalent CO_2 emissions (expressed in t_{CO_2}) in terms of their global warming potential. In order to estimate the emission factor of each fuel type, the following equation is used (RETScreen 2002):

$$e_i = \left(e_{CO_2,i} + e_{CH_4,i} \cdot 21 + e_{N_2O,i} \cdot 310\right) \cdot \frac{0.0036}{n_i \cdot (1-\lambda_i)}, \qquad (8.53)$$

8.4 Proposed TOC Incorporating Environmental Cost

where e_i is the emission factor (t_{CO_2}/MWh) for fuel type i, $e_{CO_2,i}$ is the CO_2 emission factor (kg/GJ) for fuel i, $e_{CH_4,i}$ is the CH_4 emission factor (kg/GJ) for fuel i, $e_{N_2O,i}$ is the N_2O emission factor (kg/GJ) for fuel i, n_i is the conversion efficiency (%) for fuel i, and λ_i represents the fraction (%) of electricity lost in transmission and distribution for fuel i. The factor 0.0036 in (8.53) is used so as to convert kg/GJ into t_{CO_2}/MWh. It can be seen from (8.53) that CH_4 and N_2O emissions are converted into equivalent CO_2 emissions by multiplying their emission factors by 21 and 310, respectively, since CH_4 is 21 times more powerful a greenhouse gas than CO_2 and N_2O is 310 times more powerful than CO_2 (Houghton et al. 1996).

8.4.2.2.2 Reference Transformer

The definition of the reference transformer, i.e., a transformer with reference no-load loss NLLR (W) and reference rated load loss LLR (W) is important because the NLLR and LLR are required for the computation of $\Delta PNLL$ and ΔPLL ((8.54) and (8.55), respectively) that are involved in the proposed TOCE formula (8.51). The selection of the reference transformer losses NLLR and LLR is based on the contribution of the transformer losses to the total greenhouse gas emissions of the power system of the considered electric utility and their responsibility for the violation of the maximum greenhouse gas emission values imposed by international standards or protocols concerning each country.

8.4.2.2.3 Loss Difference Between the Evaluated and the Reference Transformer

The electric utility defines in the specification the no-load loss of the reference transformer, NLLR (W), and the rated load loss of the reference transformer, LLR (W). On the other hand, the transformer manufacturer gives in the offer the no-load loss of the evaluated transformer, NLL (W), and the rated load loss of the evaluated transformer, LL (W).

The no-load loss difference between an evaluated transformer and a reference transformer, $\Delta PNLL$ (W), and the rated load loss difference between an evaluated transformer and a reference transformer, ΔPLL (W), are computed as follows:

$$\Delta PNLL = NLL - NLLR, \tag{8.54}$$

$$\Delta PLL = LL - LLR. \tag{8.55}$$

It should be noted that if $\Delta PNLL > 0$, i.e., if $NLL>NLLR$ as (8.54) implies, i.e., if the no-load loss of the evaluated transformer is greater than the no-load loss of the reference transformer, then, since A_e is always positive as implied by (8.56), the quantity $A_e \cdot \Delta PNLL$ that is added to the *TOCE* formula (8.51) is positive and consequently it increases the value of *TOCE*, thus partially affecting negatively the decision to purchase from the considered transformer manufacturer. On the other hand, if $\Delta PNLL < 0$, i.e. if the no-load loss of the evaluated transformer is smaller than the no-load loss of the reference transformer, then the quantity $A_e \cdot \Delta PNLL$ that is added to the *TOCE* formula (8.51) is negative and consequently it decreases the value of *TOCE*, thus partially affecting positively the decision to purchase from the considered transformer manufacturer. Similar conclusions can be drawn if the quantity ΔPLL takes positive or negative values.

8.4.2.2.4 Computation of Environmental Factors A_e and B_e

The no-load loss environmental factor A_e and the load loss environmental factor B_e are computed as follows:

$$A_e = \frac{LECN_e}{ET \cdot FCR \cdot IF}, \tag{8.56}$$

$$B_e = \frac{LECL_e \cdot TLF^2}{ET \cdot FCR \cdot IF}, \tag{8.57}$$

where $LECN_e$ is the levelized annual environmental cost of no-load loss (\$/kW-yr) and $LECL_e$ is the levelized annual environmental cost of load loss (\$/kW-yr) that are computed as follows:

$$LECN_e = CRF \cdot HPY \cdot AF \cdot \sum_{j=1}^{BL} C \cdot \frac{(1+EIR_e)^j}{(1+d)^j}, \tag{8.58}$$

$$LECL_e = \frac{LECN_e}{AF}, \tag{8.59}$$

where EIR_e is the annual escalation rate (%) of the current year GHG emission cost C_{cy}.

Equation 8.58 can be further simplified as follows:

$$LECN_e = CRF \cdot HPY \cdot AF \cdot C \cdot \left(\frac{1+EIR_e}{d-EIR_e}\right) \cdot \left[1-\left(\frac{1+EIR_e}{1+d}\right)^{BL}\right]. \tag{8.60}$$

8.4 Proposed TOC Incorporating Environmental Cost

Table 8.11 Electric utility electricity mix and GHG emission data

Fuel type	Coal	Diesel	Hydro	Natural gas	Wind
Indicator of fuel type, i	1	2	3	4	5
f_i (%)	69.77	7.6	7.6	15	0.03
n_i (%)	35	30	100	45	100
λ_i (%)	8	8	8	8	8
$e_{CO_2,i}$ (kg/GJ)	94.6	74.1	0	56.1	0
$e_{CH_4,i}$ (kg/GJ)	0.002	0.002	0	0.003	0
$e_{N_2O,i}$ (kg/GJ)	0.003	0.002	0	0.001	0
e_i ($/MWh)	1.0685	0.9752	0.0000	0.4911	0.0000

8.4.3 Example 8.6

The electric utility of Example 8.4 produces electricity using five fuels. Table 8.11 presents the values for the parameters f_i, n_i, and λ_i of the electricity mix for each fuel i. The emission factors per fuel and per GHG are shown in the rows six to eight of Table 8.11. The current year GHG emission cost value is $C_{cy} = 50\ \$/t_{CO_2}$. Compute the current year GHG emission cost factor C.

Solution
Equation 8.53 is used to compute the emission factor of coal (fuel type indicator $i = 1$):

$$e_1 = \left(e_{CO_2,1} + e_{CH_4,1} \cdot 21 + e_{N_2O,1} \cdot 310\right) \cdot \frac{0.0036}{n_1 \cdot (1-\lambda_1)} \Rightarrow$$

$$e_1 = (94.6 + 0.002 \cdot 21 + 0.003 \cdot 310) \cdot \frac{0.0036}{0.35 \cdot (1-0.08)} \Rightarrow$$

$$e_1 = 1.0685\ \$/MWh.$$

Similarly, the emission factors of the other four fuels are computed and the results are shown in the last row of Table 8.11.

The current year GHG emission cost factor C is computed using (8.52):

$$C = C_{cy} \cdot \sum_{i=1}^{5} f_i \cdot e_i \Rightarrow$$

$$C = 50 \cdot (0.6977 \cdot 1.0685 + 0.076 \cdot 0.9752 + 0 + 0.15 \cdot 0.4911 + 0) \Rightarrow$$

$$C = 44.66 \text{ \$/MWh}.$$

8.4.4 Example 8.7

It is given that $EIR_e = 0.035$ for the electric utility of Example 8.4. Compute the environmental factors A_e and B_e.

Solution
In Example 8.4 we found that $CRF = 0.0805864$ and $TLF = 0.44626$. In Example 8.6 we found that $C = 44.66$ \$/MWh.

Equation 8.60 is used to compute the levelized annual environmental cost of no-load loss:

$$LECN_e = CRF \cdot HPY \cdot AF \cdot C \cdot \left(\frac{1+EIR_e}{d-EIR_e}\right) \cdot \left[1 - \left(\frac{1+EIR_e}{1+d}\right)^{BL}\right] \Rightarrow$$

$$LECN_e = 0.0805864 \cdot 8760 \cdot 0.97 \cdot 44.66 \cdot \left(\frac{1+0.035}{0.07-0.035}\right) \cdot \left[1 - \left(\frac{1+0.035}{1+0.07}\right)^{30}\right] \Rightarrow$$

$$LECN_e = 570887.3 \text{ \$/MW} - \text{yr} \Rightarrow LECN_e = 570.89 \text{ \$/kW} - \text{yr}.$$

Equation 8.59 is used to compute the levelized annual environmental cost of load loss:

$$LECL_e = \frac{LECN_e}{AF} \Rightarrow LECL_e = \frac{570.89}{0.97} \Rightarrow LECL_e = 588.54 \text{ \$/kW} - \text{yr}.$$

Equation 8.56 is used to compute the no-load loss environmental factor:

$$A_e = \frac{LECN_e}{ET \cdot FCR \cdot IF} \Rightarrow A_e = \frac{570.89}{0.95 \cdot 0.10 \cdot 1} \Rightarrow A_e = 6009.4 \text{ \$/kW} \Rightarrow,$$

8.4 Proposed TOC Incorporating Environmental Cost

$A_e = 6.01 \ \$/W$.

Equation 8.57 is used to compute the load loss environmental factor:

$$B_e = \frac{LECL_e \cdot TLF^2}{ET \cdot FCR \cdot IF} \Rightarrow B_e = \frac{588.54 \cdot 0.44626^2}{0.95 \cdot 0.10 \cdot 1} \Rightarrow B_e = 1233.8 \ \$/kW \Rightarrow$$

$B_e = 1.23 \ \$/W$.

8.4.5 Example 8.8

The electric utility of Examples 8.4 to 8.7 has defined in the transformer specification that $A = 13.08 \ \$/W$, $B = 2.33 \ \$/W$, $A_e = 6.01 \ \$/W$, $B_e = 1.23 \ \$/W$, and the transformer selection will be based on the minimum $TOCE$. It is given that the reference transformer has $NLLR = 1100 \ W$ and $LLR = 10500 \ W$. The electric utility receives the three offers of Table 8.8 for 1000 kVA transformer. Compute the offer that the electric utility will select.

Solution
In Example 8.5 we computed the TOC for each offer and the results were presented in Table 8.10. In order to compute the $TOCE$, it is required to calculate the EC as (8.47) shows.

Equation 8.54 is used to compute the no-load loss difference between the evaluated transformer of offer D1 and the reference transformer:

$$\Delta PNLL_1 = NLL_1 - NLLR \Rightarrow \Delta PNLL_1 = 1700 - 1100 \Rightarrow \Delta PNLL_1 = 600 \ W .$$

Equation 8.49 is used to compute the environmental cost due to transformer no-load loss throughout the transformer lifetime for offer D1:

$$ECNLL_1 = A_e \cdot \Delta PNLL_1 \Rightarrow ECNLL_1 = 6.01 \cdot 600 \Rightarrow ECNLL_1 = \$ \ 3606 .$$

Equation 8.55 is used to compute the load loss difference between the evaluated transformer of offer D1 and the reference transformer:

$$\Delta PLL_1 = LL_1 - LLR \Rightarrow \Delta PLL_1 = 13000 - 10500 \Rightarrow \Delta PLL_1 = 2500 \ W .$$

Equation 8.50 is used to compute the environmental cost due to transformer load loss throughout the transformer lifetime for offer D1:

$$ECLL_1 = B_e \cdot \Delta PLL_1 \Rightarrow ECLL_1 = 1.23 \cdot 2500 \Rightarrow ECLL_1 = \$ \ 3075.$$

Equation 8.48 is used to compute the environmental cost due to transformer total losses throughout the transformer lifetime for offer D1:

$$EC_1 = ECNLL_1 + ECLL_1 \Rightarrow EC_1 = 3606 + 3075 \Rightarrow EC_1 = \$ \ 6681.$$

The total owning cost incorporating environmental cost of offer D1 is computed using (8.47):

$$TOCE_1 = TOC_1 + EC_1 \Rightarrow TOCE_1 = 79676 + 6681 \Rightarrow TOCE_1 = \$ \ 86357,$$

where the value of TOC_1 was retrieved from Table 8.10.
Since:

$$\frac{EC_1}{TOCE_1} = \frac{6681}{86357} = 0.077 = 7.7\%,$$

it means that the environmental cost of losses throughout the transformer lifetime is the 7.7% of the *TOCE* for Offer D1 and the remaining 92.3% of the *TOCE* is the *TOC* for Offer D1.

The above calculations are repeated for Offers D2 and D3 and the results are shown in Table 8.12 and in graphical form in Fig. 8.11. The electric utility will select Offer D3 since it has the lowest *TOCE* among the three offers, as can be seen from Table 8.12 and Fig. 8.11.

Table 8.12 *TOCE* for the three offers of Table 8.8

Parameter	Offer D1	Offer D2	Offer D3
Bid price, *BP* ($)	27150	29600	35300
Cost of no-load losses, *CNLL* ($)	22236	18312	14388
Cost of load losses, *CLL* ($)	30290	24465	22135
Cost of losses, *CL* ($)	52526	42777	36523
Total owning cost, *TOC* ($)	79676	72377	71823
Environmental cost of no-load losses, *ECNLL* ($)	3606	1803	0
Environmental cost of load losses, *ECLL* ($)	3075	0	-1230
Environmental cost of losses, *EC* ($)	6681	1803	-1230
Total owning cost with *EC*, *TOCE* ($)	86357	74180	70593
EC / *TOCE* (%)	7.7	2.4	-1.7
TOC / *TOCE* (%)	92.3	97.6	101.7

8.4 Proposed TOC Incorporating Environmental Cost

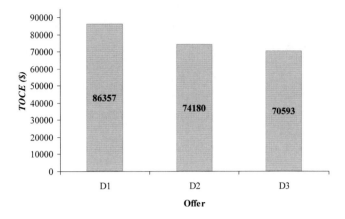

Fig. 8.11 Graphical representation of *TOCE* results of Table 8.12

8.4.6 Example 8.9

Perform a sensitivity analysis by studying the impact of parameters *d*, *CYEC*, *LIC*, *C*, *BL*, and *LDF* on *TOCE* for the three offers of Table 8.8. As base case, consider that $d = 0.07$, $CYEC = 0.084 \ \$/kWh$, $LIC = 270 \ \$/kW\text{-}yr$, $BL = 30$ yr, $C = 44.66 \ \$/MWh$, and $LDF = 0.678$. Vary each one of these six parameters from −20% to +20% of the base case value with a step of 5%. The parameters *AF*, *HPY*, *EIR*, *FCR*, *IP*, *g*, *PRF*, *IF*, and *ET* remain constant and take the values given in Example 8.4. The parameter EIR_e remains constant and takes the value given in Example 8.7. The parameters *NLLR* and *LLR* remain constant and take the values given in Example 8.8. Draw the sensitivity analysis graph of *TOCE* assuming that offer D3 is the base case offer.

Solution
The requested sensitivity analysis is implemented as follows:

1. Each time, only one parameter is varied. If, for example, the parameter *d* is varied by −20% in relation to its base case value, then the value of *d* is $d = (1 - 0.2) \cdot 0.07 \Rightarrow d = 0.056$, since the base case value of *d* is 0.07.

2. Repeating the calculations shown in Example 8.4, the values of the following parameters are computed: *CRF*, *LECN*, *LECL*, *PUL*, *LSF*, *TLF*, *A*, and *B*.

3. Repeating the calculations shown in Example 8.7, the values of the following parameters are computed: $LECN_e$, $LECL_e$, A_e, and B_e.

4. Repeating the calculations shown in Example 8.8, the value of *TOCE* is computed for each one of the three offers of Table 8.8.

The above four-step procedure is repeated 48 times, because there are six parameters that are varied (i.e., *d*, *CYEC*, *LIC*, *C*, *BL*, and *LDF*) and each parameter takes eight values (i.e., -20%, -15%, -10%, -5%, 5%, 10%, 15%, and 20% in relation to its base case value). It should be noted that the base case has been already studied in Example 8.8.

The sensitivity analysis results are shown in Tables 8.13 to 8.18. Next, assuming that offer D3 is the base case offer, Table 8.19 presents the sensitivity analysis results of *TOCE*. Figure 8.12 presents in graphical form the sensitivity analysis results of Table 8.19. Table 8.19 is computed based on the results of Tables 8.13 to 8.18 as follows. For example, the *TOCE* variation (% of base case *TOCE*) for $d = -20\%$ of base case *d* is computed as follows:

$$\Delta TOCE\Big|_{d=-20\%}^{D3} = \left[\frac{TOCE\Big|_{d=-20\%}^{D3} - TOCE\Big|_{d=0\%}^{D3}}{TOCE\Big|_{d=0\%}^{D3}} \right] \cdot 100\% \Rightarrow$$

$$\Delta TOCE\Big|_{d=-20\%}^{D3} = \left[\frac{72277 - 70593}{70593} \right] \cdot 100\% \Rightarrow \Delta TOCE\Big|_{d=-20\%}^{D3} = 2.39\%,$$

where $TOCE\Big|_{d=-20\%}^{D3}$ denotes the *TOCE* of Offer D3 for $d = -20\%$ and $TOCE\Big|_{d=0\%}^{D3}$ denotes the *TOCE* of Offer D3 for the base case value of *d*. The values of $TOCE\Big|_{d=-20\%}^{D3}$ and $TOCE\Big|_{d=0\%}^{D3}$ are retrieved from Table 8.13.

Based on Tables 8.13 to 8.19 and Fig. 8.12, the following conclusions are drawn:

1. The Offer D3 is always the optimum offer, because it has the lowest *TOCE*.

2. The parameters *CYEC* and *LDF* have large impact on *TOCE*, as can be seen from Table 8.19 and Fig. 8.12.

3. The parameters *d*, *LIC*, *C*, and *BL* have a small impact on *TOCE*, as can be seen from Table 8.19 and Fig. 8.12.

4. For the same offer, an increase of *d* results in a decrease of *TOCE*, as Table 8.13 shows.

5. For the same offer, an increase of *CYEC* results in an increase of *TOCE*, as Table 8.14 shows.

8.4 Proposed TOC Incorporating Environmental Cost

Table 8.13 *TOCE* for the three offers of Table 8.8 as a function of *d*. The base case, corresponding to $d = 0.07$, has been studied in Example 8.8

d (% of base case)	d	TOCE ($) Offer D1	TOCE ($) Offer D2	TOCE ($) Offer D3	Optimum
-20	0.056	89209	76253	72277	D3
-15	0.0595	88507	75734	71851	D3
-10	0.063	87780	75215	71435	D3
-5	0.0665	87084	74699	71009	D3
0	0.07	86357	74180	70593	D3
5	0.0735	85655	73661	70167	D3
10	0.077	84959	73145	69741	D3
15	0.0805	84410	72748	69421	D3
20	0.084	83844	72337	69090	D3

Table 8.14 *TOCE* for the three offers of Table 8.8 as a function of *CYEC*. The base case, corresponding to $CYEC = 0.084 \ \$/kWh$, has been studied in Example 8.8

CYEC (% of base case)	CYEC ($/kWh)	TOCE ($) Offer D1	TOCE ($) Offer D2	TOCE ($) Offer D3	Optimum
-20	0.0672	77412	66900	64348	D3
-15	0.0714	79579	68664	65859	D3
-10	0.0756	81876	70533	67465	D3
-5	0.0798	84060	72311	68987	D3
0	0.084	86357	74180	70593	D3
5	0.0882	88524	75944	72104	D3
10	0.0924	90821	77813	73710	D3
15	0.0966	92988	79577	75221	D3
20	0.1008	95302	81460	76838	D3

Table 8.15 *TOCE* for the three offers of Table 8.8 as a function of *LIC*. The base case, corresponding to $LIC = 270$ \$/kW-yr, has been studied in Example 8.8

LIC (% of base case)	LIC ($/kW-yr)	TOCE ($) Offer D1	TOCE ($) Offer D2	TOCE ($) Offer D3	Optimum
-20	216	84738	72857	69491	D3
-15	229.5	85106	73158	69740	D3
-10	243	85474	73459	69989	D3
-5	256.5	85989	73879	70344	D3
0	270	86357	74180	70593	D3
5	283.5	86725	74481	70842	D3
10	297	87093	74782	71091	D3
15	310.5	87461	75083	71341	D3
20	324	87846	75398	71600	D3

Table 8.16 *TOCE* for the three offers of Table 8.8 as a function of *C*. The base case, corresponding to $C = 44.66$ \$/MWh, has been studied in Example 8.8

C (% of base case)	C ($/MWh)	TOCE ($) Offer D1	TOCE ($) Offer D2	TOCE ($) Offer D3	Optimum
-20	35.728	85037	73820	70833	D3
-15	37.961	85367	73910	70773	D3
-10	40.194	85697	74000	70713	D3
-5	42.427	86027	74090	70653	D3
0	44.66	86357	74180	70593	D3
5	46.893	86712	74270	70523	D3
10	49.126	87042	74360	70463	D3
15	51.359	87372	74450	70403	D3
20	53.592	87702	74540	70343	D3

8.4 Proposed TOC Incorporating Environmental Cost

Table 8.17 *TOCE* for the three offers of Table 8.8 as a function of *BL*. The base case, corresponding to $BL = 30$ yr, has been studied in Example 8.8

BL (% of base case)	BL (yr)	TOCE ($) Offer D1	TOCE ($) Offer D2	TOCE ($) Offer D3	Optimum
-20	24	81204	70414	67524	D3
-15	25.5	82472	71344	68281	D3
-10	27	83748	72260	69017	D3
-5	28.5	84982	73162	69752	D3
0	30	86357	74180	70593	D3
5	31.5	87610	75079	71318	D3
10	33	88838	75978	72053	D3
15	34.5	90074	76863	72767	D3
20	36	91415	77853	73586	D3

Table 8.18 *TOCE* for the three offers of Table 8.8 as a function of *LDF*. The base case, corresponding to $LDF = 0.678$, has been studied in Example 8.8

LDF (% of base case)	LDF	TOCE ($) Offer D1	TOCE ($) Offer D2	TOCE ($) Offer D3	Optimum
-20	0.5424	76387	66935	64438	D3
-15	0.5763	78692	68615	65868	D3
-10	0.6102	81152	70400	67383	D3
-5	0.6441	83637	72185	68888	D3
0	0.678	86357	74180	70593	D3
5	0.7119	89127	76175	72278	D3
10	0.7458	92002	78275	74068	D3
15	0.7797	95057	80480	75933	D3
20	0.8136	98242	82790	77893	D3

Table 8.19 Sensitivity analysis: *TOCE* variation (% of base case *TOCE*) based on varying parameter values for Offer D3 of Table 8.8

Parameter variation (%)	*TOCE* (% of base case *TOCE*) when varying parameter *d* to *LDF*					
	D	CYEC	LIC	C	BL	LDF
-20	2.39	-8.85	-1.56	0.34	-4.35	-8.72
-15	1.78	-6.71	-1.21	0.25	-3.28	-6.69
-10	1.19	-4.43	-0.86	0.17	-2.23	-4.55
-5	0.59	-2.28	-0.35	0.08	-1.19	-2.42
0	0.00	0.00	0.00	0.00	0.00	0.00
5	-0.60	2.14	0.35	-0.10	1.03	2.39
10	-1.21	4.42	0.71	-0.18	2.07	4.92
15	-1.66	6.56	1.06	-0.27	3.08	7.56
20	-2.13	8.85	1.43	-0.35	4.24	10.34
Minimim	-2.13	-8.85	-1.56	-0.35	-4.35	-8.72
Maximum	2.39	8.85	1.43	0.34	4.24	10.34

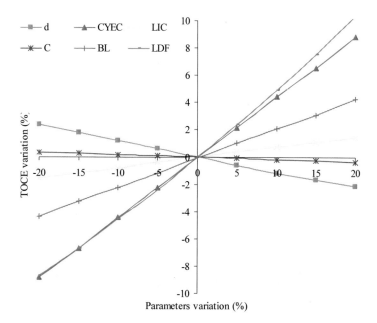

Fig. 8.12 Sensitivity analysis graph of *TOCE* to changes in each parameter. The Offer D3 is considered as the base case offer

8.4 Proposed TOC Incorporating Environmental Cost

6. For the same offer, an increase of *LIC* results in an increase of *TOCE*, as Table 8.15 shows.

7. For Offers D1 and D2, an increase of *C* results in an increase of *TOCE*, as Table 8.16 shows. On the contrary, for Offer D3, an increase of *C* results in a decrease of *TOCE*.

8. For the same offer, an increase of *BL* results in an increase of *TOCE*, as Table 8.17 shows.

9. For the same offer, an increase of *LDF* results in an increase of *TOCE*, as Table 8.18 shows.

8.4.7 Example 8.10

Compute the *TOCE* for the four different load types of Table 8.20 and for the three offers of Table 8.8. Figure 8.13 shows the daily loading profiles for the year zero for the load types of Table 8.20. It should be noted that the domestic type of load of Table 8.20 has been studied in Examples 8.4 to 8.9.

Solution
The requested computation of *TOCE* is implemented as follows:

1. Repeating the calculations shown in Example 8.4, the values of the following parameters are computed: *CRF*, *LECN*, *LECL*, *PUL*, *LSF*, *TLF*, *A*, and *B*.

2. Repeating the calculations shown in Example 8.7, the values of the following parameters are computed: $LECN_e$, $LECL_e$, A_e, and B_e.

3. Repeating the calculations shown in Example 8.8, the value of *TOCE* is computed for each one of the three offers of Table 8.8.

The *TOCE* results are shown in Table 8.21 and in Fig. 8.14. The following conclusions are drawn from Table 8.21:

1. For the same load type, Offer D3 is always the best choice, because it has the lowest *TOCE*.

2. Offer D3 is best suited to industrial loads, because it has the lowest *TOCE* ($ 64783).

Table 8.20 Parameter values for different load types

Parameter	Domestic	Industrial	Rural	Tourist
AF	0.97	0.97	0.97	0.97
HPY (h/yr)	8760	8760	8760	8760
BL (yr)	30	30	30	30
CYEC ($/kWh)	0.084	0.084	0.084	0.126
EIR	0.027	0.027	0.027	0.032
FCR	0.10	0.10	0.10	0.10
d	0.07	0.07	0.07	0.07
IP	0.480	0.580	0.565	0.450
g	0.025	0.018	0.019	0.027
PRF	0.443	0.507	0.565	0.432
LDF	0.678	0.461	0.709	0.735
IF	1	1	1	1
ET	0.95	0.95	0.95	0.95
LIC ($/kW-yr)	270	270	270	305
EIR_e	0.035	0.035	0.035	0.035
C ($/MWh)	44.66	44.66	44.66	48.00
NLLR (W)	1100	1100	1100	1100
LLR (W)	10500	10500	10500	10500

Table 8.21 *TOCE* values ($) for different load types

Offer	Domestic	Industrial	Rural	Tourist
D1	86357	76627	98072	117324
D2	74180	67250	82895	98899
D3	70593	64783	78108	91569

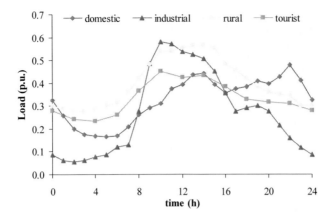

Fig. 8.13 Daily transformer loading profiles for the year zero corresponding to four load types, i.e., domestic, industrial, rural and tourist type of load. For the vertical axis, the load is expressed in per-unit (p.u.) of transformer rated power

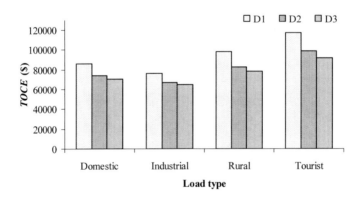

Fig. 8.14 *TOCE* as a function of load type and offer

References

Amoiralis EI (2008) Energy savings in electric power systems by development of advanced uniform models for the evaluation of transformer manufacturing and operating cost. PhD dissertation. Technical University of Crete, Chania, Greece

Amoiralis EI, Tsili MA, Georgilakis PS, Kladas AG (2007) Energy efficient transformer selection implementing life cycle costs and environmental externalities. Proc International Conference on Electrical Power Quality and Utilisation (EPQU)

Amoiralis EI, Georgilakis PS, Tsili MA, Souflaris AT (2008) Utility-based economic assessment of distribution transformers considering specific load characteristics and environmental factors. Journal of Optoelectronics and Advanced Materials 10(5):1184–1191

ANSI/IEEE (1992) IEEE loss evaluation guide for power transformers and reactors. ANSI/IEEE Standard C57.120-1991, IEEE, New York

Baldwin TL, Ykema JI, Allen CL, Langston JL (2003) Design optimization of high-temperature superconducting power transformers. IEEE Transactions on Applied Superconductivity 13(2):2344–2347

Baranowski JF, Hopkinson PJ (1992) An alternative evaluation of distribution transformers to achieve the lowest TOC. IEEE Transactions on Power Delivery 7(2):614–619

Bins DF, Crompton AB, Jaberansari A (1986a) Economic design of a 50 kVA distribution transformer. Part 1: The use of liquids of low flammability. IEE Proc Generation, Transmission, and Distribution 133(7):445–450

Bins DF, Crompton AB, Jaberansari A (1986b) Economic design of a 50 kVA distribution transformer. Part 2: Effect of different core steels and loss capitalisations. IEE Proc Generation, Transmission, and Distribution 133(7):451–456

Bode S (2006) Multi-period emissions trading in the electricity sector – winners and losers. Energy Policy 34:680–691

Chen CS, Chuang HJ, Fan LJ (2002) Unit commitment of main transformers for electrified mass rapid transit systems. IEEE Transactions on Power Delivery 17(3):747–753

Delarue E, Lamberts H, D'haeseleer W (2007) Simulating greenhouse gas (GHG) allowance cost and GHG emission reduction in Western Europe. Energy 32:1299–1309

European Commission (2005) EU action against climate change, EU emissions trading: an open system promoting global innovation

Georgilakis PS (2007) Decision support system for evaluating transformer investments in the industrial sector. Journal of Materials Processing Technology 181(1-3):307–312

Georgilakis PS, Amoiralis EI (2009) Distribution transformer cost evaluation methodology incorporating environmental cost. IET Generation, Transmission, and Distribution (submitted for publication)

Grenard S, Strbac G (2003) Effect of regulation on distribution companies investment policies in the UK. Proc IEEE Power Tech Conference 4:23–26

Hong YY, Wu JJ (2004) Determination of transformer capacities in an industrial factory with intermittent loads. IEEE Transactions on Power Delivery 19(3):1253–1258

Houghton JT, Meiro Filho LG, Callander BA, Harris N, Kattenburg A, Maskell K (1996) Climate change 1995: The science of climate change. Cambridge University Press, Cambridge, UK

Ise T, Murakami Y (1999) Design method of volt/turn for minimizing life cost of a superconducting transformer. IEEE Transactions on Applied Superconductivity 9(2):1297–1299

Kennedy BW (1998) Energy efficient transformers. McGraw-Hill, New York

Kovacs JP (1980) Economic considerations of power transformer selection and operation. IEEE Transactions on Industry Applications 16(5):595–599

Lupi S (1987) The application of amorphous magnetic alloys in induction heating medium-frequency transformers. IEEE Transactions on Magnetics 23(5):3026–3028

Mamane C (1984) Transformer loss evaluation: user–manufacturer communications. IEEE Transactions on Industry Applications 20(1):11–15

McShane CP (2001) Relative properties of the new combustion-resist vegetable-oil-based dielectric coolants for distribution and power transformers. IEEE Transactions on Industry Applications 37(4):1132–1139

Merritt S, Chaitkin S (2003) No load versus load loss. IEEE Industry Applications Magazine 9(6):21–28

References

Nickel DL, Braunstein HR (1981a) Distribution transformer loss evaluation: I – Proposed techniques. IEEE Transactions on Power Apparatus and Systems 100(2):788–797

Nickel DL, Braunstein HR (1981b) Distribution transformer loss evaluation: II – Load characteristics and system cost parameters. IEEE Transactions on Power Apparatus and Systems 100(2):798–811

Nochumson CJ (2002) Considerations in application and selection of unit substation transformers. IEEE Transactions on Industry Applications 38(3):778–787

Rasmusson PR (1984) Transformer economic evaluation. IEEE Transactions on Industry Applications 20(2):355–363

RETScreen (2002) Appendix B – GHG emission reduction analysis model. RETScreen International Renewable Energy Decision Support Centre, Ministry of Natural Resources Canada. Available: http://www.retscreen.net

Saied MM, Fetih NH, El-Shewy HM (1982) Optimal expansion of transformer substations. IEEE Transactions on Power Apparatus and Systems 101:4333–4340

Schneider KC, Hoad RF (1992) Initial transformer sizing for single-phase residential load. IEEE Transactions on Power Delivery 7(4):2074–2081

Scofield JB (1982) Selection of distribution transformer efficiency characteristics based on total levelized annual costs. IEEE Transactions on Power Apparatus and Systems 101(7):2236–2242

SEEDT (2008) Selecting energy efficient distribution transformers: a guide for achieving least-cost solutions. Report of European Commission Project No EIE/05/056/S12.419632, Leonardo Energy, European Copper Institute

Targosz R, Belmans R, Declercq J, De Keulenaer H, Furuya K, Karmarkar M, Martinez M, McDermott M, Pinkiewicz I (2005) The potential for global energy savings from high efficiency distribution transformers. Leonardo Energy, European Copper Institute

Index

A

ambient temperature, 28
Ampere's law, 5
artificial neural networks, 185
    advantages, 186
    applications to power systems, 186
    applications to transformer engineering, 186
    architecture, 190
    backpropagation training algorithm, 192
    configuration, 205
    entropy network, 210
    learning rate, 193
    momentum, 193
    multilayer perceptrons, 210
    neuron, 185, 188
    training, 191
    training functions, 206
    transfer functions, 188, 206
    types, 187, 206
automatic learning, 158
    attribute, 158
    data mining, 158
    database, 158
    example, 158
    input information, 158
    learning set, 160
    output information, 158
    supervised learning, 158
    test set, 160

B

basic insulation level, 31

bid price, 52
branch-and-bound, 239
    applications to power systems, 241
    branching, 240
    fathoming rules, 240
    mixed-integer nonlinear programming, 239

C

classification, 157, 160
    learning set, 160
    test set, 161
constraint, 52
    efficiency, 55
    flux density, 55
    heat transfer, 55
    impedance voltage, 55
    impulse voltage, 57
    induced voltage, 53, 56
    load loss, 54
    no-load current, 56
    no-load loss, 54
    tank height, 57
    tank length, 57
    tank width, 57
    temperature rise, 55
    total loss, 54
    turns ratio, 54
    voltage regulation, 56
cooling method, 40
core loss, 29
core-loss current, 12

D

data mining, 158

attribute selection, 159
interpretation and validation, 159
model selection, 159
model use, 160
representation, 159
decision trees, 162
acceptability index, 176
advantages, 163
applications to power systems, 163
applications to transformer design, 164
binary decision tree, 162
classification success rate, 178
crisp decision tree, 162
deadend node, 171
fuzzy decision tree, 163
growing, 166
inductive inference, 165
information, 170
leaf node, 170
minimum node entropy, 173
optimal splitting rule, 167
posterior classification entropy, 170
prior classification entropy, 168
pruning, 166
regression tree, 162
risk level, 173
root node, 164
score, 168, 170
stop splitting rule, 170
successor node, 164
terminal node, 164
test node, 164
top down induction, 165
distribution transformer, 40
dry type transformer, 41

E

electric current density, 125
empirical formula
average copper rise, 104
average oil rise, 106
bid price, 119
coil length, 101
cost of main materials, 117
cost of materials, 119
heat transfer, 108
impedance voltage, 99
inductive part of impedance voltage, 93
load loss of HV winding, 96
load loss of LV winding, 95
manufacturing cost, 119
mean turn length of HV winding, 96
mean turn length of LV winding, 95
minimum tank height, 102
no-load loss, 86
ohmic part of impedance voltage, 99
tank length, 102
tank width, 102
thickness of core leg, 76
total inductance, 93
transformer load loss, 96
volts per turn, 74
weight of corrugated panels, 117
weight of duct strips, 115
weight of HV winding, 99
weight of insulating materials, 114
weight of LV winding, 99
weight of magnetic material, 86
weight of mineral oil, 116
energy efficient transformers, 377
annual energy savings, 381
excitation current, 12, 32

F

Faraday's law, 9
finite element method, 128
advantages, 129
applications to power engineering, 129
applications to transformer design, 130
energy functional, 132
mesh, 129
post-processing, 129
pre-processing, 129
processing, 129
shape function, 129, 130
triangular finite element, 130
flux linkage, 9
forecasting, 157, 161
learning set, 161
test set, 162

G

genetic algorithms, 244
advantages, 244
applications to power systems, 248
applications to transformer design, 248
binary genetic algorithm, 246
chromosome, 244
continuous genetic algorithm, 246
crossover, 247

Index 425

fitness function, 246
mutation, 248
natural selection, 247
objective function, 246
parent selection, 247

H

hysteresis loop, 10

I

impedance voltage, 29
impedance voltage evaluation
    detailed modeling of windings, 309
    simplified modeling of windings, 305
    using finite element method, 315
individual core
    annealing cycle, 268
    parameters affecting no-load loss, 268
    production process, 267
individual core no-load loss
    annealing experiments, 271
    attributes, 269
    classification accuracy, 273
    classification criterion, 271
    classification success rate, 273
    classification with decision trees, 271
    classification with entropy networks, 279
    classification with multi-layer perceptrons, 276
    forecasting with hybrid multi-layer perceptrons, 296
    forecasting with multi-layer perceptrons, 294
    learning set, 271
    optimum decision tree, 273
    test set, 271
inductance, 10
input power, 4, 23

L

leakage flux, 12
load losses, 23, 30

M

magnetic field intensity, 5, 125

magnetic flux, 6, 74
magnetic flux density, 5, 125
magnetic materials, 10
    importance, 10
magnetization curve, 10
magnetizing current, 12
magnetizing flux, 12
magnetomotive force, 4
magnetostatic problems, 125
    analytical techniques, 127
    linear, 130
    magnetic vector potential, 126
    Maxwell equations, 125
    nonlinear, 146
    numerical techniques, 128
    Poisson equation, 126
multiple design method, 59

N

no-load current, 32
no-load loss classification, 266
no-load loss forecasting, 292
    accuracy, 294
    analytical methods, 293
    artificial intelligence methods, 293
    empirical methods, 293
    numerical methods, 293
no-load loss reduction, 331, 333
    exploitation of results, 342
    problem formulation, 334
    results, 341
    solution by conventional core grouping process, 333
    solution by genetic algorithm, 336
no-load losses, 23, 29

O

objective function, 50
    active part cost, 51
    active part mass, 50
    main materials cost, 51
    manufacturing cost, 51
    rated power, 52
    total owning cost, 52
oil-immersed transformer, 41
optimization, 219
    constrained, 219
    constraints, 219
    direct search methods, 220

discrete, 219
gradient search methods, 220
linear programming, 219
metaheuristic techniques, 221
mixed-integer programming, 219
nonlinear programming, 219
objective function, 219
quadratic programming, 219
unconstrained, 219
output power, 4, 23

P

permeability
ferromagnetic material, 5
free space, 6
relative, 5
per-unit load, 23, 378
power factor, 23
power transformer, 3, 40

Q

quadratic programming, 222
active set, 223
active set method, 223
applications to power systems, 225
blocking constraints, 225
convex quadratic program, 222
Karush-Kuhn-Tucker conditions, 223
step length parameter, 224
unique global optimum, 223

R

rated power, 27
recursive genetic algorithm, 368
external elitism strategy, 368
internal elitism strategy, 368
variable crossover rate, 368
variable mutation rate, 368
reluctance, 7

S

sequential quadratic programming, 231
applications to power systems, 233
feasibility error, 238
line search procedure, 232
nonlinear programming problem, 231

optimization error, 238
step length parameter, 232
short-circuit current, 32
short-circuit impedance, 29
short-circuit voltage, 29

T

temperature rise, 28
average winding, 28
top-oil, 28
tests
routine, 37
special, 39
type, 37
total loss, 54
total owning cost for electric utilities, 391
capital recovery factor, 394
cost of losses, 391
cost-effective transformer, 391
formula, 391
load loss factor B, 392
loss factor, 394
no-load loss factor A, 392
operating cost, 391
peak per-unit load, 394
transformer loading factor, 394
total owning cost for industrial users, 378
formula, 380
load loss factor B, 380
no-load loss factor A, 380
purchasing decision, 380
total owning cost with environmental cost, 402
environmental cost of load losses, 402
environmental cost of losses, 402
environmental cost of no-load losses, 402
formula, 403
importance, 401
load loss environmental factor, 406
no-load loss environmental factor, 406
transformer
approximate equivalent circuit, 13
efficiency, 23
excitation admittance, 15
excitation branch, 14, 15
maximum efficiency, 24
no-load current, 12
open-circuit test, 14
overloading, 33
parallel operation, 33

Index

427

series impedance, 15
short-circuit test, 14
standards, 35
step-down, 3
step-up, 3
T equivalent circuit, 13
tolerances, 36
voltage regulation, 18
transformer design optimization, 49, 331
    acceptable solution, 60
    characteristics, 58
    impact of current density, 364
    impact of objective function, 365
    mathematical formulation, 57
    non-acceptable solution, 60
    solution by branch-and-bound, 360
    solution by multiple design method, 60
    solution by recursive genetic algorithm, 370
transformer no-load loss
    attributes, 281
    classification criterion, 283
    classification with decision trees, 285
    classification with entropy networks, 290
    classification with multi-layer perceptrons, 290
    forecasting with hybrid multi-layer perceptrons, 299
    forecasting with multi-layer perceptrons, 298
    learning sets, 283
    test sets, 283
turns ratio, 12, 38

V

vector group, 31
voltage ratio, 12, 38

W

winding
    high voltage, 3
    low voltage, 3
    primary, 3, 12
    secondary, 3, 12
    tertiary, 3
winding material selection, 331
    creation of knowledge base, 344
    solution by adaptive trained neural network, 349
    solution by decision tree, 346